RHABDOVIRUSES

Volume III

Editor

David H. L. Bishop
Professor
Department of Microbiology
University of Alabama in Birmingham
Birmingham, Alabama

CRC PRESS, INC.
Boca Raton, Florida 33431

Library of Congress Cataloging in Publication Data
Main entry under title:

Rhabdoviruses.

Includes bibliographies and indexes.
1. Rhabdoviruses. I. Bishop, David H. L. [DNLM:
1. Rhabdoviruses. QW168.5.R4 R468]
QR415.R46 576'.6484 79-20575
ISBN 0-8493-5915-5

This book represents information obtained from authentic and highly regarded sources. Reprinted material is quoted with permission, and sources are indicated. A wide variety of references are listed. Every reasonable effort has been made to give reliable data and information, but the author and the publisher cannot assume responsibility for the validity of all materials or for the consequences of their use.

Direct all inquiries to CRC Press, 2000 N.W. 24th Street, Boca Raton, Florida, 33431.

International Standard Book Number 0-8493-5913-9 (Volume I)
International Standard Book Number 0-8493-5914-7 (Volume II)
International Standad Book Number 0-8493-5915-5 (Volume III)

Library of Congress Card Number 79-20575
Printed in the United States

FOREWORD

The Rhabdoviridae is a diverse family of enveloped RNA viruses (comprising 70 or more known serotypes). Member viruses infect homeothermic or poikilothermic vertebrates, invertebrates, or plants. Particular viruses are discussed in Volumes I and III in this series. Although some of the viruses are antigenically related to each other (so far only three genera of rhabdoviruses have been recognized), for many, no antigenic cross reactivities can be detected using the techniques commonly employed to categorize viruses.

A striking feature of the Rhabdoviridae family is the diversity of the host species that are susceptible to virus infection. Some rhabdoviruses are arboviruses while others have no arthropod host or vector (Volume I). Certain members of the Rhabdoviridae family cause severe diseases of man (e.g. rabies, see Volume I), domestic animals (e.g. bovine ephemeral fever virus, Volume III), fish (e.g. spring viremia of carp virus, Volume III), or plants (Volume III). Many other rhabdoviruses appear to be innocuous in their primary host (e.g. sigma virus, Volume III).

Due to their economic and disease importance, a considerable amount of research into the molecular biology of rhabdoviruses has been undertaken in many countries of the world over the past decade. While there are several important gaps in our understanding, what is known about virus structure (Volume I), the viral infection processes, assay systems, growth potential, molecular biology, and genetics (Volume II) of representative viruses are described in this series.

The process of interference of rhabdovirus replication by defective interfering virus particles has been a feature of rhabdoviruses which has received significant attention over the last few years. The subject is still under investigation and although the developments of the last few months are not covered in this series, background information on defective particle generation, replication and interference capabilities are discussed in Volume II. Other features of rhabdovirus infections which are discussed in this series are their ability to kill cells, form pseudotypes, and establish persistent infections (Volume III).

No book series on rhabdoviruses would be complete without a discussion of virus vaccines and other possible therapeutic processes. The advancements made in rabies vaccine development over the past years are described in Volume III. The development of vaccines for other rhabdoviruses will presumably be governed by the importance of the diseases they cause.

The goal of this book series has been to provide an overview of rhabdovirology as a whole (including an appraisal of current research findings), suitable for students, teachers, and research workers. To realize this goal I asked many of the research leaders in the different disciplines of rhabdovirology to contribute chapters. Only a few were not able to particpate due to prior commitments; most of those asked responded with articles which I believe do justice to what is known about the subject.

A final point, the four International Rhabdovirus Symposia that have been held every two years since 1973 have brought together rhabdovirologists from different parts of the world and from different disciplines of the field; the Symposia have been a major factor in the development of the subject and the communication that exists among the research scientists. This book series is therefore dedicated to those who had the foresight to initiate these Symposia, the participants of the meetings, and to the contributors who have given their time and energies to the compilation of this work.

David H. L. Bishop
October 1979

THE EDITOR

David H. L. Bishop, Ph.D., is Professor of Microbiology in the Medical Center of the University of Alabama in Birmingham, Alabama. He is also a Senior Scientist in the Medical School Comprehensive Cancer Center and Diabetes Research and Training Center.

Dr. Bishop was graduated from the University of Liverpool, England, with a B.Sc. (Hons.) degree in Biochemistry in 1959. He received a Ph.D. (Biochemistry) in 1962 also from the University of Liverpool. After a postdoctoral year sponsored by a Research Fellowship at the Centre Nationale de la Recherche Scientifique, Gif-sur-Yvette, France, Dr. Bishop was a Research Associate from 1963 to 1966 in the Department of Zoology, Edinburgh University, a Research Fellow from 1966 to 1969 in the Department of Microbiology, University of Illinois, Assistant Professor (1969 to 1970) then Associate Professor (1970 to 1971) at Columbia University College of Physicians and Surgeons, Department of Human Genetics and Development. Before joining the faculty as a Professor of Microbiology at the Medical Center of the University of Alabama in Birmingham in the fall of 1975, Dr. Bishop was an Associate Professor (1971 to 1975), then Professor (1975) at the Waksman Institue of Microbiology, Rutgers University.

Dr. Bishop has published more than 100 research papers over his career and has been on the Editorial Board of the *Journal of Virology,* since 1974, and *Virology* since 1979. In addition to being active in the American Society of Microbiology and the American Society of Tropical Medicine and Hygiene, Dr. Bishop is Chairman of the Bunyaviridae Study Group of the International Committee for Taxonomy of Viruses as well as a member of its Rhabdoviridae Study Group. He has also been a member of the National Cancer Institute Scientific Review Committee from 1975 to 1979.

CONTRIBUTORS

Gilbert Brun
Directeur du laboratoire de Genetique des Virus
Laboratoire de Génétique des Virus CNRS
Gif sur Yvette
France

A. J. Della-Porta
Research Scientist
CSIRO Division of Animal Health
Animal Health Research Laboratory
Parkville, Victoria
Australia

Contamine Didier
Attaché de Recherche
Laboratoire de Génétique des Virus CNRS
Gif sur Yvette
France

R. I. B. Francki
Reader in Pathology
Plant Pathology Department
Waite Agricultural Research Institute
University of Adelaide
South Australia

Philip I. Marcus
Professor of Biology
Microbiology Section U-44
Biological Sciences Group
University of Connecticut
Storrs, Connecticut

Thomas P. Monath
Director, Vector-Borne Diseases Division
Bureau of Laboratories
Center for Disease Control
Public Health Service
U.S. Department of Health, Education, and Welfare
Fort Collins, Colorado

J. W. Randles
Senior Lecturer in Plant Pathology
Plant Pathology Department
Waite Agricultural Research Institute
University of Adelaide
South Australia

Polly Roy
Assistant Professor
Department of Public Health
University of Alabama in Birmingham
Birmingham, Alabama

Margaret J. Sekellick
Graduate Student
Microbiology Section U-44
Biological Sciences Group
University of Connecticut
Storrs, Connecticut

W. A. Snowdon
Officer-in-Charge
CSIRO Division of Animal Health
Australian National Animal Health Laboratory
Melbourne, Victoria
Australia

Danielle Teninges
Chargée de Recherche au CNRS France
Laboratoire de Génétique des Virus CNRS
Gif sur Yvette
France

Robin A. Weiss
Doctor
Imperial Cancer Research Fund Laboratories
Lincoln's Inn Fields
London, England

Tadeusz J. Wiktor
Professor
The Wistar Institute
Philadelphia, Pennsylvania

TABLE OF CONTENTS

Volume I

TABLE OF CONTENTS

Volume II

TABLE OF CONTENTS

Volume III

Chapter 1

THE INTERFERON SYSTEM AND RHABDOVIRUSES

Philip I. Marcus

TABLE OF CONTENTS

I. INTRODUCTION

The interferon system appears as a formidable means of preventing the dire consequences of infection by rhabdoviruses,[1-3] attesting to the acute sensitivity of this family of viruses to interferon action. The prototype rhabdovirus, vesicular stomatitis virus (VSV), with its ubiquitous range of hosts and host cells, is used as an agent *par excellance* for the detection and biological assay of interferon.[4] The rhabdoviruses are perhaps less well recognized for their prowess as inducers of interferon, despite the significant levels of interferon observed in animals injected with rabies virus or certain rabies vaccines,[1,5] or in rabies virus-infected cultures of cells competent for the interferon system.[6] The capacity of rhabdoviruses to induce interferon seems worthy of reexamination in light of these observations and the recent discovery that defective interfering particles of VSV which contain covalently linked, self-complementary (±) RNA[7,8] are extraordinarily efficient inducers of interferon,[9,10] as are certain *ts* mutants and wild-type revertants of this virus.[11-14,31]

In this chapter, we will examine the role of the interferon system in rhabdovirus infections, considering, in turn, rhabdoviruses as inducers of interferon and as infectious agents sensitive to its action. Studies concerned with the molecular basis of interferon induction and of interferon action encompass many virus-cell systems; we will consider only briefly those concerned with rhabdoviruses. Finally, we will examine the prospects of interferon as an antiviral agent in diseases of rhabdovirus etiology and consider the concept that the prophylactic efficacy of vaccines for rabies may depend, in part, on their content of interferon-inducing particles.

II. VESICULAR STOMATITIS VIRUS

A. Interferon Induction: In Vitro

Vesicular stomatitis virus is generally considered a poor inducer of interferon.[15-17] Either an interferon inducer moiety in sufficient amount is not produced during virus replication and sensed by the cell, or the rapid inhibition of cellular macromolecular synthesis — a hallmark of VSV infection — precludes transcription of interferon mRNA and/or its translation.[17-22,30] However, Wertz and Youngner[17] observed that low levels of interferon can be induced by stocks of VSV under certain conditions. Thus, while stocks of a large-plaque wild-type VSV produced no interferon in mouse L cells, a small-plaque mutant produced significant levels. Stocks of both small- and large-plaque virus induced interferon in monolayers of chick embryo cells, albeit at low levels. These studies provided evidence that actual production of interferon may be related inversely to the efficiency of the viruses at inhibiting cellular macromolecular synthesis. Presumably, if viral-induced inhibition of cellular macromolecular synthesis is delayed long enough, interferon production will ensue. Clearly, the potential to induce interferon resides in stocks of VSV, but generally is not expressed. Is this potential in the form of a preformed inducer moiety intrinsic to the virion of VSV, or must a new molecular species be synthesized? Studies with a special type of defective-interfering (DI) particle of VSV provided part of the answer.

Lazzarini et al.[7] described a defective-interfering (DI-011) particle of VSV (IND) which contained a single strand of covalently linked self-complementary (±) RNA which, upon deproteinization, assumed a helical form (double-stranded, dsRNA) which was 90% resistant to ribonuclease. DI particles with (±) RNA are widely distributed in nature,[8,23] increasing the probability that they may be present as adventitious agents in some, if not many, stocks of wild-type VSV. Marcus and Sekellick[9] tested their capacity to induce interferon, and discovered that a single (±) RNA DI particle, under nonreplicating conditions, could induce a quantum yield of interferon — thus defining their activity as an interferon-inducing particle (IFP). The IFP capacity of (±) RNA DI-011 particles is very resistant to inactivation by heat (50°C) or UV radiation,[9] an attribute consistent with the preexistance of an interferon-inducer moiety in the particle. This inducer is presumed to be a single molecule of dsRNA formed, at least in part, upon entry of the DI particle into the cell.[9]

The reports of interferon production in cells persistently infected with VSV,[24] but presumably free of, or low in, DI particles, prompted us to measure the interferon-inducing particle capacity of VSV *ts* mutants; especially since Youngner and colleagues[25] had shown that *ts* mutants were the predominant species selected for during the establishment of persistent infection. We found that all *ts* mutants of VSV which failed to inhibit cellular protein synthesis (ts^{PSI-}) or kill cells (ts^{CKP-}) at a nonpermissive temperature (40°C)[21] were excellent inducers of interferon — a single particle usually sufficed to induce a quantum yield of interferon[11-13] (see Table in Chapter 4 of this volume)[24] — such mutants were designated phenotypically as $ts^{PSI-,CKP-,IFP+}$. Some ts^{IFP+} mutants induced over 25,000 VSV-PR_{50} units of interferon when monolayers containing 10^7 "aged" chick-embryo cells[27,39] were infected with an average of one plaque-forming particle(PFP) per cell. Under these same conditions wild-type VSV induced only 300 units. In contrast to the characteristics of (±) RNA DI-011 particles with their preformed inducer, upon heating (50°C), the interferon-inducing particle activity of the ts^{IFP+} mutant G11(I) is lost at the same rate as infectivity (plaque-forming particle activity)[31] — a rate set by the lability of the virion transcriptase,[28] and an indication that primary transcription is a requisite event for IFP activity. Our preliminary results with UV inactivation of *ts*G11(I) *ifp* activity suggest that perhaps only one or two genes need be expressed to form the interferon inducer.[31] If subsequent experiments confirm these preliminary findings, then the genes and virion functions required for expression of the IFP phenotype may be the same as those necessary for the expression of cell-protein synthesis inhibition and killing.[21,28,29]

Most VSV mutants which turned off cell-protein synthesis and killed cells rapidly at 40°C produced less than 1% the level of interferon as did the ts^{IFP+} mutants, and, hence, qualified as $ts^{PSI+,CKP+,IFP-}$[10-12] However, the third category of mutants was lethal to cells, but nonetheless induced significant levels of interferon. Conceivably, these mutants turn off cell macromolecular synthesis slow enough to permit expression of the *ifp* phenotype. Also in this category is the VSV non-*ts* revertant, R1, of Stanners et al.,[30] which appears to manifest a delay in the expression of the *psi* phenotype, such that early after infection, the rate of cellular protein synthesis is high, relative to that in cells infected with wildtype virus. We have found that revertant R1 induces high levels of interferon,[31] confirming a report by Francoeur et al.,[14] and, hence, would classify that virus and these mutants as psi^+ (delayed expression), ckp^+ (delayed expression), and ifp^+. Francoeur et al.[14] also noted that interferon induction by revertant R1 appears to be responsible for the small plaques produced by this virus in hamster embryo fibroblasts[29] — an explanation seemingly equally applicable (perhaps, in general, to many small-plaque formers) to the production of small plaques by VSV mutants described earlier by Wagner et al[16] and by Wertz and Youngner.[17] In support of this thesis, Stanners[75] and Sekellick and Marcus[31] both have observed that the R1 revertant of VSV produces only large plaques on GMK-Vero cells — a line genetically incompetent in its response to inducers of interferon.[9,10,24,32] In addition, Vilček et al.[26] showed that preparations of wild-type VSV (Ind.) were capable of inducing, in human cells, significant levels of interferon, as assayed on the hyperresponsive trisomic 21 GM-258 cells.

These observations provide convincing evidence to support the thesis advanced by Wertz and Youngner[17] that differences in the rate or efficiency of cell macromolecular synthesis determine (at least in part) whether interferon is produced in a given virus-cell encounter. We conclude that VSV is potentially an extremely efficient inducer of interferon and that that potential is realized only when the virus is defective or delayed in its expression of cell-protein synthesis inhibition or killing.[21,31] As discussed in a section below, rabies virus inherently may possess these same attributes.

B. Interferon Induction: In Vivo

In view of the efficient interferon-inducing capacity of certain DI particles, *ts* mutants, and revertants of VSV revealed by in vitro studies, it perhaps is not surprising (in retrospect) to read reports of interferon induction by VSV in vivo. Youngner and Wertz[33] found that wild-type VSV and a small-plaque mutant both induced several thousand units of newly synthesized interferon in mice within 12 hr after intravenous inoculation. The higher the inoculum, the higher the amount of interferon produced. Significantly, the wildtype virus was as good, or better, an inducer in vivo than the small-plaque mutant, the latter being the better inducer in vitro.[33] The induction of interferon in vivo by viruses which are relatively poor inducers in vitro may be more the rule than the exception. The studies of Gresser et al.[34] provide another example: they showed that wild-type VSV (and other viruses, some which were relatively poor inducers of interferon in vitro), when inoculated into mice treated with anti-interferon serum, produced an exaggerated expression of disease and shortened onset of death — indicating that these "poor" inducers were producing biologically significant levels of interferon in vivo. It is equally clear from other studies of Gresser et al.[35] that exogenously added interferon can protect mice from the lethal effects of VSV or delay their demise, even when administered some 4 days after infection.

It is not immediately apparent why stocks of wild-type virus should be more efficient in inducing interferon in vivo than they are in vitro. Conceivably, for cells in vivo, the rate of macromolecular synthesis inhibition by some viruses is significantly less than that observed in vitro, thus permitting the interferon-inducer moiety of VSV to be expressed. It is also possible that small numbers of interferon-inducing DI particles[9,10] may be expressed in vivo more readily as the large inocula of wild-type virus are distributed (diluted out) within the body. Clearly, the situation is complex and requires more experimentation for its resolution.

Two other studies in animals warrant consideration with regard to the possible role of interferon in the course of rhabdovirus (VSV) infection. Doyle and Holland[36] described the prophylactic use of VSV DI particles in mice to protect against intracerebral infection by wild-type VSV. They showed that large numbers of purified DI particles (5×10^{10}) could protect mice from the lethal action of hundreds of infectious particles inoculated at the same time. These DI particles failed to protect mice against the lethal action of a heterologous VSV(NJ) and NWS influenza virus, suggesting that only homotypic interference was operative under their test conditions. In keeping with the apparent absence of interferon-mediated heterologous interference, we note that the purified DI particles used by Doyle and Holland[36] contained an RNA strand about one third the size of the wild-type (IND) genome and, hence, presumbly represented what we have termed conventional (−)RNA DI particles.[9] Using similar purified preparations of a one-third size (−)RNA DI particle, we showed that they do not induce interferon ($\leqslant$ 10 units per 10^7 cells) in vitro under conditions where a one-fifth genome size (±)RNA DI-011 particle induces over 30,000 units.[9]

The results of Doyle and Holland[36] contrast significantly with those from a related study by Crick and Brown.[37] The latter investigators demonstrated that VSV (IND) DI particles induced a heterologous interference [against VSV(NJ) and (Brazil), rabies, and foot and mouth disease virus]. However, from our view, the two studies differed in at least one important aspect: Crick and Brown used DI particles which had been treated with acetyl ethyleneimine (AEI) to render the preparation biologically inactive for homotypic interference. Full-size wild-type virions inactivated similarly and inoculated in equally large numbers (10^{10} per mouse), also induced a heterologous interference of lasting duration.[37] In spite of the absence of demonstrable levels of circulating interferon in mice, it is conceivable that the AEI-inactivated preparations of DI or infectious particles contained enough (±)RNA DI particles to induce significantly high levels of interferon locally to produce an antiviral state, yet go undetected in the circulation. In this context, we know that IF → DI particles can produce very high levels of a heterologous antiviral state (in human FS4 cells) in the absence of detectable interferon.[9] Indeed, Dianzani et al.[38] have presented evidence that the local concentration of interferon in a tissue mass may be very much higher than that measured in the circulation. Also, we know that the homotypic interfering capacity of (±)RNA DI-011 particles can be inactivated with high doses of UV radiation which leave virtually unaffected the IFP activity of the preparation.[9] Therefore, to account for the different results in these two studies, we suggest that the DI particles used by Doyle and Holland[36] were of the (−)RNA type *(ifp⁻* phenotype), and the protection observed was due to homotypic interference; whereas those used by Crick and Brown[37] may have contained some (±)RNA type *(ifp⁺* phenotype) and, hence, could produce a heterologous interferon-mediated interference and protection.

Questions concerning the possible role of interferon in animal protection experiments, which show heterologous interference induced by inactive virus, might be resolved readily through the use of specific antiinterferon serum[34] and a test of the inactivated virus for its IFP activity in a hyperresponsive system, such as primary chick embryo cells "aged" in vitro.[9,27,39]

C. Sensitivity to Interferon

There is general agreement that VSV is relatively sensitive to interferon action, ranking perhaps close to the highly sensitive togaviruses, and on a par with vaccinia virus in many test systems.[40] However, the host cell may influence the sensitivity; thus, two strains of VSV were inhibited to different degrees by interferon acting in mouse L cells, but showed the same sensitivity in chicken cells.[41] Furthermore, individual clones from an uncloned population of wildtype VSV (IND) were observed to vary as much as seven fold in sensitivity, with the absolute sensitivity also dependent upon the host cell used for assay, but not the method of assay.[42]

D. Interferon Action

The sensitivity of VSV to interferon action and its broad host range, coupled with recent insights into its molecular biology — attested to by the contributions in this monograph, make VSV an excellent subject for the study of interferon-mediated interference. We will consider only briefly its contribution to our understanding of interferon action.

Following the discovery of a viron-associated transcriptase in VSV and its demonstration in vitro by Baltimore et al.,[43] Marcus et al.[44] described an assay for the in vivo accumulation of transcription products in which inhibitors of protein synthesis, such as cycloheximide, were used to restrict transcripts to those originating from the parental genome, i.e., primary transcripts. With that procedure, they demonstrated that chick embryo cells manifesting interferon-mediated interference accumulated, in a dose-dependent manner, fewer viral transcripts than mock-treated cells also infected in the presence of cycloheximide.[44] These observations were confirmed and extended by Manders et al.,[45] using VSV in human muscle skin fibroblasts. They showed that full-size transcripts accumulated in the interferon-treated cells, albeit fewer in number. Under the same conditions, i.e., when cycloheximide was present, Repik et al.[46] also observed a significant decline in the rate of VSV transcript accumulation in mouse L cells treated with interferon, as did Baxt et al.[47] in LLC-MK2 monkey cells and U human amnion cells. Results similar to the earlier study by Marcus et al.,[44] with chick embryo cells have also been obtained by Thacore[48] in human cells. Thacore[48] has also presented evidence that the species of host cell may be important in dictating the extent to which VSV primary transcription is affected by interferon action. Thus, in different laboratories, using several cell species, interferon action was shown to have an adverse effect on the accumulation of VSV virion-derived transcripts in vivo, as measured in the presence of cycloheximide. Recognizing that cycloheximide might produce aberrant effects on the regulation of viral transcription in different virus-cell systems, Marcus and Sekellick[49] turned to the use of a temperature-sensitive mutant, VSV *ts*G41 (IV),[50,51] to examine the effect of interferon action on primary transcription in green monkey kidney-Vero cells, in the absence of inhibitors of protein synthesis — a condition made possible by the failure of this mutant to amplify RNA synthesis at a non-permissive temperature (40°C), thus restricting transcription solely to that associated with the virion transcriptase, i.e., to primary transcription. Their studies showed that in Vero cells, the initial rate of virion-associated (primary) transcription increased linearly for 1 to 2 hr after infection, and that interferon acted to reduce this rate (approximately four fold with 50 units per milliliter interferon), irrespective of the presence or absence of cycloheximide.[49] In addition, they showed that the VSV mRNA transcripts synthesized in mock- or interferon-treated cells were equal in size[45] and had an equivalent half-life of 17 hr at 40°C. Thus, it seems likely that once transcription is initiated in interferon-treated cells, it is completed successfully. Furthermore, since interferon reduces the rate of VSV primary transcript synthesis to below that achieved in the presence of cycloheximide, Marcus and Sekellick[49] concluded that interferon had an effect on transcription beyond that attributable solely to protein synthesis inhibition, and postulated that interferon decreased the probability of initiating viral transcription. They also pointed out that viral mRNA escaping this facet of interferon action may then encounter other facets, such as post-transcriptional modification and/or inhibition of translation.[47,48,52] However, it is important to note that the mandatory sequence of primary transcription → primary translation for negative-strand viruses like VSV (and other rhabdoviruses) dictates that the overall inhibitory effect of interferon on translation would derive in part from a prior inhibition of transcription.[49]

A more complete discussion of interferon action as it relates to the transcription and translation of VSV mRNA is considered in papers by Marcus and Sekellick[49] and by Oxman.[53]

III. RABIES VIRUS

A. Interferon Induction: In Vitro

Although early in vitro studies appear equivocal regarding rabies virus as an inducer of interferon,[6] it now seems evident that most genetically competent cell types can produce interferon upon overt infection with most preparations of rabies virus. For example, Wiktor and Clark[54] demonstrated the production of a few hundred units of interferon from primary rabbit-kidney cell cultures infected at plaque-forming particle multiplicities (m_{PFP}) $\geqslant 10$ of the HEP (Flury) strain of rabies virus. At $m_{PFP} \simeq 1$, less than ten units of interferon were produced, but the monolayer of infected cells challenged after an overnight incubation period was capable of reducing by ten fold the plaquing efficiency of VSV, suggesting an interferon-mediated antiviral state was extant — even in the absence of readily demonstrable interferon. In this study, these same preparations of rabies virus, inactivated by high doses of UV radiation or β-propiolactone, were essentially negative as interferon-inducers. Rabies virus (HEP) did induce significant levels of interferon in primary human embryonic fibroblasts, but, as with primary rabbit kidney cells, only active preparations were effective.

In addition to interferon induction in cells overtly infected with rabies virus, cultures of interferon-competent cells persistently infected with rabies also appear to produce interferon and/or develop heterologous interference, presumably reflecting an interferon-mediated antiviral state.[54] Thus, Wiktor and Clark[54] reported that a line of embryonic hamster cells (Nil-2) persistently infected with rabies virus (Pasteur or HEP - Flury strains) acquired a cyclical heterologous antiviral state, with peaks of resistance to challenge by VSV occurring about 1 day after rabies virus reached maximal levels of infectivity. Interferon, characterized as such, was demonstrable at 100 units or more during the time of maximal resistance to VSV. Viper cell cultures infected with rabies virus also developed a cyclical appearance of infectious virus and resistance to superinfection with VSV, and, although no interferon was demonstrable, the heterologous nature of the interference suggests an interferon-mediated antiviral state may have been induced.[54] Wiktor and Clark[54] also reported that BHK-21 cells persistently infected with rabies virus (HEP) manifested a heterologous antiviral state (as measured by resistance to VSV) which persisted over many cell passages, whether low levels of interferon were demonstrable or not. These results contrast with those of Kawai et al.,[55] who found no evidence for a heterologous antiviral state (as measured by resistance to VSV) or the presence of interferon in their cultures of BHK persistently infected with HEP-Flury rabies virus. Instead, Kawai and his colleagues accounted for the cyclical appearance of rabies virus and/or antigen by the presence of homologous DI particles[56] with the special property of suppressing rabies-virus induced cytopathic effects (CPE).[57] Perhaps the BHK-21 line used by Wiktor and Clark[54] is unusual, with respect to its capacity to respond to inducers of interferon; since most investigators fail to demonstrate interferon production or the development of a heterologous antiviral state in their lines of BHK cells,[55,58,59] even when highly efficient inducers of interferon are used.[9] Perhaps these studies emphasize the value of working with cell systems fully competent for the interferon system when extrapolation to animal systems is desired, and where interferon competence is the rule rather than the exception. Further studies seem warranted to place the interferon-inducing capacity of rabies virus and its defective particles on as firm a quantitative basis as that established for VSV.[9-11,17,26,31]

B. Interferon Induction: In Vivo

An impressive record has been amassed as to the effectiveness of rabies virus and certain rabies vaccines in inducing interferon in vivo, and to the sensitivity of the virus to interferon action in animals.[1,3,5,60] We will assume, with others,[61] that the early presence, and/or stimulation and appearance of significant levels of rabies-immune antibody may be a critical determinant in the final outcome of any prophylactic treatment of rabies, and we might add, particularly so in the absence of the interferon system. In this context, we will attempt to place in perspective a special role that the interferon system may assume in rabies-virus infection and prophylaxis. In a section below, we will emphasize, in particular, a view propounded perhaps first by Stewart and Sulkin,[62] that "—interferon production may be involved in the classical post-exposure antirabies prophylaxis (Pasteur treatment)."

C. Sensitivity to Interferon

Rabies virus is sensitive to the action of interferon as tested in vitro. Thus, treatment with homologous exogenous interferon has been shown to protect the following cell species from rabies virus infection: human,[63] canine,[64] mouse,[65] hamster,[54,63] rabbit,[63] and chicken.[66] With the finding of approximately equal sensitivities of VSV and rabies virus to interferon action,[63] we might hope that other rhabdoviruses of concern as disease entities in humans and animals will prove equally sensitive.

IV. PERSPECTIVES: THE INTERFERON SYSTEM AS AN ANTIVIRAL AGENT IN RABIES PROPHYLAXIS

It is clear that the interferon system can prevent death from infection by rabies virus. Several types of animals, including primates, can be protected fully from the lethal effects of rabies virus if homologous interferon or interferon inducers are administered shortly before or after the time of infection.[1,2,5] Indeed, the protective effect of some rabies vaccines may be related to their interferon-inducing capacity.[63,67-69] Furthermore, vaccines lacking this capacity can be made more efficacious if interferon or an interferon-inducer is included as part of the prophylactic treatment.[69,70] However, one study with monkeys indicates that interferon may not be the only factor acting to enhance the effectiveness of rabies vaccines.[68] Two observations from this study by Wiktor et al.[68] warrant further consideration: (1) the level of virus-neutralizing antibody stimulated by the vaccine, i.e., its antigenicity, did not correlate with the protection observed and (2) the vaccine used, and which induced interferon, had been inactivated with β-propiolactone. In another animal study where the interferon-inducing capacity of live, and UV- and β-propiolactone-inactivated, virus were compared, active virus induced the highest levels of serum interferon, but the two preparations of inactive virus also induced significant levels.[63] An important experiment in that study by Wiktor et al.[63] showed that live virus did not induce interferon in animals vaccinated 2 weeks earlier and already immune to rabies. These results indicate that the interferon-inducer (actual or potential) resides in a particle with the specificity of rabies virus, and that its expression can be blocked by neutralization with antibody. (Anti-VSV serum acted similarly to block interferon induction in vitro by the VSV (±)RNA IF → DI particle).[9]

Interferon production in animals injected with UV- or β-propiolactone-inactivated rabies virus raises an interesting question regarding the nature of the interferon inducer in rabies vaccines, and in this context, two findings from our studies with VSV seem pertinent. We have preliminary data which show that some degree of primary transcription from the virion of VSV is required to produce an interferon-inducer moiety[31] and UV- and heat-inactivated (±)RNA IF → DI-011 particles still function as excellent inducers of interferon.[9] These results, taken in concert with the observations of interferon induction by some rabies vaccines, suggest that the interferon-inducing moiety in live-virus vaccines may be produced in cells in vivo following virion-associated (primary) transcription; whereas inactivated vaccines may already contain preformed (potential) inducer in the population of particles making up the vaccine. Since the existing evidence points to the induction of interferon by rabies vaccine as having a beneficial effect on prophylaxis, it behooves us to identify the interferon-inducing moiety in vaccines and enrich for it. Based on the discovery of a (±)RNA IF → DI particle in VSV stocks,[9] we suggest a similar particle may be responsible for, or contribute to, the interferon-inducing capacity of both active or inactivated rabies vaccines. This suggestion requires a qualification, since it is conceivable that a sufficient portion of the rabies virus genome in full-size particles survives functionally, following inactivation by UV radiation or β-propiolactone, and can provide the primary transcripts required to form an interferon-inducer moiety.[11-13,21,28,71] The discovery of rabies virus DI particles[56,57,72,73] provides a possible candidate for the inactivation-resistant interferon-inducer moiety in some rabies vaccines. The presence of 18S RNA material extractable from purified preparations of the Pittman-Moore strain of virus[74] suggests the candidate is real. In this context, we note that the rabies virus DI particles isolated by Wiktor et al.,[73] and those used in the studies of Kawai et al.,[55] appeared to give negative results with respect to their capacity to induce heterologous interference (against VSV). However, those results may lack significance, since it is not clear whether the host cells (BHK) used in the experiments were competent for the interferon system.

Finally, we note that the development of subunit rabies vaccines may serve to eliminate the interferon-inducing capacity of the vaccines by removing the viral RNA and its potential to function as an interferon-inducer, presumably in the form of double-stranded RNA.[9] On the other hand, subunit vaccines in conjunction with effective interferon inducers may provide the optimal regimen for prophylactic treatment. However, until the exact role of "heterologous" inducers of interferon is resolved,[68] it may be prudent to search specifically for (±)RNA IF → DI particles of rabies and evaluate their efficacy in vaccines. Such an approach would not sacrifice the antigenic potency of the vaccine and might, through its interferon-inducing capacity, render it a superior prophylactic agent.

ACKNOWLEDGMENTS

I thank my associate Margaret J. Sekellick for a critical reading and discussion of this chapter. The research from the author's laboratory referred to in this chapter was aided in part by National Science Foundation grant No. PCM 77-24875, National Cancer Institute grant No. CA 20882, and benefited from use of the Cell Culture Facility supported by National Cancer Institute grant No. CA 14733.

REFERENCES

1. **Baer, G. M.,** Antiviral action of interferon in animal systems: effect of interferon on rabies infections of animals, in *The Interferon System, Tex. Rep. Biol. Med.,* 35, 461, 1977.
2. **Krim, M. and Sanders, F. K.,** Prophylaxis and therapy with interferons, in *Interferons and Their Actions,* Stewart, W. E., II, Ed., CRC Press, Cleveland, 1977, 153.
3. **Murphy, F. A.,** Rabies pathogenesis, *Arch. Virol.,* 54, 279, 1977.
4. **Finter, N. B.,** The assay and standardization of interferon and interferon inducers, in *Interferon and Interferon Inducers,* Finter, N. B., Ed., North-Holland, Amsterdam, 1973, 135.
5. **Sulkin, S. E. and Allen, R.,** Interferon and rabies virus infection, in *The Natural History of Rabies,* Vol. 1, Baer, G. M., Ed., Academic Press, New York, 1975, 355.
6. **Wiktor, T. J. and Clark, H F.,** Growth of rabies virus in cell culture, in *The Natural History of Rabies,* Vol. 1, Baer, G. M., Ed., Academic Press, New York, 1975, 155.
7. **Lazzarini, R. A., Weber, G. H., Johnson, L. D., and Stamminger G. M.,** Covalently-linked message and anti-message (genomic) RNA from a defective vesicular stomatitis virus particle, *J. Mol. Biol.,* 97, 289, 1975.
8. **Perrault, J.,** Cross-linked double-stranded RNA from a defective vesicular stomatitis virus particle, *Virology,* 70, 360, 1976.
9. **Marcus, P. I. and Sekellick, M. J.,** Defective interfering particles with covalently linked (±)RNA induce interferon, *Nature (London),* 266, 815, 1977.
10. **Sekellick, M. J. and Marcus, P. I.,** Persistent infection. I. Interferon-inducing defective-interfering particles as mediators of cell sparing: possible role in persistent infection by vesicular stomatitis virus, *Virology,* 85, 175, 1978.
11. **Sekellick, M. J. and Marcus, P. I.,** Persistent infection by vesicular stomatitis virus: interferon induction by *ts*-mutants, and cell sparing, *J. Supramol. Struct.,* Suppl. 2, 246, 1978.
12. **Sekellick, M. J. and Marcus, P. I.,** Interferon induction by vesicular stomatitis virus, *Am. Soc. Microbiol., Abstr.,* 264, 1978.

13. **Sekellick, M. J. and Marcus, P. I.,** Interferon-inducing defective-interfering particles and temperature-sensitive mutants as mediators of cell-sparing: their possible role in persistent infection by VSV, *4th Int. Cong. Virol., Abstr.*, 1978, 241.
14. **Francoeur, A. M., Lam, T., and Stanners, C. P.,** The role of interferon in persistent infection with T1026, *J. Supramol. Struct.*, Suppl. 2, 245, 1978.
15. **Paucker, K., Skurska, Z., and Henle, W.,** Quantitative studies on viral interference in suspended L cells. I. Growth characteristics and interfering activities of vesicular stomatitis virus, Newcastle disease virus, and influenza A virus, *Virology*, 17, 301, 1962.
16. **Wagner, R. R., Levy, A. H., Snyder, R. M., Ratcliff, G. A., Jr., and Hyatt, D. F.,** Biological properties of two plaque variants of vesicular stomatitis virus (Indiana serotype), *J. Immunol.*, 91, 112, 1963.
17. **Wertz, G. W. and Younger, J. S.,** Interferon production and inhibition of host synthesis in cells infected with vesicular stomatitis virus, *J. Virol.*, 4, 476, 1970.
18. **Wagner, R. R. and Huang, A. S.,** Inhibition of RNA and interferon synthesis in Krebs-2 cells infected with vesicular stomatitis virus, *Virology*, 28, 1, 1966.
19. **Yamazaki, S. and Wagner, R. R.,** Action of interferon: kinetics and differential effects on viral functions, *J. Virol.*, 6, 421, 1970.
20. **Weck, P. K. and Wagner, R. R.,** Inhibition of RNA synthesis in mouse myeloma cells infected with vesicular stomatitis virus, *J. Virol.*, 25, 770, 1978.
21. **Marvaldi, J. L., Lucas-Lenard, J., Sekellick, M. J., and Marcus, P. I.,** Cell killing by viruses. IV. Cell killing and protein synthesis inhibition by VSV require the same gene functions, *Virology*, 79, 267, 1977.
22. **Marvaldi, J., Sekellick, M. J., Marcus, P. I. and Lucas-Lenard, J.,** Inhibition of mouse L-cell protein synthesis by ultraviolet-irradiated vesicular stomatitis virus requires viral transcription, *Virology*, 84, 127, 1978.
23. **Perrault, J. and Leavitt, R. W.,** Characterization of snap-back RNAs in vesicular stomatitis defective interfering virus particles, *J. Gen. Virol.*, 38, 21, 1978.
24. **Sekellick, M. J. and Marcus, P. I.,** Persistent infections by rhabdoviruses, in *Rhabdoviruses*, Vol. III, Bishop, D. H. L., Ed., CRC Press, Boca Raton, Fla., 1979, chap. 4.
25. **Youngner, J. S., Dubovi, E. J., Quagliana, D. O., Kelly, M., and Preble, O. T.,** Role of temperature-sensitive mutants in persistent infections initiated with vesicular stomatitis virus, *J. Virol.*, 19, 90, 1976.
26. **Vilček, J., Yamazaki, S., and Havell, E. A.,** Interferon induction by vesicular stomatitis virus and its role in virus replication, *Infect. Immun.*, 18, 863, 1977.
27. **Carver, D. H. and Marcus, P. I.,** Enhanced interferon production from chick embryo cells aged in vitro, *Virology*, 32, 247, 1967.
28. **Marcus, P. I. and Sekellick, M. J.,** Cell killing by vesicular stomatitis virus: a requirement for virion-derived transcription, *Virology*, 63, 176, 1975.
29. **Marcus, P. I. and Sekellick, M. J.,** Cell killing by vesicular stomatitis virus: The prototype rhabdovirus, in *Rhabdoviruses*, Vol. III, Bishop, D. H. L., Ed., CRC Press, Boca Raton, Fla., 1979, chap. 2.
30. **Stanners, C. P., Francoeur, A. M., and Lam, T.,** Analysis of vesicular stomatitis virus mutants with attenuated cytopathogenicity: mutation in viral function, P, for inhibition of protein synthesis, *Cell*, 11, 273, 1977.
31. **Sekellick, M. J. and Marcus, P. I.,** Persistent infection. II. Interferon-inducing *ts*-temperature-sensitive mutants as mediators of cell-sparing: possible role in persistent infection by vesicular stomatitis virus, *Virology*, 95, 36, 1979.
32. **Desmyter, J., Melnick, J. L., and Rawls, W. E.,** Defectiveness of interferon production and of rubella virus interference in a line of African green monkey kidney cells (Vero), *J. Virol.*, 2, 955, 1968.
33. **Youngner, J. S. and Wertz, G.,** Interferon production in mice by vesicular stomatitis virus, *J. Virol.*, 2, 1360, 1968.
34. **Gresser, I., Tovey, M. G., Maury, C., and Bandu, M. T.,** Role of interferon in the pathogenesis of virus diseases in mice as demonstrated by the use of anti-interferon serum. II. Studies with herpes simplex, Moloney sarcoma, vesicular stomatitis, Newcastle disease, and influenza viruses, *J. Exp. Med.*, 144, 1316, 1976.
35. **Gresser, I., Tovey, M. G., and Bourali-Maury, C.,** Efficacy of exogeneous interferon treatment initiated after onset of multiplication of vesicular stomatitis virus in the brains of mice, *J. Gen. Virol.*, 27, 395, 1975.
36. **Doyle, M. and Holland, J. J.,** Prophylaxis and immunization in mice by use of virus-free defective T particles to protect against intracerebral infection by vesicular stomatitis virus, *Proc. Natl. Acad. Sci. U.S.A.*, 70, 2105, 1973.

37. **Crick, J. and Brown, F.,** In vivo interference in vesicular stomatitis virus infection, *Infect. Immun.,* 15, 354, 1977.
38. **Dianzani, F., Gullino, P., and Baron, S.,** Rapid activation of the interferon system in vivo, *Infect. Immun.,* 20, 55, 1978.
39. **Lockart, R. Z., Jr.,** Viral interference in aged cultures of chick embryo cells, in *Medical and Applied Virology,* Sanders, M. and Lennette, E. H., Eds., Warren Green, Inc. St. Louis, 1968, 45.
40. **Lockart, R. Z., Jr.,** Criteria for acceptance of a viral inhibitor as an interferon and a general description of the biological properties of known interferons, in *Interferons and Interferon Inducers,* Finter, N. B., Ed., North-Holland, Amsterdam, 1973, 11.
41. **Wagner, R. R., Levy, A. H., Snyder, R. M., Ratcliff, G. A., Jr., and Hyatt, D. F.,** Biological properties of two plaque variants of vesicular stomatitis virus (Indiana serotype), *J. Immunol.,* 91, 112, 1963.
42. **Ito, Y. and Montagnier, L.,** Heterogeneity of the sensitivity of vesicular stomatitis virus to interferons, *Infect. Immun.,* 18, 23, 1977.
43. **Baltimore, D., Huang, A. S., and Stampfer, M.,** Ribonucleic acid synthesis of vesicular stomatitis virus. II. An RNA polymerase in the virion, *Proc. Natl. Acad. Sci. U.S.A.,* 66, 572, 1970.
44. **Marcus, P. I., Engelhardt, D. L., Hunt, J. M., and Sekellick, M. J.,** Interferon action: Inhibition of vesicular stomatitis virus RNA synthesis induced by virion-bound polymerase, *Science,* 174, 593, 1971.
45. **Manders, E. K., Tilles, J. G., and Huang, A. S.,** Interferon-mediated inhibition of virion-directed transcription, *Virology,* 49, 573, 1972.
46. **Repik, P., Flamand, A., and Bishop, D. H. L.,** Effect of interferon upon the primary and secondary transcription of VSV and influenza viruses, *J. Virol.,* 14, 1169, 1974.
47. **Baxt, B., Sonnabend, J. A., and Bablanian, R.,** Effects of interferon on vesicular stomatitis virus transcription and translation, *J. Gen. Virol.,* 35, 325, 1977.
48. **Thacore, H. R.,** Effect of interferon on transcription and translation of vesicular stomatitis virus in human and simian cell cultures, *J. Gen. Virol.,* 41, 421, 1978.
49. **Marcus, P. I. and Sekellick, M. J.,** Interferon action. III. The rate of primary transcription of vesicular stomatitis virus is inhibited by interferon action, *J. Gen. Virol.,* 38, 391, 1978.
50. **Pringle, C. R.,** The induction and genetic characterization of conditional-lethal mutants of vesicular stomatitis virus, in *The Biology of Large RNA Viruses,* Barry, R. D. and Mahy, B. W. J., Eds., Academic Press, New York, 1970, 567.
51. **Unger, J. T. and Reichmann, M. E.,** RNA synthesis in temperature-sensitive mutants of vesicular stomatitis virus, *J. Virol.,* 12, 570, 1973.
52. **Marcus, P. I., Terry, T. M., and Levine, S.,** Interferon action. II. Membrane-bound alkaline ribonuclease activity in chick embryo cells manifesting interferon-mediated interference, *Proc. Natl. Acad. Sci. U.S.A.,* 72, 182, 1975.
53. **Oxman, M. N.,** Molecular mechanisms of the antiviral action of interferon: Effects of interferon on the transcription of viral messenger RNA, in *The Interferon System, Tex. Rep. Biol. Med.,* 33, 230, 1977.
54. **Wiktor, T. J. and Clark, H F.,** Chronic rabies infection of cell cultures, *Infect. Immun.* 6, 988, 1972.
55. **Kawai, A., Matsumoto, S., and Tanabe, K.,** Characterization of rabies viruses recovered from persistently infected BHK cells, *Virology,* 67, 520, 1975.
56. **Crick, J. and Brown, F.,** An interfering component of rabies virus which contains RNA, *J. Gen. Virol.,* 22, 147, 1974.
57. **Kawai, A. and Matsumoto, S.,** Interfering and non-interfering defective particles generated by a rabies small plaque variant virus, *Virology,* 76, 60, 1977.
58. **Holland, J. J., Villarreal, L. P., Welsh, R. M., Oldstone, M. B. A., Kohne, D., Lazzarini, R., and Scolnick, E.,** Long-term persistent vesicular stomatitis virus and rabies virus infection of cells in vitro, *J. Gen. Virol.,* 33, 193, 1976.
59. **Morgan, M. J.,** The production and action of interferon in Chinese hamsters cells, *J. Gen. Virol.,* 33, 351, 1976.
60. **Clark, H F., Wiktor, T. J., and Koprowski, H.,** Human vaccination against rabies, in *Natural History of Rabies,* Vol. 2, Baer, G. M., Ed., Academic Press, New York, 1975, 341.
61. **Baer, G. M., Shaddock, J. H., Moore, S. A., Yager, P. A., Baron, S. S., and Levy, H. B.,** Successful prophylaxis against rabies in mice and rhesus monkeys: the interferon system and vaccine, *J. Infect. Dis.,* 136, 286, 1977.
62. **Stewart, W. E., II and Sulkin, S. E.,** Interferon production in hamsters experimentally infected with rabies virus, *Proc. Soc. Exp. Biol. Med.,* 123, 650, 1966.
63. **Wiktor, T. J., Postic, B., Ho, M., and Koprowski, H.,** Role of interferon induction in the protective activity of rabies vaccines, *J. Infect. Dis.,* 126, 408, 1972.
64. **Depoux, R.,** Virus virologie-rabique fixe et interferon, *C. R. Acad. Sci. (Paris),* 260, 354, 1965.

65. **Barroeta, M. and Atanasiu, P.,** Action inhibitrice de l'interferon sur le developement du virion rabique en culture cellulaire, *C. R. Acad. Sci. (Paris),* 269, 1353, 1969.
66. **Yoshino, K., Taniguchi, S., and Arai, K.,** Autointerference of rabies virus in chick embryo fibroblasts, *Proc. Soc. Exp. Biol. Med.,* 123, 387, 1966.
67. **Baer, G. M. and Cleary, W. F.,** A model in mice for the pathogenesis and treatment of rabies, *J. Infect. Dis.,* 125, 520, 1972.
68. **Wiktor, T. J., Koprowski, H., Mitchell, J. R., and Merigan, T. C.,** Role of interferon in prophylaxis of rabies after exposure, *J. Infect. Dis.,* 133 (Suppl. A), 260, 1976.
69. **Baer, G. M. and Yager, P. A.,** A mouse model for post-exposure rabies prophylaxis: the comparative efficacy of two vaccines and of antiserum administration, *J. Gen. Virol.,* 36, 51, 1977.
70. **Harmon, M. W. and Janis, B.,** Therapy of murine rabies after exposure: efficacy of polyriboinosinic-polyribocytidylic acid alone and in combination with three rabies vaccines, *J. Infect. Dis.,* 132, 241, 1975.
71. **Ball, L. A.,** Transcriptional mapping of VSV in vivo, *J. Virol.,* 21, 411, 1977.
72. **Holland, J. J. and Villarreal, L. P.,** Purification of defective interfering T particles of vesicular stomatitis and rabies viruses generated in vivo in brains of newborn mice, *Virology,* 67, 438, 1975.
73. **Wiktor, T. J., Dietzshold, B., Leamnson, R. N., and Koprowski, H.,** Induction and biological properties of defective interfering particles of rabies virus, *J. Virol.,* 21, 626, 1977.
74. **Aalestad, H. G. and Urbano, C.,** Nucleotide composition of the ribonucleic acid of rabies virus, *J. Virol.,* 8, 922, 1971.
75. **Stanners, C. P.,** personal communication.

Chapter 2

CELL KILLING BY VESICULAR STOMATITIS VIRUS: THE PROTOTYPE RHABDOVIRUS

Philip I. Marcus and Margaret J. Sekellick

TABLE OF CONTENTS

I. INTRODUCTION

The demise of the host represents one extreme in the expression of rhabdovirus infection — the ultimate consequence of an initial encounter of a single cell with a lethal virus. In spite of the detailed *catalogue raisonné* of events which characterize death of the host or host cell, the molecular reactions which result in cell killing by animal viruses, or its prevention through antiviral agents or viral interference, still pose an enigma for the molecular virologist. Quantification of an event like cell killing by viruses invariably proceeds its definition at the molecular level. With the development of single-cell plating procedures for animal cells, it became possible to detect and quantitate the lethal action of animal viruses on cells and, through single-cell survival curve analysis, to obtain a precise measure of the capacity of a virus to kill cells, i.e., prevent colony (clone) formation — thereby defining virions in terms of cell killing (CK) particle activity.[1-4]

The measurement of infectious virus requires that all viral gene functions be expressed and that sufficient new virus be produced, so that multiple cycles of infection may generate some readily detected amplified response, for example, plaque formation or death of the host. In contrast, the detection of CK particles requires only that the initial infecting virion prevent cell reproduction, as scored by a loss in the plating efficiency of single cells. New infectious virus need not be produced to express cell killing. Thus, virions which kill cells can be infectious or noninfectious. Virions which lack infectivity but nonetheless kill cells are termed defective cell-killing particles. Such particles may be intrinsically toxic, i.e., contain a sufficient quantity of a cell-killing factor to bring about cell death, or they may require the expression of some viral gene functions, short of the total required for infectivity. Finally, the expression of cell killing may be aborted, as in persistent infection, representing the other extreme in rhabdovirus-cell interaction, and a subject considered separately in this monograph.[5]

This communication is concerned primarily with our studies of the past 5 years, which have led to a definition of the viral genes and functions required for the expression of cell killing and protein synthesis inhibition by vesicular stomatitis virus (VSV) — the prototype rhabdovirus. Our experimental approach is based on earlier studies[1-4] and is basic enough for use in the resolution of similar questions in other lytic virus-cell systems. Several reasons commend the use of VSV in such studies: large numbers of temperature-sensitive *(ts)* virus mutants are available, representing the five complementation groups and structural proteins reported for the Indiana serotype of this virus, and assignment of the biochemical functions of specific polypeptides to each of these five groups is virtually complete.[6,7] In addition, subvirion-sized structures which contain all the polypeptides of the complete virus are available in the form of defective-interfering (DI) particles;[8] one type is capable of primary transcription,[9,10] while others also function as extraordinarily efficient interferon-inducing (IF → DI) particles.[11] Furthermore, details in the strategy of replication of VSV are becoming increasingly clear.[6,7,12] Thus, we are in an excellent position to determine which viral genes and proteins are required for the expression of cell killing. We can ask whether cell killing results directly from the action of a preexisting structural element of the virion or whether synthetic reactions attending viral development are required. The experiments described below represent our attempts to define this requirement for cell killing by VSV. Also, we will consider briefly how these model studies with VSV might enhance our understanding of the viral and host factors that regulate the expression of virulence (neurovirulence) or persistent infection by rhabdoviruses.

II. THE VIRUS

Since several chapters in this monograph will address in detail the structure, genetics, and replication strategy of VSV and its defective particles, only a brief description is warranted here — enough to provide continuity and perspective with respect to the lethal action of this virus on cells.

Vesicular stomatitis virus is a bullet-shaped virion which contains five structural proteins and consists of a ribonucleocapsid core surrounded by a membrane of lipoprotein. The membrane contains about a thousand molecules of matrix (M) protein buried within the lipid complex which derives from the host cell and about 200 molecules of a glycoprotein (G) in the peplomers that constitute the receptor molecules which stud the surface of the virion. Beneath the lipoprotein membrane lies a transcribing ribonucleoprotein complex which consists of one genome molecule of single-stranded (−)RNA ($\simeq 3.8 \times 10^6$ daltons), contained within about 2000 molecules of viral-specific nucleoprotein (N), and associated with about 40 molecules of a "nonstructural" (NS) protein and some 50 molecules of a large molecular weight ($\simeq$ 190,000 daltons) L protein.[7] Three of the viral proteins, N, NS, and L, along with the RNA genome as template, are required for primary transcription, i.e., mRNA synthesis from the parental strand.[7] The L protein, the transcriptase of the virion, under appropriate conditions, presumably may function as a replicase.[7] The single, negative-strand genome of RNA contains the genes for these five polypeptides of the Indiana serotype VSV, and all are found both in the virion and infected cell. Transcription has been established as initiating at a single site, presumably utilizing a leader sequence at or near the 3′ terminus of the (−)RNA genome,[12] and occurring according to the gene sequence 3′−N−NS−M−G−L−5′.[12] Temperature-sensitive mutants representing the five structural proteins of the Indiana serotype comprise five complementation groups; mutations affecting the transcriptase (L protein) are located in Group I.[7,9] The other complementation groups and their respective products are (in the order of gene transcription): protein N (Group IV), NS (II), M (III), and G (V).[7,9,13]

III. CELL-KILLING PARTICLE ASSAY: THE SINGLE-CELL SURVIVAL TEST

Populations of VSV were tested for their capacity to kill green monkey kidney (GMK)-Vero cells by the single-cell survival procedure, as first described by Marcus and Puck[1] and Marcus,[3,4] and modified by Marcus and Sekellick,[14,15] to permit tests of cell killing by *ts*-mutants at permissive and nonpermissive temperatures. The general procedure is outlined schematically in Figure 1. Briefly, monolayers of cells are infected with various dilutions of virus to achieve a range of multiplicities. Following

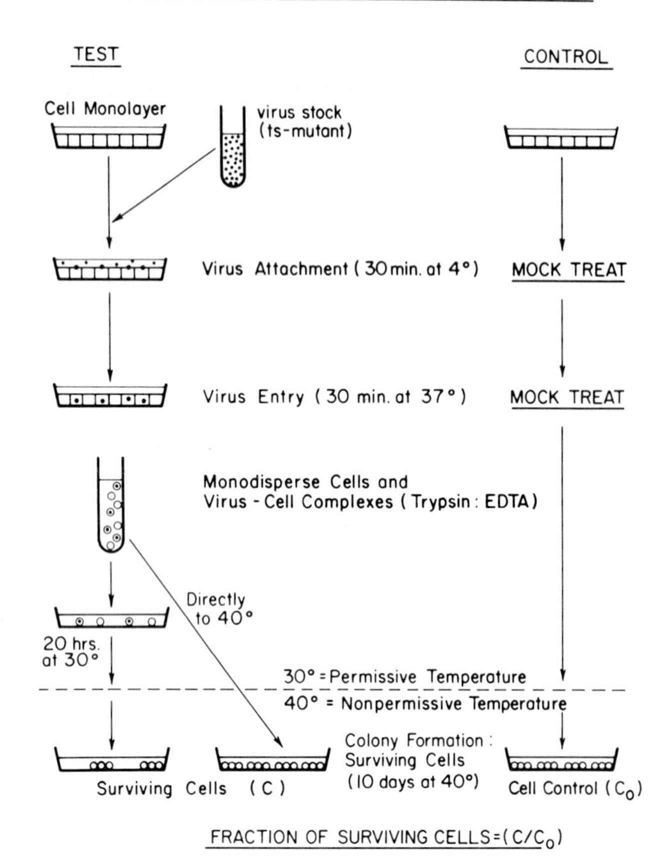

FIGURE 1. Cell-killing particle assay: single-cell survival procedure for ts-virus mutants. (From Marcus, P. I., Cancer biology IV, in *Advances in Pathobiology*, Vol. 6, Stratton Intercontinental, New York, 1977, 192. With permission.)

virus attachment and entry, infected cell monolayers are treated with a trypsin: EDTA solution, to produce a monodisperse suspension consisting of single, uninfected cells and virus-cell complexes. These are plated in duplicate sets, with one set held at a permissive temperature (30°C in this case), for 20 hr. We have established that this "pulse infection" at 30°C suffices to score all virions with cell-killing particle potential.[15] Following the 20 hr "pulse infection" at 30°C, this set of plates is placed at 40°C for the development of colonies from surviving cells. The second set of plates is incubated directly at the nonpermissive temperature (40°C) to determine whether the viral function, which is defective at the restrictive temperature, is required for the expresion of cell killing. Cross-infection of cells during the incubation period for cell survivors is prevented by inclusion of VSV antiserum in the growth medium. Control (uninfected) cells are carried through these same temperature manipulations which have no adverse effect on the high plating efficiency of GMK-Vero cells. Colonies from surviving cells are fixed, stained, and counted after 10 days incubation at 40°.* A typical survival curve generated from data obtained in this manner is illustrated in

* The surviving colonies obtained from a typical experiment with Vero cells infected with wildtype VSV are like those illustrated in Figure 12.

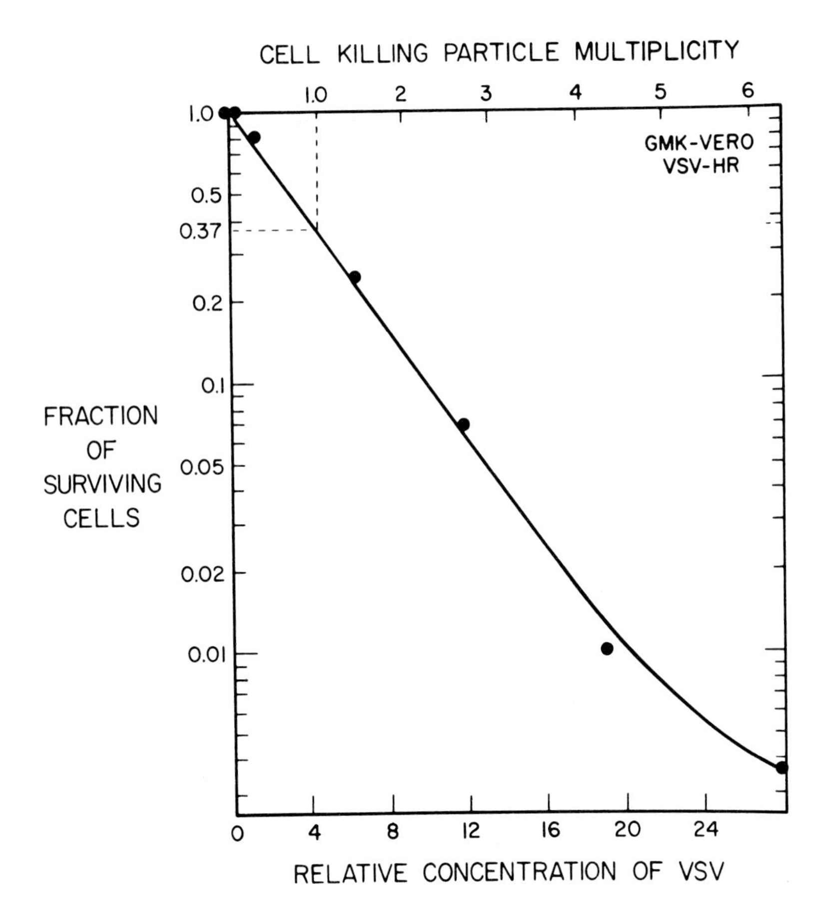

FIGURE 2. Survival curve of GMK-Vero cells exposed to a crude preparation of VSV-HR. Cell monolayers in 60 mm dishes were infected with various dilutions of virus in 0.3 mℓ of attachment solution with DEAE-dextran (10 μg/mℓ), and adsorption was carried out for 30 to 40 min at 4°C. The infected cell monolayer was freed of unattached virus by aspiration and incubated for 1 hr at 37°C in 1 mℓ attachment solution to permit VSV entry into the cells. These infected cell monolayers then were dispersed to single-cell suspensions by incubating for 15 min at 37°C with 2 mℓ of trypsin-EDTA solution and plating at various cell concentrations in the appropriate growth medium. Incubation for colony formation was carried out at 37.5°C for 9 to 11 days. Colonies were fixed with 10% formalin and stained with Giemsa prior to counting. The upper abscissa represents the CKP multiplicities calculated from the 37% survival level. (From Marcus, P. I. and Sekellick, M. J., *Virology*, 57, 321, 1974. With permission.)

Figure 2. The one-particle-to-kill nature of the survival curves generated at low multiplicities identifies a viral activity, attributable to a cell-killing (CK) particle, through its capacity to prevent colony formation from single cells.[14] These data demonstrate that a cell encounter with a single particle of VSV suffices to initiate the molecular events that lead to cell death. As resolved by velocity sedimentation on linear sucrose gradients (Figure 3), CK particles are located predominantly in the peak of infectious (plaque-forming) particles, but they present a significantly broader distribution. Cell-killing particle assays of several crude, or once gradient, purified stocks of VSV, revealed an average ratio of cell killing: plaque-forming particles ≃ 5, with extremes of 1 and 16. The high ratios show that stocks of VSV may contain, in addition to infectious particles that kill cells (plaque particles), killing particles that are noninfectious. By analogy with defective-interfering particles, they may be termed defective CK particles.[14]

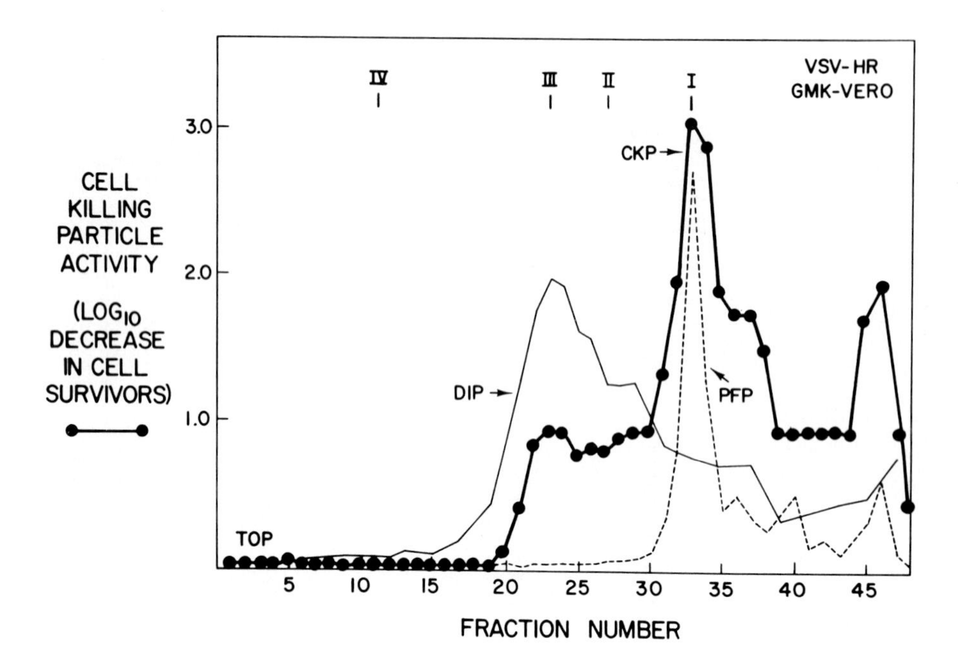

FIGURE 3. VSV-HR profiles of cell-killing (CKP), plaque-forming (PFP), and defective-interfering (DIP) particle activity from a 5 to 40% linear sucrose gradient following velocity sedimentation. Cell-killing particle activity was scored by adding 0.1 mℓ of each 0.25 mℓ fraction to a monolayer of Vero-GMK cells (2×10^6 cells) containing 0.2 mℓ attachment solution with DEAE-dextran (10 μg/mℓ). After infection, cells were dispersed with trypsin-EDTA and plated for cell survivors as described in the legend to Figure 2. Infectivity was measured as PFP and plotted as total activity per fraction. Defective-interfering particles were scored by adding 0.1 mℓ of each 0.25 mℓ fraction to a monolayer of Vero-GMK cells containing 0.2 mℓ of attachment solution with DEAE-dextran and an $m_{PFP} = 0.1$ of thrice-gradient purified B-particles (peak VSV-I). Yield reduction of PFP was calculated from plaque assays on supernatant material of control and test plates collected 12 hr after infection. (From Marcus, P. I. and Sekellick, M. J., *Virology*, 57, 321, 1974. With permission.)

IV. DEFECTIVE CELL-KILLING PARTICLES

The discovery of defective CK particles[14] which cosediment with complete infectious virus, through their excess over plaque-forming particles in standard preparations of VSV, demonstrates that a single virion of VSV can kill a cell in the absence of complete viral replication. Although this observation appears to indicate that the virions of VSV may be intrinsically "toxic", i.e., contain all the preformed components required for cell killing, as discussed below, further studies proved this was not the case.[14-19]

V. CELL KILLING BY ULTRAVIOLET (UV)- AND HEAT (50°C)-INACTIVATED VSV

If the expression of cell killing by VSV requires all viral gene functions, as presumably does the expression of infectivity, the survival curves for the inactivation of these two functions by UV radiation should be identical. Should the lethal action of VSV require fewer than all five gene functions, or if the virions were intrinsically toxic, then cell-killing particle activity would be significantly more resistant to UV radiation than infectivity. The following experiment was performed to distinguish between these possibilities: wild-type VSV was exposed to different doses of UV radiation, and the fraction of surviving activity of both infectivity (as plaque-forming particles) and cell

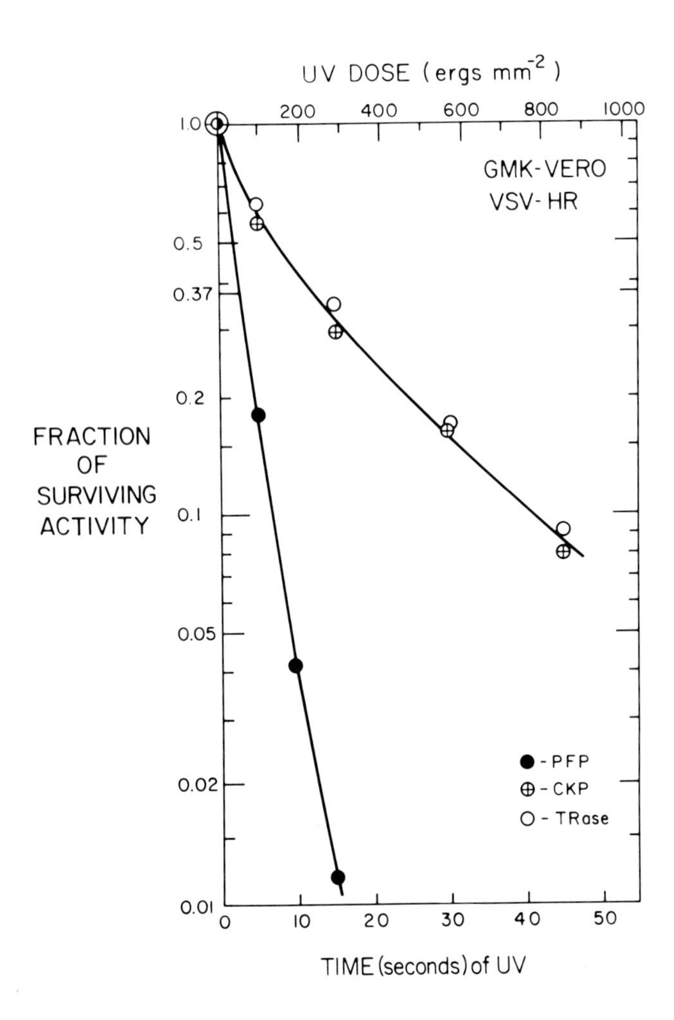

FIGURE 4. Surviving activity curves for GMK-Vero cells infected with VSV-HR previously exposed to UV radiation for the time intervals shown on the abscissa. Survival curves for PFP, CKP, and in vivo virion-associated transcriptase (TRase) activity were determined as previously described[15] and were used as the basis of calculations for surviving activity of each function at different doses of UV radiation. For VSV-PFP, D^o ($1/_e$ dose) = 52.3 ergs mm^{-2} when calibrated against Newcastle disease virus (D^o = 42.5 ergs mm^{-2}) as a biological actinometer. For CKP and in vivo transcriptase activity, D^o = 262 ergs mm^{-2}. (From Marcus, P. I. and Sekellick, M. J., *Virology*, 63, 176, 1975. With permission.)

killing (as CK particles) was determined as a function of UV dose. The results presented in Figure 4 reveal that the capacity of VSV to kill cells is about five times more resistant to UV radiation than is infectivity.[15] It follows from these results that UV radiation can convert a significant fraction of the virions in a VSV preparation into defective CK particles, providing additional evidence that cell killing can be expressed in the absence of complete viral replication, a conclusion already deduced from the high ratio of CK particles: plaque-forming particles in crude stocks and gradient purified preparations.[14] Figure 4 also shows that the virion-associated transcriptase activity of VSV, measured by primary transcript accumulation in cells infected with VSV in the presence of cycloheximide,[20] is also lost at the same rate as CK particle activity. (The value of D_o = 262 ergs mm^{-2} for UV inactivation of virion transcriptase[15] has been confirmed by Dubovi and Youngner[21]). Kinetic studies such as those represented by the UV-surviving activity curves cannot be interpreted rigorously, but they suggest that only about one fifth of the viral genome must remain undamaged and be transcribed in order to express cell killing. These data do not permit a distinction between the need for transcription of a specific or a random sequence of the genome. (We will address this question in the section on cell killing by *ts*-mutants). The absence of any multihit component to the UV-survival curves for CK particle activity tends to preclude a model in which the VSV genome contains more than a single sequence of information that can be expressed (transcribed) independently to produce a putative cell-killing factor. Also, as noted, the capacity of VSV to accumulate primary transcripts in vivo following UV radiation is lost at the same rate as cell-killing particle activity. Because of the relatively low doses of UV radiation used in these experiments, we favor as the inactivation event (hit), a subtle damge to the VSV genome, most likely the formation of uracil dimers, resulting in a block to transcription beyond the site of damage.[22]

These experiments provided the first clue that virion-derived primary transcription might be a required, though not necessarily sufficient, reaction for the expression of cell killing by VSV. Were this the case, and primary transcription the rate-limiting step in cell killing, inactivation of virion-associated transcriptase (L protein) by heat should lead to the concomitant loss of infectivity, cell-killing particle activity, and transcript accumulation. In Figure 5, the upper curve shows that for wild-type virus, these three activities indeed are lost at equivalent rates as a function of time exposed to heat (50°C). (We have determined that this treatment does not affect virus attachment or entry into cells[11]). Were the inactivation of virion-derived primary transcription the rate-limiting synthetic event for cell killing (and in this case for infectivity, too), then heat inactivation of a mutant with an extremely labile virion-associated transcriptase should generate surviving activity curves in which these same three functions are lost at an equal rate, but one significantly faster than that of wildtype virus. VSV tsG114(I) is such a mutant — it is both *ts* and *tl* (thermolabile): the virion-associated transcriptase (L protein) of this mutant is so sensitive to heat that its genetic-based biochemical defect is attributable to failure of the virion-associated enzyme to function at nonpermissive temperature.[6,23] Indeed, the lower curve in Figure 5 demonstrates that these three functions are lost at a rate 6.5 times faster than that observed for wildtype virus (the upper curve).

These results with UV- and heat-inactivated virus provide significant evidence that primary transcription is a requisite, though not necessarily sufficient, event for cell killing. Predictably then, virions or subvirions of VSV deficient in transcribing capacity would not kill cells. This possibility was tested experimentally, as described below.

VI. NONTRANSCRIBING DEFECTIVE-INTERFERING PARTICLES ARE NOT INTRINSICALLY TOXIC

Almost all isolets of DI particles are reported to be deficient in virion-bound transcriptase activity[24] and fail to produce detectable primary transcripts in cells.[25] (An important exception is considered separately below). Nonetheless, these nonreplicating subvirions contain the same five polypeptides present in the complete virion, albeit, proportionately fewer molecules per DI particle, and hence, permit a direct test for intrinsic toxicity of the infecting virion. In Figure 6, the upper curve demonstrates that preparations of thrice-gradient purified conventional DI particles do not kill cells, as tested up to a multiplicity of 70, a result consistent with our hypothesis that transcription is a requisite event for cell killing by VSV. The observation that cells exposed to nontranscribing DI particles are able to divide and produce macroscopically visible colonies, indistinguishable in both efficiency of plating and size from uninfected con-

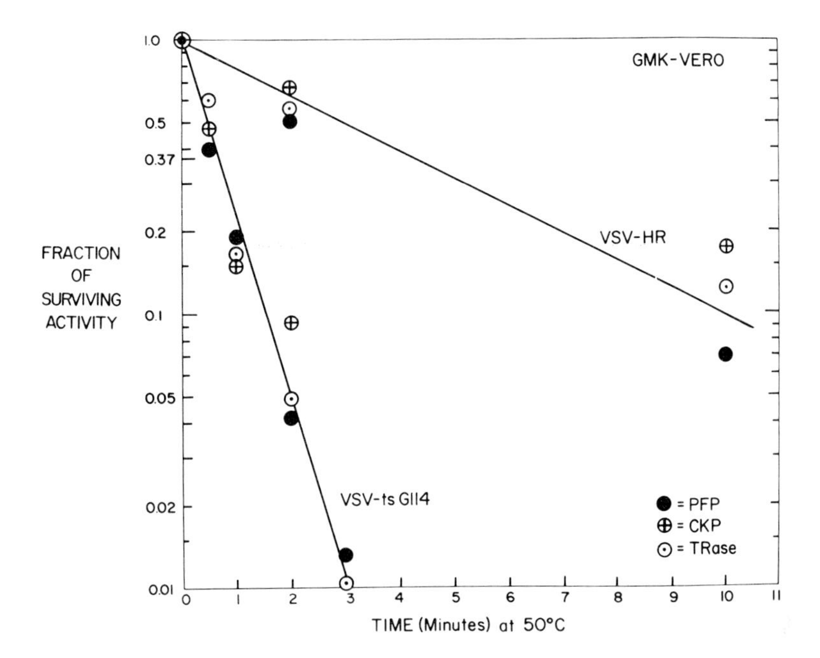

FIGURE 5. Surviving activity curves for GMK-Vero cells infected with VSV-HR (upper curve) or VSV *ts* G114 (lower curve) previously held at 50°C for various intervals. Survival curves for PFP, CKP, and in vivo virion-associated transcriptase (TRase) activity were carried out as previously described[15,20] and were used to calculate surviving activity for each function after exposure of virions to 50°C for various periods of time. All assays for activity were carried out at 30°C. (From Marcus, P. I. and Sekellick, M. J., *Virology*, 63, 176, 1975. With permission.)

trol cells, demonstrates conclusively that these DI particles are not intrinsically lethal to cells. Cooper and Bellett[26] had arrived at a similar conclusion in 1959 based on the absence of a cytopathic effect with VSV stocks prepared from "undiluted" passage of the New Jersey strain. Reports of toxic effects of DI particle preparations have diluted the significance of these earlier findings. The presence of defective CK particles in DI particle preparations at possibly $\geqslant 5$ times the concentration of infectious virus (see Figure 3) would appear to reconcile these disparate observations. Because a *single* defective CK particle suffices to kill a cell, even in the presence of DI particles,[14, 15] it is important to use highly purified populations of DI particles to test for their intrinsic toxicity. Using such preparations, Doyle and Holland[27] could add $\simeq 10^5$ DI particles to a BHK cell with very little deleterious effect! Although a ten-fold increase in multiplicity produced inhibition of cell-protein synthesis and a cytopathic reaction, the significance of this "toxic" effect seems muted, in view of the extraordinarily high multiplicity ($m \simeq 10^6$). Even so, some proponents of intrinsic toxicity for the VSV virion consider these results as supportive evidence.[28,29]

We tested the cell-killing activity of another nontranscribing DI particle — one unique in its potential to form double-stranded *(ds)* RNA within the cell. This particle, DI-011, described by Lazzarini et al.,[30] contains a molecule of covalently linked, self-complementary (±)RNA within the subvirion as a single strand of 8×10^5 daltons in a ribonucleoprotein (RNP) complex. On deproteinization, 80 to 90% of the RNA self-anneals into *ds*RNA. There is good evidence that the single-stranded (±)RNA of the DI-011 particle can achieve some measure of helical structure within the cell, as deduced from the exceptional efficiency of the particles at inducing interferon; a single particle induces a quantum yield of interferon in primary chick-embryo cells aged in vitro;[11] (−)RNA DI particles, like those used in the experiments illustrated in Figure

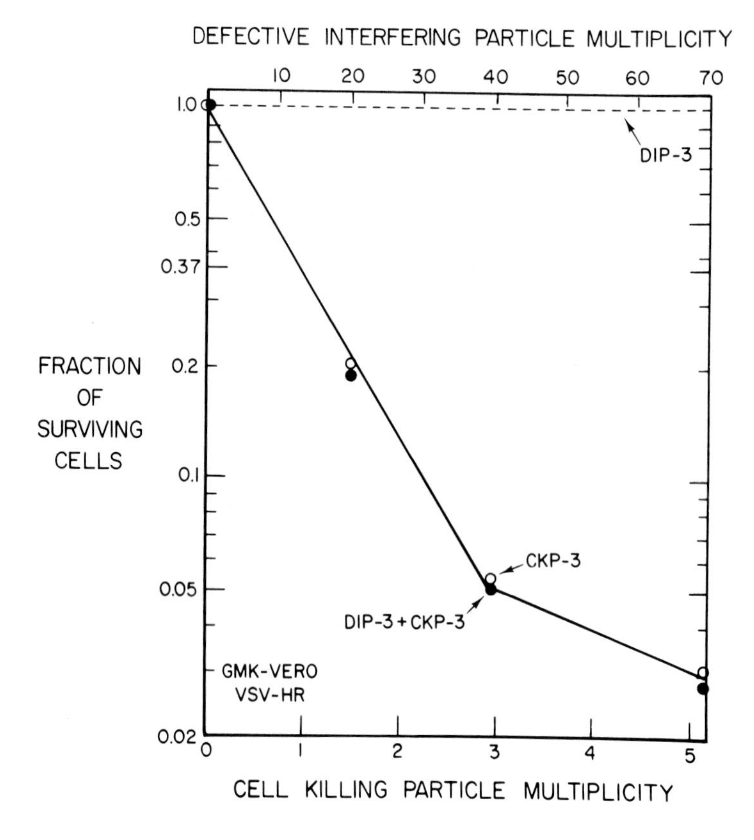

FIGURE 6. Survival curves of GMK-Vero cells exposed to thrice-gradient purified cell-killing particles (CKP-3 at $m_{CKP-3} = 3$), defective-interfering particles (DIP-3 at $m_{DIP-3} = 100$), and a mixture of the two ($m_{DIP-3} = 100 + m_{CKP-3} = 3$). The bottom abscissa applies to the two survival curves generated when CKP are present. The top abscissa relates to the survival curve produced when only DIP-3 is added to the cell population. (From Marcus, P. I. and Sekellick, M. J., *Virology*, 57, 321, 1974. With permission.)

6, did not induce interferon.[11] If some form of *ds*RNA constitutes the putative cell-killing factor of VSV, as normally produced by transcribing virions, then it is conceivable that the (±)RNA DI-011 particles would be intrinsically toxic, i.e., kill cells in the absence of primary transcription. We tested thrice-gradient purified preparations of DI-011 particles and found that they did not affect the plating efficiency of single GMK-Vero or chick-embryo cells, as tested up to $m = 50$.[11] Thus, they behaved like conventional (−)RNA DI particles.[14] It is possible that more than 50 molecules of *ds*RNA are required to kill a cell, even though a single molecule, presumed to form on infection with DI-011, can induce interferon.[11] In some preparations, multiplicities in excess of 50 DI-011 particles resulted in a measureable decrease in plating efficiency and the appearance of a cytopathic effect in monolayer cultures of Vero cells. It is not clear whether these results reflect a cumulative "toxic" effect of these particles, the presence of contaminating defective CK particles, or the cooperative action of complementary (+) and (−) RNAs. We are continuing this study.

Certain conclusions can be derived from the nonlethal nature of individual subvirions of these two kinds of nontranscribing DI particles. Since the cell-killing potential of VSV can reside in a single noninfectious virion (defective CK or UV-inactivated particles), and we can assume that the polypeptidey composition of DI particles is the same as in complete parental (helper) virions,[31,32] it follows that virion proteins per se (and their attendant cellular lipids) are not lethal for cells, as delivered in this multiplicity range, and hence, cannot account for the lethal action of a defective CK particle.

Although DI particles are known to block totally the replication of complete virus,[26,33] they do not interfere with the expression of CK particle activity by coinfecting helper virions. This fact is documented in Figure 6, which reveals that identical cell-survival curves were obtained when Vero cell populations were exposed to CK particles alone, or in the presence of DI particles, conferring homotypic interference. These results are a confirmation in quantitative terms of a conclusion reached by Cooper and Bellett[26] from observing the cytopathic effect of a mixture of DI and plaque-forming particles on chick embryo cells. The expression of cell killing by a single particle of VSV in the presence of homotypic interference takes on added significance when we realize that virion-derived transcription from the coinfecting helper virus goes on unabated during homotypic interference,[34,35] thus satisfying our hypothesis that primary transcription is a requisite synthetic event for cell killing by VSV.[14,15]

Not all DI particles may be inactive with respect to their failure to block the expression of cell killing by coinfecting helper virus. We will consider later a report on the cytopathic effect (CPE)-suppressing activity of rabies DI particles.[36]

VII. CELL KILLING BY *ts*-MUTANTS OF VSV

Integrating our data on the lack of cell killing by nontranscribing DI particles with that from heat-inactivated virions, which implicated virion-associated transcription, and the UV inactivation results, which defined a target for CK particle activity, we deduced that only one fifth of the VSV genome need be intact and transcribed to produce a putative cell-killing factor and express cell killing. In order to determine if the undamaged portion of the genome represented a particular set of genes and hence, a requirement for specific viral proteins, we measured the cell-killing particle capacity of *ts*-mutants from the five complementation groups which define the five structural proteins of VSV (Ind.). We reasoned that if the function of a viral protein was required for cell killing, then *ts*-mutants of complementation groups representing that particular protein would not kill cells at a nonpermissive temperature. On the other hand, if the function of this protein was not critical to the formation of cell-killing factor, then cell-killing particle activity at permissive and nonpermissive temperatures should be essentially equivalent. Figure 7 presents the survival curves of GMK-Vero cells exposed to wild-type (HR-C) VSV or to each of 14 *ts*-mutants at permissive (30°C) and nonpermissive (40°) temperatures. Several conclusions are apparent from these results and are considered below.

A. Wild-type Virus

In Figure 7, the survival curves for Vero cells infected with wild-type VSV, the HR-C strain,[14] show that cell-killing particle activity is almost the same at permissive and nonpermissive temperatures. Over several experiments,[15] we observed that the slight decrease in the slope of the survival curve at the higher temperature represents less than a 25% loss of CK particle activity. Clearly, wild-type virus can kill cells with almost equal efficiency at 30°C and 40°C.

B. Complementation Group I (L protein)

Four *ts*-mutants from complementation Group I were tested: *ts* G114, *ts* G11, *ts* T1026, and *ts* 05. As already mentioned, the virion-associated transcriptase of mutant *ts* G114 is extremely heat labile, to the extent that primary transcripts do not accumulate in the infected cell held at nonpermissive temperature.[15,37,38] Therefore, phenotyp-

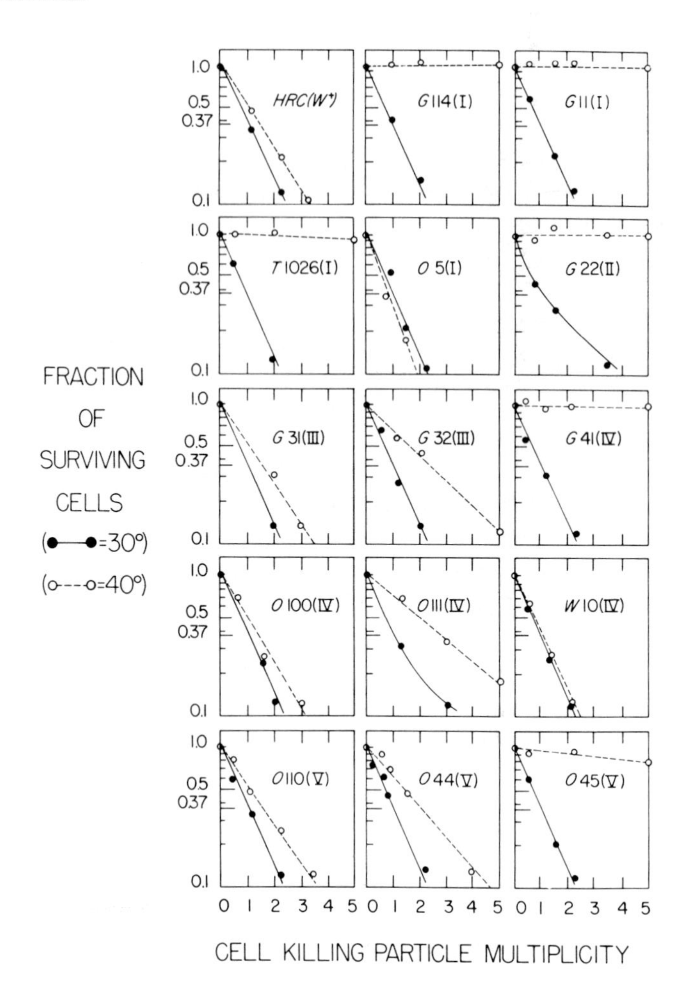

FIGURE 7. Survival curves of GMK-Vero cells exposed to low multiplicities of ts-mutants representing the five complementation groups and wildtype VSV *(Ind.)*. Experimental details are outlined in Figure 1 and described in the legend to Figure 2. (From Marvaldi, J. L., Lucas-Lenard, J., Sekellick, M. J., and Marcus, P. I., *Virology*, 79, 267, 1977. With permission.)

ically, *ts* G114 qualifies as primary transcription-negative, a condition we will denote as *tra*⁻. In contrast, mutants *ts* 05, *ts* G11, and *ts* T1026 appear to be phenotypically *tra*⁺, producing measurable amounts of primary transcripts in vivo at nonpermissive temperature.[38-40] Figure 7 shows that individual virions of mutants *ts* G114, *ts* G11, and *ts* T1026 do not kill GMK-Vero cells at 40°, whereas those of mutant *ts* 05 do. Similar results were obtained with mouse L cells as host.

These results appear to represent an anomalous situation: first, because they indicate that L protein (the transcriptase), coded for by one half of the viral genome, may be required for cell killing — even though the UV-target determination clearly indicated that only one fifth of the genome need be intact; and secondly, because with respect to CK particle (CKP) activity, different mutants from the same complementation group reveal contrasting phenotypes at nonpermissive temperature, i.e., *ckp*⁺ (*ts* 05) and *ckp*⁻ (*ts* G114, G11, and T1026). We can resolve these apparent anomalies if we assume that the requirement for L protein reflects the need for virion-associated molecules which are minimally functional, rather than a need for newly synthesized protein. Thus, *ts* G114 is not lethal, because its virion-associated L protein is nonfunctional *(tra⁻)* at 40°C (this predicts that all mutants which display the tra⁻ phenotype will also be *ckp⁻*). Because *ts* G11 and T1026 are phenotypically both *tra⁺* and *ckp⁻*, we conclude that primary transcription per se does not suffice to express the *ckp⁺* phenotype, and that the virion-associated L protein of these two mutants lacks some minimal function (beyond that of transcriptase activity) required to express the *ckp⁺* phenotype. However, the *tra⁺*, *ckp⁺* phenotype of mutant *ts* 05 demonstrates that its virion-associated L protein possesses that minimal function (termed the 05L function) required for cell killing. Even though the L protein of *ts* 05 is still defective with respect to a function needed for complete replication *(rep⁺)* and production of plaque-forming particles *(pfp⁺)*, it is minimally functional (05L⁺) with respect to the *ckp* phenotype. This concept will be considered in greater depth in a discussion of cell killing by the *ts* -mutants summarized in a section below.

C. Complementation Groups II (NS Protein) and IV (N Protein)

Figure 7 shows that the Group II *tra⁺* mutant *ts* G22 does not kill cells at 40°C, thus demonstrating that functional NS protein is required for expression of the *ckp* phenotype. In contrast, Group II mutant *ts* 052 did kill cells at 40°,[127] indicating that the NS protein of *ts* 052 possessed a minimal function (the 052NS function) required for cell killing.

Group IV mutant *ts* G41 does not kill cells at 40°, thus identifying functional N protein as required for cell-killing particle activity. However, the N protein of three other Group IV mutants, *ts* W10, 0100, and 0111, appears to possess the appropriate configuration to function phenotypically as *tra⁺*, *ckp⁺*. We have termed this function W10N, after the first Group IV mutant to display it. In addition to identifying minimally functional proteins N and NS as essential for the expression of cell killing by VSV, these studies permit genotypic resolution within a complementation group, in terms of the *ckp* phenotype. Also, we note that genes N and NS constitute about one fifth of the coding capacity of the VSV genome and hence, would be equivalent to the portion of the genome which must remain undamaged for the expression of cell killing. This interpretation gains added significance, because genes N and NS constitute the two coding sequences most proximal to the manditory initiation site for transcription at the 3′ terminus of the genome, and hence, the only two genes whose combined radiation target would equal one fifth of the genome.[41,42]

The N protein is synthesized as a soluble protein, assembled into nucleocapsids which accumulate in the cytoplasm, and eventually bind to M- and G-containing membranes in the budding process.[93] Knipe et al.[98] showed that the N protein encoded by the *psi⁻*, *ckp⁻* mutant *ts* G41(IV) was degraded much faster than the wildtype protein. It would be of interest to know whether the N protein of the *psi⁺*, *ckp⁺* mutant *ts* W10(IV) is stable at nonpermissive temperatures. However, since *ts* G41(IV) and *ts* W10(IV) both fail to amplify RNA synthesis,[15,39,59,99] i.e., are *rep⁻*, the formation of cell-killing factor and the expression of the *psi⁺*, *ckp⁺* phenotype does not appear to require RNA replication.

D. Complementation Groups III (M Protein) and V (G Protein)

Group III (M protein) mutants *ts* G31 and G32 both kill cells with a significantly high efficiency at 40°, indicating that the matrix protein M need not be functional for the expression of cell killing (Figure 7). Group V (G protein) mutants *ts* 0110 and 044 also kill cells at 40°, indicating that cell killing does not require functional glycoprotein G (Figure 7). However, the almost complete reduction in CK particle activity of mutant *ts* 045(V) at 40° remains anomalous; it might simply be due to the thermolabile (tl) character of the G protein defect[43] and failure of individual virions to achieve proper attachment or entry under our experimental conditions where $m_{PFP} \leqslant 5$ (See Figure 1).

This explanation seems creditable in view of the observations by Knipe et al.[98] who reported that another Group V mutant, *ts* M502, appeared to synthesize very little RNA or protein at a nonpermissive temperature — even after a few hours at 31°. These studies too were carried out at relatively low multiplicities (m = 10).

Cell killing by a single particle of UV-irradiated VSV which has sustained one or more hits, with respect to expression (transcription) of the G protein gene (but none to genes N and NS),[15,41] demonstrates that newly synthesized G protein is not required for the formation and action of cell-killing factor. (The same consideration applies to the M protein). It follows that the presence of new G protein in intracellular membranes, or its dynamic flow in the membrane systems of the cells,[93,94] is not necessary for expression of the *psi, ckp* phenotype. Conceivably, the approximately 200 molecules of G protein in the infecting virion may seed the plasma membrane[95] and somehow serve a modulating role in cell killing; however, if so, it is difficult to understand why the equivalent of about 10^6 molecules per cell has little or no effect on host protein synthesis, as reported by McSharry and Choppin.[96] In this context, McSharry and Choppin[96] point out that some *ts*-mutants of VSV which inhibit protein synthesis and kill cells at nonpermissive temperatures also produce large amounts of G protein. They suggest that these results are compatible with some role for G protein in the inhibition of cellular macromolecular synthesis and cytopathic effects. However, two additional observations tend to blunt this argument: (1) some *ts*-mutants which do not inhibit protein synthesis or kill cells nonetheless produce large amounts of viral G (and all other) protein(s)[97] and (2) two mutants with *psi*$^+$, *ckp*$^+$ phenotypes, *ts* 05(I) and *ts* W10(IV), synthesize very small amounts of G protein compared to wild-type or other *psi*$^+$, *ckp*$^+$ mutants[17] (Figure 7). We think the bulk of the evidence indicates that newly synthesized G protein is not required to produce the pleiotypic effects attributable to VSV infection. Whether cell killing requires the 200 molecules of G protein associated with the infecting virion is still a moot point. However, the absence of the budding process and new virion producton in cells infected with *ts* G31(III)[39] or a similar mutant, *ts* M301(III),[94] suggests that the *psi*$^+$, *ckp*$^+$ phenotype does not require these maturation processes for expression. Nucleocapsid formation appears normal for these mutants.[39,94]

VIII. PROTEIN SYNTHESIS INHIBITION BY *ts*-MUTANTS OF VSV

Inhibition of macromolecular synthesis is a hallmark of VSV action on host cells.[44] In particular, cellular protein synthesis inhibition *(psi)* has been well documented.[17,37,45-49,60] Conceivably, a *psi*$^+$ phenotype might be related to the expression of cell killing, although protein synthesis inhibition per se through drug action certainly does not result in the precipitous cytopathic effect which characterizes infection by VSV. We determined the *psi* phenotypes for representative mutants from each complementation group for comparison with the *ckp* phenotype. Mouse L cells were infected at m_{PFP} = 100 for 4.5 hr at 30° or 40° and then labeled isotopically with a ^{3}H-amino acid mixture for 30 min at those respective temperatures. Proteins were extracted from the infected cells, processed by polyacrylamide slab-gel electrophoresis, and analyzed following densitometer scans of fluorographic patterns. All *ts*-mutants incubated with cells at a permissive temperature, 30°, brought about inhibition of cellular protein synthesis in a manner indistinguishable from that of wild-type virus (data not shown[17]), and all produced patterns of virus-specific components identifiable on the scanning profiles as proteins M, N + NS, G, and L, by comparison with wildtype marker proteins (See Figure 8, the bottom left-hand panel). In contrast, incubation of mutant virus-cell complexes at a nonpermissive temperature, 40°, revealed clear differences in the expression of the *psi* phenotype. Typical results are illustrated in Figure

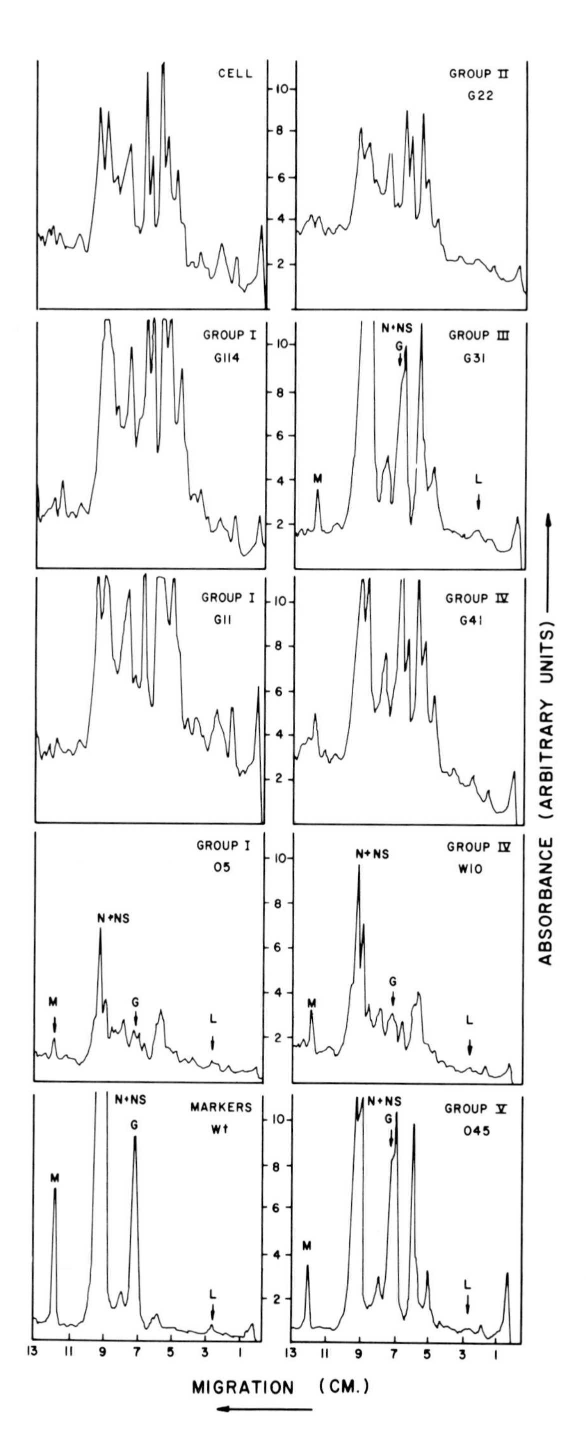

FIGURE 8. Gel electrophoresis of proteins from mouse L cells infected with ts-mutants representing the five complementation groups and wildtype VSV (IND). The scanning profiles of fluorographic patterns represent protein synthesis during a 30 min period 4.5 hr postinfection at m_{PFP} = 100 and incubation at nonpermissive temperature, 40°C. In all tests at permissive temperature, 30°C, the scanning profiles of all the ts-mutants were the same as that shown here for W⁺ virus at 40°C. (From Marvaldi, J. L., Lucas-Lenard, J., Sekellick, M. J., and Marcus, P. I., *Virology*, 79, 267, 1977. With permission.)

8, including a control of uninfected cells. Several mutants, *ts* 05(I), G31(III), W10(IV) and 045(V), displayed fluorographic patterns like those obtained from cells infected with wild-type virus and, hence, qualified as ts^{PSI+}. As expected, the *psi*$^+$ mutants which were defective in RNA replication *(rep*$^-$*)*, i.e., *ts* 05(I), and W10(IV), produced smaller amounts of viral proteins than wild-type VSV or the *rep*$^+$ mutants *ts* G31(III) and *ts* 045(V). The remaining mutants, *ts* G114(I), G11(I), G22(II), and G41(IV), generated fluorographic profiles indistinguishable from uninfected cells, i.e., failed to inhibit host protein synthesis and hence, qualified as ts^{PSI-}.

For every mutant tested, with the reservation for *ts* 045(V) noted in the preceding section, there was an exact correlation between cellular protein synthesis inhibition, measured at m_{PFP} = 100, and CK particle activity scored by the action of single virions. This correlation extended to a subdivision of the complementation groups and revealed that the *psi* and *ckp* phenotypes are conjointly expressed.

IX. CELL KILLING AND PROTEIN-SYNTHESIS INHIBITION BY VSV REQUIRE THE SAME GENE FUNCTIONS

The data in Table 1 summarize our studies on cell killing and protein-synthesis inhibition by *ts*-mutants of VSV, and in addition, include information on *psi* phenotypes from a study by McAllister and Wagner.[37] Also identified in the table are mutants which Dubovi and Youngner report inhibit the replication of pseudorabies virus.[21] Although the mechanism underlying this interesting form of viral interference is not understood, the interference with pseudorabies virus replication (yield) by VSV appears to correlate with the *psi*$^+$, *ckp*$^+$ phenotype. These data reveal that the *psi*$^+$, *ckp*$^+$ phenotypes of VSV are expressed coordinately — even within a complementation group divisible by these two functions. Thus far, all *ts*-mutants may be classified either as $ts^{PSI+,CKP+}$ or $ts^{PSI-,CKP-}$, allowing us to conclude that the expression of cell killing and protein-synthesis inhibition require the same gene functions — identified as the products of genes N, NS, and L. Since the small target size for CK particle activity (Figure 4) appeared to rule out a requirement for newly synthesized L protein, this function was relegated to the molecules formed at permissive temperature and contained within the virion. This conclusion was proved correct in experiments described in the next section.

X. CELL KILLING BY A TRANSCRIBING DEFECTIVE-INTERFERING PARTICLE, DI-LT

Although conventional DI particles of VSV contain a (−)RNA molecule and the same polypeptides as wild-type virus,[31,32] including the transcriptase,[52] they are phenotypically transcription-negative *(tra*$^-$*)*; and, in keeping with our postulate that primary transcription is required for cell killing,[14,15] they are phenotypically *ckp*$^-$ (Figure 6). However, the availability of a transcribing *(tra*$^+$*)* DI particle would permit a more direct test of the requirement for transcription and, depending upon the portion of the genome deleted, a confirmation of the genes involved in cell killing. Such a DI particle was recently described by Colonno et al.[9] and Johnson and Lazzarini.[10] These workers established that a long DI particle of VSV *(Ind.)*, DI-LT, was capable of transcribing leader RNA[53] and the four mRNAs coding for proteins, N, NS, M, and G. Furthermore, since DI-LT particles lack most of the gene for L protein,[54,55] no new L protein could be synthesized, and only virion-derived molecules would be available.

We compared the CK particle activity of purified preparations of the *tra*$^+$ DI-LT particle with that of wild-type virus and two *tra*$^-$ DI particles: a conventional (−)RNA DI-HR and the unusual interferon-inducing (±)RNA DI-011 which contains covalently linked self-complementary RNA.[11,18,30] The photomicrographs in Figure 9 illustrate the

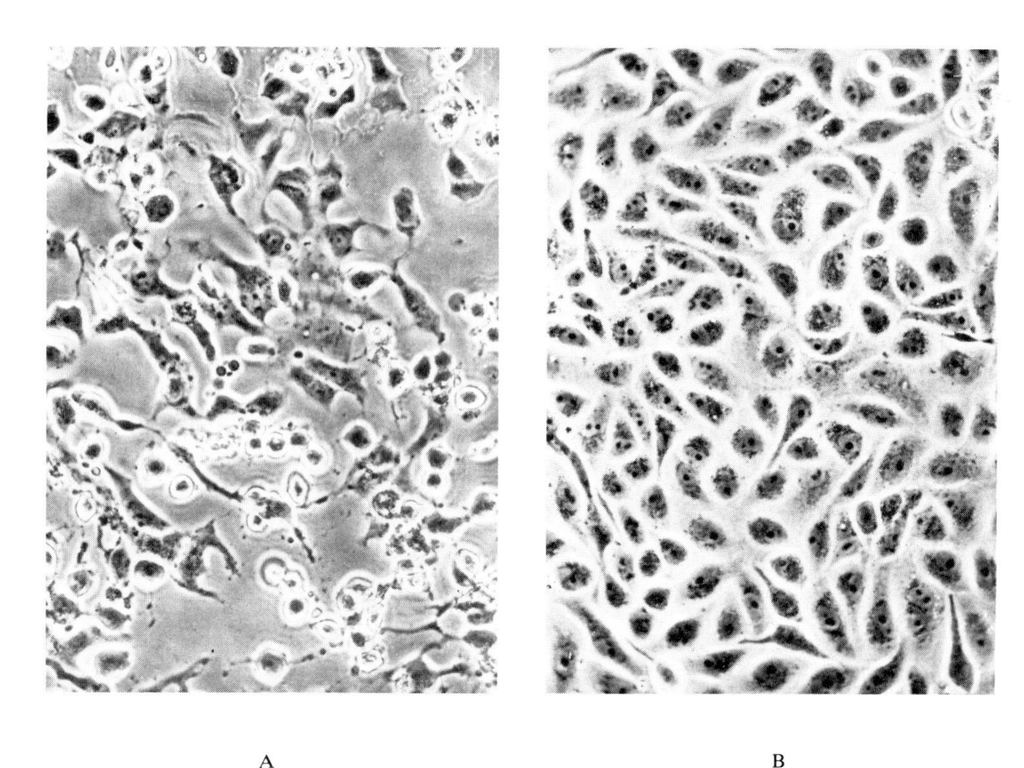

A B

FIGURE 9. Photomicrographs showing the cytopathic effect produced in GMK-Vero cells following infection with (A) the heat-resistant (HR) wild-type VSV at m_{PFP} = 8; (B) DI-HR particles at m_{DIP} = 50; (C) DI-011 particles at m_{DIP} = 50; or (D) DI-LT particles at m_{DIP} = 50, the equivalent of m_{CKP} = 8. The plaque-forming particle (PFP) (= cell-killing particle [CKP]) multiplicity of the wildtype-HR preparation was chosen to be equivalent to the cell-killing particle multiplicity in the DI-LT particle preparations when the defective-interfering particle (DIP) multiplicity was 50. Infected cells were incubated at 37°C for 21 hr prior to photomicrography. Uninfected cells (not shown) were indistinguishable in appearance from those shown in (B) or (C). (From Marcus, P. I., Sekellick, M. J., Johnson, L. D., and Lazzarini, R. A., *Virology*, 82, 242, 1977. With permission.)

cytopathic effect (CPE) brought about by infection of GMK-Vero cells with wild-type VSV and these three different kinds of DI particles. When antiserum to VSV *(Ind.)* was added to the medium to restrict infection solely to that of input virus, the CPE developed at essentially equivalent rates in cells infected with either wild-type virus or DI-LT subvirions, but not at all in cells infected with the *tra*⁻ DI-HR or DI-011 particles. This experiment graphically demonstrates that there are particles in the DI-LT preparation not present in *tra*⁻ DI particle stocks, which can induce a cytopathic effect indistinguishable in its appearance and rate of development from that of a fully replicating wild-type virus. We conclude that DI-LT particles must contain those molecules and functions required to produce putative cell-killing factor.[18]

The stock of purified DI-LT particles was tested quantitatively by the single-cell survival procedure for its capacity to kill cells. The results of two such experiments are presented in Figure 10 and compared with experiments in which the two different *tra*⁻ DI particles were used to challenge cells. As revealed by the zero slope of the single-cell survival curves, the two *tra*⁻ particles (DI-HR and DI-011) do not kill cells — as tested up to a multiplicity of about 100 DI particles per cell. In marked contrast, DI-LT particle interaction with Vero cells generates a survival curve characteristic of one-particle-to-kill kinetics.[18]

Comparative assays of defective-interfering, plaque-forming, and physical particles with cell-killing particle activity indicate that about 60% of the DI-LT physical parti-

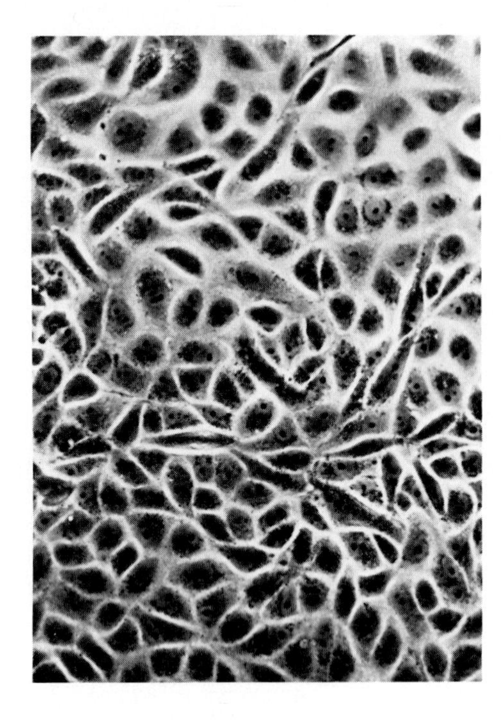

FIGURE 9C

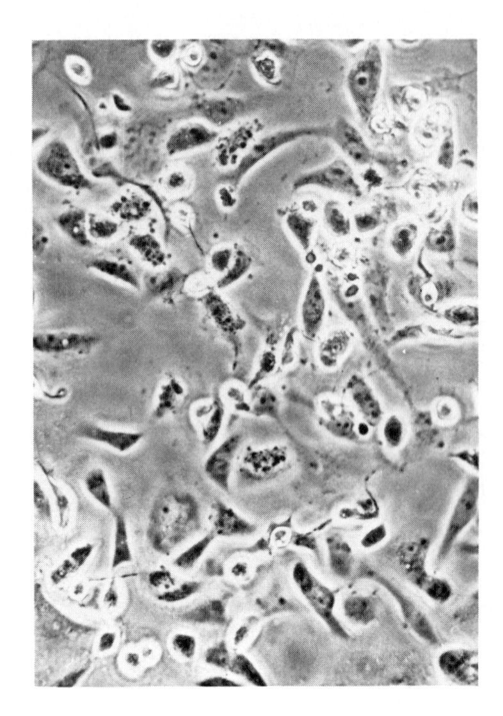

FIGURE 9D

cles are biologically functional as DI particles; whereas about 10% appear to possess cell-killing particle activity. This suggests that only one of every six biologically active DI-LT particles can kill cells, or that all particles are active, but there is only a one in six chance that a given DI-LT particle-cell interaction results in cell killing. We favor the former view, however it is possible that every DI-LT subvirion possesses the potential to function as a defective CK particle, but the absence of newly synthesized L protein acts to limit the formation of putative cell-killing factor. Since, on a particle basis, the rate of primary transcription is the same as for wild-type VSV,[9,10] we infer that the 50 or so molecules of virion-associated L protein function with normal efficiency for transcription, but that expression of the 05L function may be rate limiting and thereby, account for the observed low titer of CK particle activity. Regardless of the actual situation, these data demonstrate that a single DI-LT particle, capable of transcribing genes N, NS, M, and G, but missing most of gene L, is capable of functioning as a cell-killing particle. The absence of an intact gene for L protein in DI-LT particles provides direct proof that newly synthesized L protein is not required for the expression of cell killing by single virions. Most likely, the 50 or so molecules of L protein in the virion suffice to form cell-killing factor, but the addition of newly synthesized L protein may serve to increase the probability of formation to one.

XI. A MODEL FOR THE FORMATION OF A PUTATIVE CELL-KILLING FACTOR BY VESICULAR STOMATITIS VIRUS

A. The Experimental Basis

Conclusions drawn from the data offered above, and the more detailed presentations of experiments in previous publications that relate to cell killing by VSV[11,14-19,56] are summarized below, along with a brief statement describing the basis for the conclusion. A model for the formation of a putative cell-killing factor follows thereafter.

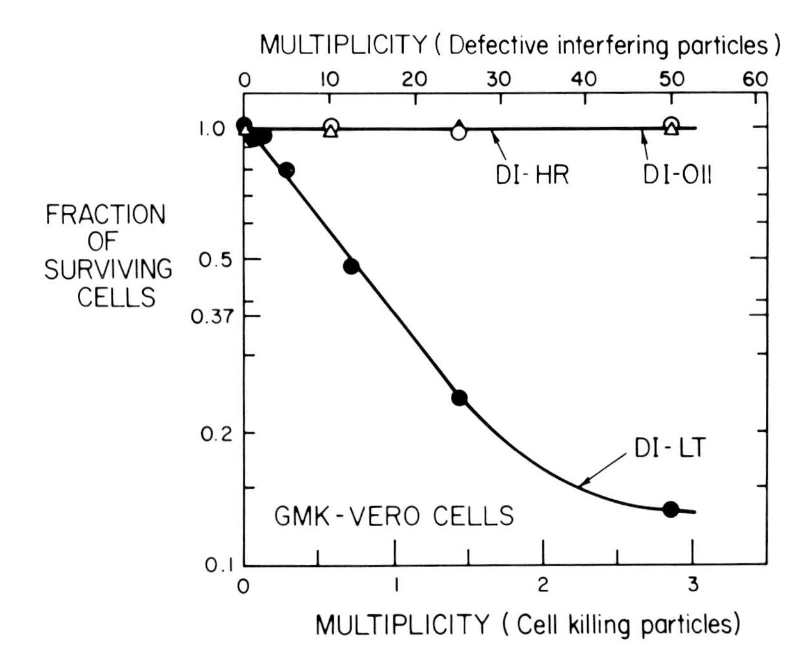

FIGURE 10. Survival curves of GMK-Vero cells infected with DI-HR, DI-011 or DI-LT particles. The top abscissa refers to the survival curves produced when DI-HR or DI-011 particles were used, and the bottom abscissa relates to the survival curve produced when DI-LT particles were used. A 1:3 × 10^4 dilution of stock virus on a monolayer of 1.7 × 10^6 cells contained, on the average, one cell-killing particle. The experimental points represent the average of two experiments. Individual points were within ±20% of the mean value for the fraction of surviving cells (colony counts) for each determination. Colony counts were carried out in duplicate, representing on the average about 80 surviving colonies per plate. (From Marcus, P. I., Sekellick, M. J., Johnson, L. D., and Lazzarini, R. A., *Virology*, 82, 242, 1977. With permission.)

Cell killing by a single particle of VSV does not depend upon infectivity or complete viral replication — Single-cell survival tests reveal that VSV stock preparations frequently contain a five- to ten-fold excess of cell-killing to plaque-forming particles. Noninfectious particles capable of killing cells have been termed defective CK particles.[14] UV radiation can create defective CK particles from infectious CK particles.[15]

Nontranscribing DI particles do not kill cells — Analyses of DI, CK, and PF particle activity profiles in velocity sedimentation gradients of VSV reveal that standard preparations of DI particles can be contaminated with defective CK particles, presumably accounting for the so-called "toxic" effect of DI particles. Successive gradient purifications rid the DI preparations of these cell-killing particles and result in a population of DI particles which does not kill cells, yet shows no dimunition in its biologic capacity as measured by homotypic interference.[14]

Primary transcription is a requisite, though insufficient, event for the expression of cell killing — Cell-killing particle activity and transcript accumulation in vivo are both lost at essentially equivalent rates upon subjecting the virion to: (1) heat at 50°C, or (2) UV radiation, (3) following infection of interferon-treated cells, or (4) by mutants with a thermolabile virion transcriptase. Furthermore, cell killing by VSV is expressed in viral interference under two different conditions where primary transcription is extant: (1) in DI particle-induced homotypic interference[14] and (2) in the heterotypic intrinsic interference,[57] but not under conditions where virion-directed transcription is inhibited significantly by the interferon system.[16,20,58,59] Finally, since several transcribing *tra*$^+$) *ts*-mutants do not kill cells (Table 1), primary transcription per se does not result in the formation of cell-killing factor.

TABLE 1

Comparison of Phenotypic Activities of Vesicular Stomatitis Virus *ts*-Mutants

Complementation group	*ts* Mutant designation	Phenotype (at nonpermissive temperature)[a]						Ref.
		tra[b]	*psi*[c]	*rni*[d]	*pho*[e]	*psr*[f]	*ckp*[g]	
Wild-type	W^+ *(Ind.)*	+	+	+	+	+	+	6, 7
I	G114	−	−	−	−	(−)	−	6, 7
(L protein)								
	G11	+	−	−	(−)	−	−	
	T1026[h]	+	−	−	(−)	(−)	−	
	G12	+	(−)	(−)	(−)	(−)	−	
	G13	+	−	(−)	(−)	−	−	
	G16	(+)	(−)	(−)	(−)	−	(−)	
	05	+	+	+	(+)	+		
II	G22	+	−	−	−	(−)	−	6
(NS protein)								
	052	+	+	(+)	(+)	+	+	
III	G31	+	+	+	(+)	(+)	+	6
(M protein)								
	G32	+	(+)	(+)	(+)	(+)	+	
	023	+	(+)	(+)	(+)	+	+	
	T54	+	(+)	(+)	(+)	(+)	+	
IV	G41[i]	+	−	−	−	−	−	13
(N protein)								
	W10	+	+	+	+	(+)	+	
	0100	+	(+)	(+)	(+)	(+)	+	
	O111	+	(+)	(+)	(+)	(+)	+	
	W16B	+	(+)	(+)	(+)	(+)	+	
	G44	+	+	(+)	(+)	(+)	+	
V	O44	+	(+)	(+)	(+)	(+)	+	6
(G protein)								
	O110	+	(+)	(+)	(+)	(+)	+	
	O45[j]	+	+	(+)	(+)	+	+	

Note: The + and − designations represent our own evaluation of data obtained from the following sources: *tra*,[6,15,16,20,23,37-40,52,59,113] *psi*,[17,37] *rni*,[133] *pho*,[85] *psr*,[21] and *ckp*.[14-18,134] The parentheses around the + and − designations signify these phenotypes were not determined, and that we predict the characteristic shown.

[a] The + and − designations denote the presence or absence, respectively, of measurable activity for each virus at a nonpermissive temperature.

[b] *tra* = primary transcription[20]

[c] *psi* = protein synthesis inhibition[17]

[d] *rni* = RNA inhibition[133]

[e] *pho* = phosphorylation (protein kinase activity)[85]

[f] *psr* = pseudorabies virus inhibition[21]

[g] *ckp* = cell-killing particle activity[14-18,126]

[h] tsT1026(I) appears to be a double mutant, with a nontemperature-sensitive defect for the *psi* function — termed P function by Stanners et al.[60]

[i] tsG41(IV) appears to leak at multiplicities exceeding ten plaque-forming particles, especially at temperatures < 40.0°C

[j] See Section VII.D. for a discussion of the anomalous behavior of ts045(V) with respect to the *ckp* phenotype.

Only one fifth of the viral genome is required for the expression of cell killing — Surviving activity curves for CK particles following UV irradiation demonstrate that the capacity of VSV to kill cells is about five times more resistent to UV radiation than is infectivity. It follows that about one fifth of the viral genome must remain undamaged, presumably in order to produce primary transcripts.

Transcription of genes N and NS, representing one fifth of the genome, and their translation into minimally functional proteins N and NS are required for cell killing — VSV genes N and NS not only constitute one fifth of the genome length, but they are the two genes most proximate to the single initiation site at the 3′ end of the genome and therefore constitute the only sequence of two genes that would present a target size proportional to their combined molecular weights — as was observed.[41,42] The absence of cell killing by some mutants of complementation Groups II (NS protein) and IV (N protein) at nonpermissive temperature indicates that the mRNAs of these two viral genes must produce polypeptides that are minimally functional in order to bring about cell killing (Table 1).[15,17]

Newly synthesized L protein is not required for cell killing: virion-associated molecules suffice — The transcription-positive *(tra⁺)* DI-LT particles contain intact nucleotide sequences for leader RNA, and for genes N, NS, M, and G. All four genes are transcribed in the infected cell, and presumably produce functional proteins.[9,10] However, although DI-LT particles lack almost all of the gene for L protein, a single particle can nonetheless kill cells (Figure 10).[18] We conclude that newly synthesized L protein is not required for the formation of putative cell-killing factor and that the virion-associated L protein can successfully fulfill this role.

Cell killing requires, in addition to transcriptase activity (tra$^+$), a second function of virion-associated L protein, the 05L function (05L$^+$) — The *tra*$^+$ mutants *ts* G11 and *ts* T1026 of complementation Group I (L protein) do not kill cells or inhibit host protein synthesis at nonpermissive temperatures,[17,60] implying that at least one additional function of L protein is required beyond that of transcribing genes N and NS. The lethal action of Group I mutant *ts* 05 at 40°C reveals that activity and demonstrates that a second function of L protein is required for cell killing, namely that resident in the *ts* 05 L protein — termed the 05L function. Since no new L protein is synthesized by the cell-killing DI-LT particles, we conclude that the two functions, *tra*$^+$ and 05L$^+$, need reside only in the molecules of L protein associated with the virion.

Detectable levels of viral RNA replication are not required for cell killing — Since mutant *ts*05(I), and possibly *ts*W10[IV] and 052[II], kills cells at 40°C, and yet appears to be deficient in viral [−] RNA synthesis,[38] we conclude that the expression of cell killing does not require viral replication (replicase activity; *rep* function) — at least not as defined by presently *detectable* levels of RNA synthesis.[38] This caveat is introduced out of respect for the capacity of a single molecule of dsRNA to induce a quantum yield of interferon.[11]

The viral [−] RNA genome — Thus far, no experimental condition has eliminated the need for the viral [−] RNA genome in the expression of cell killing by VSV. Indeed, it is required, at the minimum, to code for newly synthesized transcripts of proteins N and NS.[15,17] The cell-killing capacity of DI-LT particles[18] proves that deletion of almost one half of the genome at the 5′-end is not a deterrent to their lethal action. Possibly, conservation of a minimal portion (negligible UV target) of the 5′-end of the genome may be required for recognition by the 05L-N-NS complex. Certainly the leader RNA sequence at the 3′-end (a very small target for UV radiation[61]) is required to initiate transcription of genes N and NS.

B. A Model for Cell-Killing Factor Formation

We postulate that inhibition of cell macromolecular synthesis, the progressive development of a cytopathic effect, and the eventual demise of the cell following infection by vesicular stomatitis virus (and perhaps other rhabdoviruses) results from the formation of a putative cell-killing factor and its subsequent action through pleiotypic effects. The sequence of events that result in the formation of this putative cell-killing factor are presented below. In concert, they represent a model based on our present knowledge, and constitute a working hypothesis against which the results of others and future results can be tested. The identification of cell-killing factor in molecular terms and its mode of action await further study. Cell-Killing Factor Formation by a Single Particle of Vesicular Stomatitis Virus Requires:

1. The [−] RNA viral genome — with intact sequences for leader RNA at the 3′-end, genes N and NS, and an intact 5′-end
2. Virion-associated L protein — as a functional transcriptase *(tra⁺)*
3. Transcription of genes N and NS and the synthesis of minimally functional proteins N and NS
4. Interaction of proteins N and NS with a second activity of virion-associated L protein, the 05L⁺ function; normal viral RNA replication (the replicase function, *rep⁺)* not required
5. Proximate or actual cell-killing factor may consist of a complex of proteins N + NS + 05L, a product of their interaction with the 5′-end of the genome (−)RNA (namely dsRNA) or reflect enhanced protein kinase activity and hyperphosphorylation of cellular proteins
6. The formation and subsequent action of cell-killing factor presumably triggers the pleiotypic effects characteristic of VSV infection and culminates in cell killing

XII. DOUBLE-STRANDED RNA AS PROXIMATE CELL-KILLING FACTOR

It is clear from our results from several $ts^{PSI+,CKP+,REP-}$ mutants that cell-killing factor can be produced in the "absence" of viral RNA replication or the appearance of replicative intermediates, at least as assessed by presently detectable levels of synthesis.[38] Why then pursue a case for *ds*RNA as proximate cell-killing factor? From our view there are several reasons. Perhaps the most compelling derives from the recent demonstration that the formation in a cell of but a *single* molecule of *ds*RNA results in a quantum yield of interferon.[11] Clearly, cells are exquisitely tuned to respond to helical RNA. Conceivably then, input [−]RNA genome and [+] RNA transcripts synthesized from it could interact in the presence of a 05L + N + NS complex to produce an aberrant or abortive replication event, thereby creating one or a few molecules of *ds*RNA. Although these few molecules may escape detection by physical means, their presence may be inferred from the capacity of a single virion to induce a quantum yield of interferon under appropriate conditions.[11] Using the exquisitely sensitive induction system of primary chick-embryo cells aged in vitro,[62] we have obtained evidence for the intracellular formation of *ds*RNA by VSV. We have observed that at nonpermissive temperatures *ts*-mutants of VSV that are positive for primary transcription, but deficient in protein synthesis inhibition, cell killing, and replication ($ts^{TRA+,PSI-,CKP-,REP-}$), are nonetheless extremely efficient inducers of interferon; a single virion suffices to induce a quantum yield of interferon.[63] This latent capacity of a potentially infectious virion to function as an interferon-inducing particle *(ifp⁺* phenotype) is revealed only when the *psi* and *ckp* phenotypes are not expressed, or are delayed in expression. We interpret these results to mean that under the conditions of interferon induction by ts^{IFP+} mutants, at least a single molecule of *ds*RNA was formed — enough to trigger the induction of interferon mRNA synthesis, but not enough to constitute the threshold amount we postulate is required to activate the progressive cascade of effects that ultimately result in inhibition of host macromolecular synthesis and cell killing. There are other reasons that recommend *ds*RNA as a candidate molecule for proximate cell-killing factor. Double-stranded RNA has been shown to have a profound effect on protein synthesis both in vitro (lysates from reticulocytes[64,65] or interferon-treated cells[66,67]) and in vivo,[68] and has been implicated in the activation of an endonuclease in interferon-treated cells.[69] Also, there are clear demonstrations of overt cytotoxic effects of *ds*RNA, albeit at quite high concentrations, when added to the cell externally in the form of naked RNA[70,71] and, in the special case of enhanced cytotoxicity to *ds*RNA, in certain interferon-treated cells.[72] The myriad effects of *ds*RNA on biological systems has been recognized by Carter and De Clercq.[73] We support their view that *ds*RNA may assume a modulatory role in the virus-infected cell, and would add that the recognition-response system of the cells for *ds*RNA formed within the cell may be orders more sensitive than heretofore suspected.[11]

In this context, the "absence" of *ds*RNA in virus-infected cells under conditions where macromolecular synthesis may be impaired significantly, or CPE overtly manifested, may be more apparent than real, and simply reflect our present limitations of detection. Thus, the approximately 150 molecules of viral *ds*RNA that Collins and Roberts[74] calculate may be present under particular restrictive conditions of mengovirus replication, but where host-cell RNA and protein synthesis are inhibited, loom large when viewed against the extraordinary effectiveness of one molecule of *ds*RNA to induce interferon.[11] Furthermore, there is reason to conclude from the experiments of Marcus and Sekellick[11] that a cell can recognize (count?) the number of *ds*RNA molecules introduced into it and respond accordingly.

The examples cited above point out the pleiotypic nature of the response of the cells to *ds*RNA, suggesting that critical (threshold) numbers of molecules may dictate the quality of the response of the cells and provide the basis for our proposing this class of molecules as a prime candidate for proximate cell-killing factor.

XIII. PROTEIN PHOSPHORYLATION AS CELL-KILLING FACTOR

By relegating the role of proximate cell-killing factor to *ds*RNA, we are left with the task of identifying a molecule(s) more immediately responsible for the pleiotypic changes observed in the virus-infected cell. As a working hypothesis, we assign this role to the action of cellular protein kinases, enhanced in activity by, or through, *ds*RNA action. The myriad changes in cellular functions, both of enzymatic and structural significance, attributable to phosphorylation-dephosphorylation reactions,[83,103] lead us to favor an important role for virus (*ds*RNA?)-induced phosphorylation in cell killing and its precedent events, inhibition of host macromolecular synthesis, and CPE. Apparently, the rapid inhibition of cellular macromolecular synthesis by viruses cannot of itself account for the rapid and characteristic expression of CPE or cell killing. Thus, the exposure of most types of cells to protein synthesis inhibitors and/or actinomycin D does not begin to mimic viral CPE or cell killing until many hours after treatment with the drug(s).[44,128] Indeed, enucleated cells remain morphologically intact for many hours,[121,122,125] although they do succumb to the CPE of virus infection (see below).[122] These observations suggest that inhibition of host macromolecular synthesis may not be responsible for the major cellular malfunctions behind CPE or cell killing. Rather, we would assign that role to the extensive phosphorylation observed, for example, in VSV-infected cells,[129] and the subsequent dysfunction of preexisting cellular proteins and organelles vital for the integrity of the cells. Certainly, further studies are warranted in a search for evidence that viral-induced cellular protein kinases[83-85,103,120] may act as a cell-killing factor by adversely affecting the function of proteins through phosphorylation.

XIV. ON THE SO-CALLED "TOXICITY" OF VESICULAR STOMATITIS VIRUS

The literature is replete with reports purporting to demonstrate that the virions of VSV are intrinsically "toxic", i.e., capable of inhibiting host macromolecular synthesis and expressing cell killing in the absence of newly synthesized viral components. This conclusion is based primarily on the repeated observations that stock preparations of VSV that have been inactivated by UV radiation and high multiplicities of "purified" DI particles both display these capabilities.[48,75-80] These earlier studies, predating as they did the discovery of virion-associated transcriptase[81] and its expression in vivo,[20] could not have anticipated this synthetic capacity indigenous to the VSV virion, nor the capacity of UV-irradiated virions for limited transcription[15] — up to the point of UV damage.[41,42] Furthermore, these earlier studies antedate the discovery of defective CK particles[14] and hence, could not have taken into account the lethal action of non-infectious CK particles as contaminants of standard preparations of VSV DI particles.[14] With the wisdom of hindsight, we have reexamined this question of "toxicity", especially since reports from two laboratories continue to support the concept of an inhibitory component preexisting in the virion of VSV.[21,28,29]

It is clear that VSV exposed to increasing doses of UV radiation progressively loses its capacity to inhibit cellular protein synthesis[48] and to kill cells.[15] However, infection with high multiplicities of VSV appears to compensate for this loss of activity,[48,75-80] suggesting that structural elements of the infecting virion may suffice to bring about protein synthesis inhibition if they are present in high-enough concentration. By this reasoning, inhibition of cellular protein synthesis would require amplification of viral components only at low multiplicities of infection,[48] a view reiterated by Baxt and Bablanian[28,49] and by Dubovi and Youngner.[21,29] However, we have shown that to kill a cell, only about one fifth of the VSV genome need remain intact in order to produce transcripts of genes N and NS for subsequent translation into minimally functional proteins N and NS.[15,18] Because the *ckp* and *psi* phenotypes of VSV appear to be expressed conjointly,[17] we favor the possibility that the inhibition of cellular protein synthesis (and cell killing), observed with UV-irradiated virus at high multiplicities of infection, reflects the survival of genes N and NS and their expression through transcription and translation in the formation of cell-killing factor, rather than the direct action of preexisting components of the virion, as others have suggested.[48,75-80]

To test the predictions of our model for cell-killing factor formation, we first sought to calculate whether any viral synthetic activity could be expected in cells infected with high multiplicities of UV-irradiated virus and secondly, we sought experimentally to demonstrate synthesis of protein from the two critical genes, N and NS. The results presented in Figure 11 suffice to provide a basis for our calculations and discussion; however, the articles by Marvaldi et al.[17,19] should be consulted for full experimental details and data derivation. We note that at $m_{PFP} = 100$, each control monolayer of 2×10^7 mouse L cells was infected with 2×10^9 plaque-forming (PF) particles. Following UV irradiation for 160 sec (total dose = 1046 ergs mm^{-2}, the equivalent of 20 lethal hits to infectivity but only 4 lethal hits to CK particle[15] and transcription activity[15,21]), the number of PF particles was reduced to four ($[2 \times 10^9] \times [2 \times 10^{-9}$ survival level$]$) per plate — an effective $m_{PFP} = 2 \times 10^{-7}$. Yet, in spite of the virtually complete inactivation of *infectivity,* there was no dimunition in the capacity of the UV-irradiated VSV to inhibit cell protein synthesis (compare experimental points at 0 and 160 sec). These results would appear to represent a clear example of virion "toxicity", simply because the VSV preparation has been reduced in titer from 2×10^9 to 4 PF particles without any reduction in the expression of cell protein synthesis inhibition. However, further calculations make it equally clear that, in terms of *CK particle* activity, this same prep-

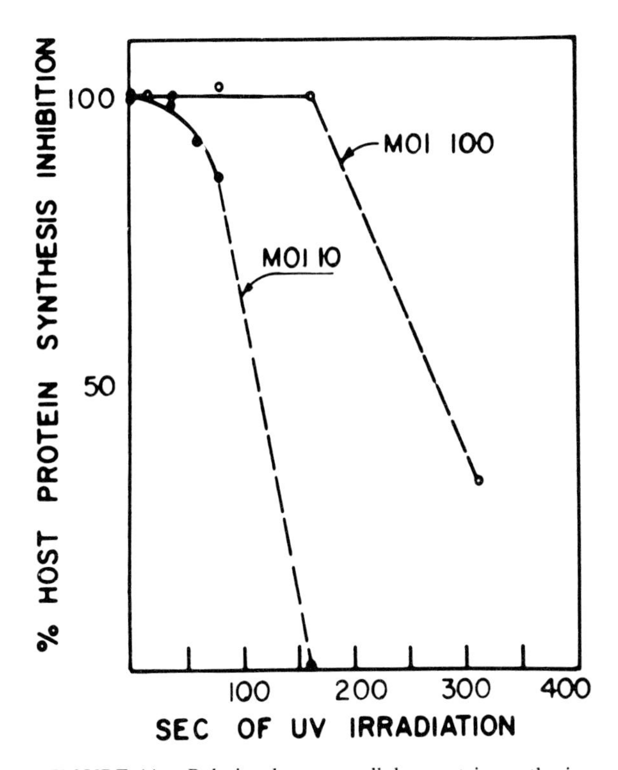

FIGURE 11. Relation between cellular protein synthesis inhibition and multiplicity of infection by ultraviolet (UV)-irradiated VSV. The dashed line indicates a degree of uncertainty because of insufficient points on the curve. The percent host (mouse L cell) protein synthesis inhibition was experimentally determined and calculated. (From Marvaldi, J. L., Sekellick, M. J., Marcus, P. I., and Lucas-Lenard, J., *Virology*, 84, 127, 1978. With permission.)

aration delivers not 4, but 3.6×10^7 functional particles to the monolayer of 2×10^7 cells ($[2 \times 10^9] \times [1.8 \times 10^{-2}$ survival level])! Furthermore, since stock preparations of VSV may contain about five times more defective CK particles than PF particles,[14] the total number of CK particles on the cell monolayer may actually have been closer to 18×10^7, the equivalent of $m_{CKP} \simeq 9$, insuring that essentially all of the cells were infected with a functional CK particle. Clealy, under conditions where UV destroys infectious particle activity virtually totally, large numbers of CK particles may survive. These calculations suggest a more tenable explanation for the so-called "toxic" effects of VSV, namely that genes N and NS, because of their small size and location proximal to the manditory single initiation site at the 3′ end of the genome,[41,42] "survive" large doses of UV radiation, and that, at high-multiplicity infection, each cell receives at least one functional set of these two genes. In turn, these gene products would interact with virion L protein and the genome RNA to express the *ckp* and *psi* characters, as previously proposed[15,17,18] and more thoroughly detailed in the model presented above. We have obtained evidence that at least N protein (and possibly NS) appears in cells infected with UV-irradiated VSV under the experimental conditions just discussed (see Reference 19 and Figure 2, m = 100, UV = 160 sec). Our failure to convincingly demonstrate NS protein on polyacrylamide-gel electropherograms, under conditions where N protein is demonstrable and cellular protein synthesis inhibition is extant, may simply reflect the limits of detection of our system, coupled with the predictably preferential inactivation of the larger target of the NS gene relative to the gene for N protein.[41,42] In this regard, we note that the polyacrylamide-gel electropherograms of Baxt and Bablanian[28] also indicate that viral protein N (possibly NS, too) may have been synthesized in cells infected with UV-irradiated VSV at super-high multiplicities (m = 3000 to 4000 physical particles). On the other hand, it is conceivable that the synthesis of NS protein may be dispensed with, in part, at high multiplicities. This is certainly not the case at low multiplicities,[15,17-19] nor does it appear to be applicable with respect to the requirement for N protein.[15,17,18] The rescue at nonpermissive temperature of VSV *ts*-mutants by UV-irradiated virus[82] is taken as further evidence that "surviving" viral genes may produce functional proteins.

Similar calculations, using the Poisson distribution,[4] revealed that after 320 sec of UV-irradiation (40 and 8 lethal hits to PF and CK particle activity, respectively) and infection at m_{UV-PFP} = 100, about 16% of the cells would receive at least one CK particle and by our model, would undergo protein synthesis inhibition. This calculated value agrees well, within the limits of the precision of the assay for protein synthesis inhibition,[19] with the observed value of 30% protein synthesis inhibition (Figure 11). In view of these results, and until such time as a level of transcription sufficient to produce functional transcripts solely of genes N and NS at one set per cell can be ruled out at high-multiplicity infections with "inactivated" VSV, we think the bulk of the evidence refutes the concept of intrinsic "toxicity" for the VSV virion and points to a definite requirement for synthesis of viral proteins (N and NS) in order to express cell killing or protein synthesis inhibition (Table 1).

XV. CELL MACROMOLECULAR SYNTHESIS INHIBITION AND THE CKP PHENOTYPE

The ubiquitous occurrence of protein kinases and the emerging role of phosphorylation in the regulation of protein function in both normal and virus-infected cells[83,84,103] invites a closer examination of this activity in lytic infection. Using VSV *ts* -mutants with defined *psi*$^+$ and *psi*$^-$ phenotypes,[17,37] Marvaldi and Lucas-Lenard[85] examined infected cells for changes in lhosphorylation patterns, specifically those associated with ribosomes, reasoning that changes in those organelles brought about by the virus infection might be linked to cellular protein synthesis inhibition. Infected mouse L cells were incubated in the presence of (^{32}P) orthophosphate, and proteins were extracted from ribosomes or their subunits with high salt and analyzed by two-dimensional electrophoresis. Two findings emerged: wild-type virus infection resulted in (1) the appearance of a new or modified nonphosphorylated protein predominantly on the 40S subunit, termed SLX, and (2) the appearance of five distinct additional (^{32}P) labeled-spots, termed Pi1-Pi5. Although the nature of these new components is still under investigation, they do not appear to be viral constituents (protein or RNA).[85] Infection at m_{PFP} = 10 resolved the *ts*-mutants and wild-type virus into *pho* (phosphorylation: protein kinase activity) phenotypes which correlated exactly with their *psi* and *ckp* characters.[17] Thus, Marvaldi and Lucas-Lenard[85] found that infection at nonpermissive temperature with mutant *ts*W10(IV) (*pho*$^+$, *psi*$^+$, *ckp*$^+$), like wild-type, resulted in the appearance of these new components, whereas mutants *ts* G114(I), G22(II), and G41(IV) (all *pho*$^-$, *psi*$^-$, *ckp*$^-$) did not (cf. Table 1). Although the *pho*$^+$, *psi*$^+$, *ckp*$^+$ mutant, *ts*W10(IV), was able to promote the appearance of these new ribosome-associated components just as effectively as wild-type virus,[85] *ts*W10 accumulates less than one tenth the amount of viral transcripts as the *pho*$^-$, *psi*$^-$, *ckp*$^-$ mutant, *ts* G41(IV).[59] These results emphasize again[15] that the quality, not the quantity, of the viral transcripts is the important element in determining the *ckp* associated characters.

Cellular RNA synthesis, like protein synthesis, also is inhibited rapidly by lytic viruses, including VSV,[44,47,77,78,123] an event Weck and Wagner[123] have shown is attributable solely to the inhibition of cellular polymerase activity, with the α-amanitin-sensitive polymerase II proving to be especially susceptible. Nonetheless, we have observed that GMK-Vero cells treated with α-amanitin (enough to protect them from the lethal action of pseudorabies virus, which presumably requires polymerase II for transcription and the subsequent formation of cell killing factor) were very slow to express the cytotoxic effects finally manifested by this potent drug.[130] Consequently, we conclude that RNA synthesis inhibition alone, like protein synthesis inhibition, by VSV, is not directly responsible for CPE or cell killing by this virus. As discussed above, we view the inhibition of cellular RNA synthesis by VSV as part of a general pleiotypic effect brought about, perhaps, by extensive protein phosphorylation, in this case, possibly of nuclear polymerases and/or histones. Regardless of the actual molecular events responsible for cellular RNA synthesis inhibition (an rni^+ phenotype), its expression appears to require the same gene functions needed to express the psi^+ and ckp^+ phenotypes (Data in Table 1 by permission of J. Lucas-Lenard and F-S. Wu). Thus, cellular macromolecular synthesis inhibition and cell killing by VSV appears to require in common, the formation of cell-killing factor. We find further support for this conclusion from the results of Weck and Wagner,[123] who reported a level of RNA synthesis inhibition in excess of that expected from the actual fraction of cells infected at m_{PFP} = 1. We infer from their data that the stocks of VSV contained an excess of defective CK particles over infectious CK (plaque-forming) particles, as has been reported,[14] and that the defective CK particles are as capable of inhibiting host-macromolecular synthesis as they are of killing cells.

As Marvaldi and Lucas-Lenard pointed out,[85] their results did not distinguish between cause or effect, i.e., whether protein SLX and/or the phosphorylated components were responsible for the psi^+, ckp^+ character, or a consequence of it. This dilemma is not unique to the problem of phosphorylated ribosomal components; similar questions remain unresolved with respect to the role of lysosomes,[50,51,86] membrane formation and integrity,[87] and ion flux and concentration[88,89] in the myriad effects which characterize lytic infection. However, since infection with psi^+, ckp^+ *ts*-mutants of VSV induces extensive phosphorylation of cellular proteins in general, and this correlates with the rni^+ character,[131] we regard it as premature to summarily dismiss such effects as noncausal. For the present, they must be catalogued as yet another of the pleiotypic responses of cells to lytic virus infection, but perhaps, for reasons discussed earlier, an important one for cell killing.

In the case of lysosomes, the lack of an absolute correlation with lysosomal enzyme activation and cytopathic effects[50,51] suggests that, on balance, release of lysosomal enzymes may not be a cause, but rather an effect, of lytic infection. As inferred by the stability of cellular[90,123] and viral[59,91] mRNA in VSV-infected cells, the role of lysosomal nuclease activity seems insignificant.

A general mechanism for cell protein synthesis inhibition by viruses was proposed recently, which presumes that viruses may induce ionophore-like changes in the plasma membrane following insertion of viral structural proteins.[88] Although an attractive proposal, Francoeur and Stanners[89] could find no experimental basis to support it. They infected mouse L cells with wild-type VSV and a mutant which is significantly delayed in its expression of the *psi* phenotype[60] and concluded that the inhibition of protein synthesis by VSV cannot be ascribed to viral-induced changes in the intracellular concentration of the monovalent ion, K^+.[89]

In spite of the adverse effects of VSV infection on the rate and extent of cellular RNA[44,123] and protein synthesis inhibition,[17,19,37,44,49,60,89] VSV appears to fall into the category of viruses that do not produce significant degradation of host mRNA.[90,123] Indeed, some special species of cell message may compete successfully and be translated unabated in the infected cell while essentially all other species of cell mRNA are inhibited.[90,92] These results connote a fine degree of regulation of protein synthesis in the VSV-infected cell — a situation more in keeping with the fact that viral mRNA is translated well into the infectious cycle and long after the synthesis of most species of cellular proteins has ceased.

XVI. CONTROL OVER CELL KILLING

The myriad changes catalogued for a rhabdovirus-infected cell in the course of its ultimate demise are matched in magnitude by the absolute control over cell killing exhibited by: (1) the interferon system,[5,16,46,56,100] (2) CPE-suppressing DI particles,[36,104] and (3) certain host cells.[105-107,116] The influence of each of these is considered below. The survival of cells during persistent infection is of particular interest, combining perhaps all of these elements, and is the subject of a separate chapter.[5]

A. The Interferon System

Cells manifesting interferon-mediated interference can survive the lethal action of VSV.[5,16,46,56,100] We have established the precise relationship between interferon dose and the inhibition of CK particle activity through use of the single-cell survival procedure (Figure 1) and "pulse-infection" with a *ts* -mutant of VSV.[16] The survival of single cells treated with interferon and infected with a CK particle is illustrated in Figure 12A to C. Figure 12A shows colonies that arose from plating 200 GMK-Vero cells previously treated for 24 hr with 54 PR_{50} (VSV) units per milliliter of human interferon. Essentially all of the cells survived to form visible colonies; the plate in Figure 12B received 200 cells from a monolayer infected with VSV mutant *ts* G114(I), and were held for 20 hr at 30°. Over 95% of the cells were killed, i.e., failed to form colonies; the plate in Figure 12(C) shows the results of combining these two experimental conditions, i.e., the cells were treated first with interferon and then challenged with VSV. Virtually all of the 200 interferon-treated cells survived the lethal action of VSV, presumably because viral primary transcription[20,58,59,102] and protein synthesis[101,102] were inhibited and cell-killing factor could not be formed.

Data from experiments of the type illustrated in Figure 12A to C allowed us to determine the exact relationship between interferon (or poly[rI]·poly[rC]) dose and the surviving activity of CK particles. In two different cell systems (GMK-Vero and mouse L cells), we found that about four to five times more interferon (or poly[rI]·poly[rC]) was required to achieve a 50% reduction in CK particle activity than was needed to reduce infectivity (by plaque assay), comparably.[16] Furthermore, accumulation of VSV primary transcripts was inhibited in unison with the loss of CK particle activity,[16] once again suggesting that viral transcription may constitute the rate-limiting reaction in the formation of cell-killing factor. These results find their parallel in the action of UV radiation on these same functions, in that the dose-response curves for the two dissimilar agents both define "one-hit-to-inactivate" kinetics in the low-dose range.[15,16] The exact interpretation of these results must await definition of interferon action in more precise terms. Regardless, interferon action stands as the singularly most effective means of preventing cell killing by viruses, and rhabdoviruses are particularly sensitive, both in cell culture[16,46,56,58,59,100-102] and in the animal.[108,109] Furthermore, recent experiments reveal the effectiveness of some VSV *ts*-mutants[5,63,110] and (±)RNA DI particles[11] (and of rabies virus[111,112]) to induce interferon, thereby activating the interferon system and providing a mechanism for cell sparing (survival) in situations like persistent infection[5] or survival of the host animal.[108,109]

B. CPE-Suppressing Particles

Kawai and his colleagues[36,104] discovered that DI particles recovered from BHK cells persistently infected with rabies virus were capable of inhibiting the cytopathic effect

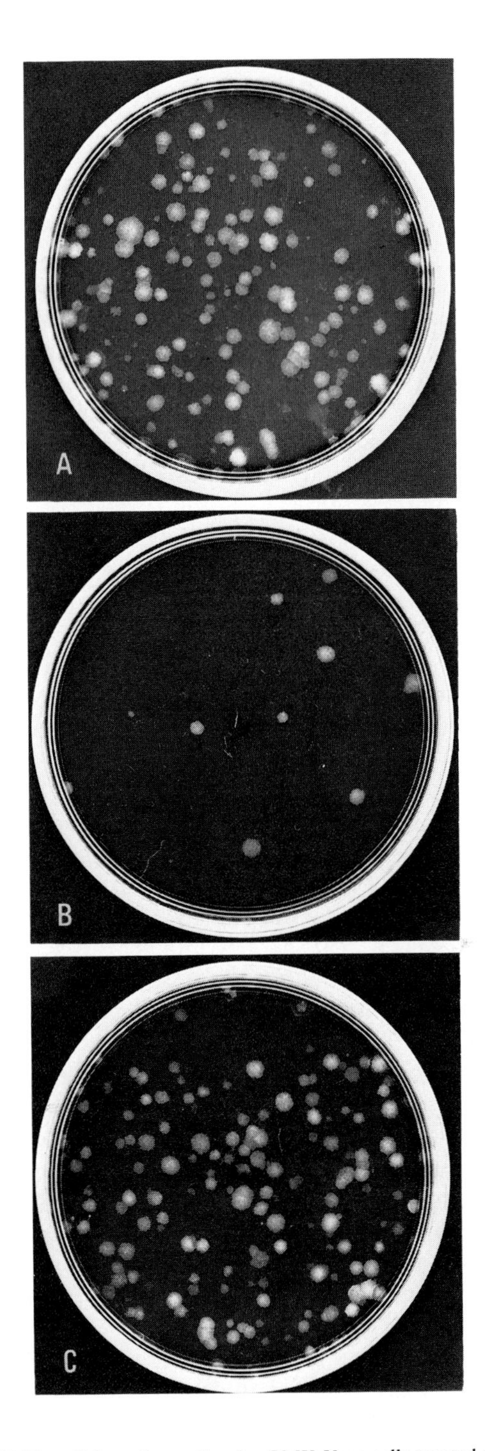

FIGURE 12. Colony formation by GMK-Vero cells treated with human interferon and challenged with VSV tsG114(I). Photographs show colonies from surviving cells incubated for 20 hr at 30.0°C (pulse-infection) and 8 days at 40.0°C. All petri dishes (50 mm) were inoculated with 200 cells from monolayers treated with interferon and challenged with virus as follows: (A) 54 units per milliliter of interferon, no virus; (B) no interferon, m_{ckp} = 6; (C) 54 units per milliliter interferon, m_{ckp} = 6. (From Marcus, P. I. and Sekellick, M. J., *Virology*, 69, 378, 1976. With permission.)

(CPE) produced by standard rabies virus. This observation is particularly important, because it provides a mechanism for sparing cells from the lethal action of a virus which does not appear to involve the interferon system.[5] The behavior of these so-called CPE-suppressing DI particles of rabies virus contrasts sharply wth our own observations that conventional (−)RNA DI particles of VSV, although nonlethal themselves, do not protect cells from the lethal action of that virus, even though homotypic interference is extant.[14] Under these conditions, primary transcription of the standard helper virus is unaffected by the presence of coinfecting conventional (−)RNA DI particles of VSV,[34,35] consistent with the requirement for cell-killing factor formation.[14-18] It will be of interest to learn whether the CPE-suppressing DI particles of rabies virus are capable of inhibiting primary transcription,[113] and indeed, whether cell killing by rabies requires such transcription. Also, it is not yet clear whether the differences reported for the cell-sparing capacity of rabies and VSV DI particles reflects a real difference between these two rhabdoviruses or a quantitative difference in the number of DI particles used to coinfect the cells with standard virus.

The paucity of quantitative data on cell killing by purified preparations of rabies virus and the lack of *ts* -mutants preclude a direct test on rabies virus of our VSV-derived model for cell killing. Furthermore, the biochemical differences in growth between rabies virus and VSV, as inferred from the inhibition of replication of the former, but not the latter, by cytosine arabinoside[114] and in the enucleated cell,[115] dictates that experimentation, not extrapolation, will be required in any serious comparative study of cell killing by these viruses.

C. The Host Cell

1. Invertebrate Cells

The ubiquitous host range of the rhabdoviruses is impressive. In vertebrate cell cultures, VSV replicates to high titer and produces characteristic cytopathic effects and cell killing, even at temperatures suboptimal for cell growth. For example, 5 hr after exposure of GMK-Vero cells to VSV at 30°C, one half of the infected cell population has acquired a lethal dose of cell-killing factor;[15] yet, at that time, less than 10% of the virus had been produced and released from the cells. Since there appear to be no reports of vertebrate cells inherently resistant to the lethal effects of VSV, excluding those exposed to interferon action,[16,46,100] CPE-suppressing particles,[36,104] or conditions of persistent infection,[5] we assume that cell-killing factor is both produced and expressed in generally all vertebrate host cells. The situation with invertebrate cells may be different. The infection of insect cells (fly, moth, mosquito) at temperatures near normal for the host, about 28°, are, in general, characterized by a singular lack of cytopathic effects.[105-107,116] Since significant yields of VSV can be obtained from invertebrate cells in culture, and virtually all cells can be shown to contain viral antigen and produce budding virus,[105] we cannot attribute the lack of CPE to the absence of virus replication. Either cell-killing factor is not produced in significant amounts, or it is produced but not expressed (recognized?) in these cells; in either case, the end result is host control over cell killing. However, this control does not appear to be immutable. Sarver and Stollar[116] have isolated a clone of *Aedes albopictus* cells which did express CPE when infected with VSV at 28° or 34°C. Other clones, resembling the parental population, did not display CPE at 28°C, but did at 34°C. As Sarver and Stollar point out, the availability of clones of CPE-susceptible and CPE-resistant mosquito cells should prove useful in probing the molecular biology of cell killing by viruses. In this context, we note that leafhopper cells undergo fusion and a cytopathic effect when infected by a plant rhabdovirus.[117]

Since invertebrate cells appear to lack the interferon system,[118] conceivably, they may also lack a recognition system for cell killing factor — a speculative view predicated on a model in which the inducer moiety of interferon and proximate cell-killing factor share similar molecular characteristics, as in the form of *ds*RNA.[11,68,70,71,119]

2. The Enucleated Vertebrate Cell

The enucleated vertebrate cell may be viewed as a unique type of host — one devoid of all potential responses of the nucleus to virus infection. If control over the expression of CPE or cell killing resides directly in the nucleus, or involves inactivation of nuclear elements, then the absence of the nucleus might have profound effects. VSV can replicate to almost normal titer in enucleated cells.[115,122] In one of these studies, Follett and his colleagues[122] observed that, "The only difference noted between the effects of VSV in nucleate and enucleate cells (GMK-BSC-1) was that a delay in cytopathic effect of about 2 hr was consistently observed in enucleate cells." It follows that new gene products, or short-lived factors, are not required for the expression of CPE. Furthermore, the stability of enucleated cell fragments[121,122,125] and their partial biochemical equivalent, actinomycin D-treated cells (most types), relative to that of a VSV-infected cell (intact or enucleated),[122] clearly demonstrates that the characteristic CPE of this virus is not due to the inhibition of cellular RNA synthesis per se. More likely, the molecular events controlling the expression of CPE and cell killing depend solely on the response of cytoplasmic constituents to cell-killing factor. In this context, a study of phosphorylation in the enucleated cell infected with VSV should provide a definitive test of our hypothesis concerning the role of protein phosphorylation in the expression of CPE and cell killing.

XVII. PERSPECTIVES ON CELL KILLING

We have reviewed cell killing by rhabdoviruses, in particular that due to the prototype, vesicular stomatitis virus (VSV). Studies based primarily on the analyses of single-cell survival curves and the activity of cell-killing (CK) particles revealed that the expression of cell killing by VSV required a consistent array of viral gene functions. Virion-associated L protein was needed for two purposes: (1) to produce transcripts of genes N and NS for subsequent translation into minimally functional proteins N and NS and (2) to interact with these proteins and part of the viral genome to produce a putative, proximate cell-killing factor. We present evidence that this factor might be viral-coded *ds*RNA and speculate that its presence in the cell may activate (or induce) protein kinases and extensive phosphorylation, producing gross malfunction of cellular proteins and organelles, which in turn would precipitate the pleiotypic effects that characterize virus infection: inhibition of cellular macromolecular synthesis (the *psi*$^+$, *rni*$^+$ phenotypes), cytopathic effects, and, ultimately, cell killing (the *ckp*$^+$ phenotype). Whether or not this view of the events leading to cell killing by VSV is verified by experimentation, it is clear that the viral gene functions required to produce the *psi*$^+$, and *ckp*$^+$ phenotypes are identical, suggesting that they may share a common biochemical event — one we have identified as the production of proximate or actual cell-killing factor.

From our own studies and our analyses of experiments purporting to demonstrate the "toxic" effect of VSV, we conclude that the virion of this virus is not intrinsically toxic. In fact, we have presented evidence that the so-called "toxic" effects of "inactive" preparations of VSV can be attributed to the survival of genes N and NS and the eventual production and action of cell-killing factor.

Important to an understanding of cell killing by VSV is our discovery that in many preparations of this virus most virions can function as CK particles even though they are noninfectious.[14] The presence of these so-called defective CK particles in VSV stocks provides a simple explanation for many reports in which the "toxic" or functional activity of a virus preparation seems excessive, relative to the amount of demonstrable infectious virus. For example, Dubovi and Youngner[21] clearly demonstrated that the number of VSV particles capable of inhibiting pseudorabies (PSR) virus replication exceeded the number of plaque-forming particles by about 50-fold, and that only one such particle per cell sufficed to produce the inhibition. We interpret their results to mean that defective CK particles constitute the predominant member of the virus population and that they are responsible for inhibition of PSR virus replication through the action of cell-killing factor. Thus, only *ts*-mutants which we know to be *ckp*$^{+}$ (and *psi*$^{+}$) are capable of inhibiting PSR virus at a nonpermissive temperature (Table 1).[21] Furthermore, since VSV inhibits cellular polymerase II,[123] and PSR virus requires that enzyme to express cell killing and for replication,[132] we conclude that the inhibition of PSR virus replication by VSV,[21] most likely reflects the inhibition of host polymerases through the action of cell-killing factor produced by defective cell-killing particles.

Some types of cells have already bested rhabdoviruses in the arena of cell killing. Either cell-killing factor is not produced, or it is not recognized as such by many insect cells, presumably an accommodation worked out over eons of coexistence, and one which might serve us in perspective. It seems paradoxical that VSV, usually so lethal to vertebrate cells in culture, produces a self-limiting disease in natural infection, while rabies virus, usually relatively innocuous in vitro, produces a highly lethal disease. Yet, continued investigation of the lethal action of rhadoviruses at the cellular level may provide the insights needed to understand and control this same action in humans — witness the recent discovery that highly virulent genetically stable rabies virus could be obtained with one or a few passages in murine or human neuroblastoma cells in vitro.[124]

Finally, we note that cell-killing factor, proximate or actual, has yet to be identified in molecular terms. Once identified, its mode(s) of action must be defined, and once defined, we must seek ways to ameliorate its expression. These challenges remain to test the acumen of our investigational skills.

ACKNOWLEDGMENTS

The research from our laboratory referred to in this chapter was aided in part by National Science Foundation grant No. PCM-76-00467, National Cancer Institute grant No. CA 20882, and benefited from use of the Cell Culture Facility, supported by National Cancer Institute grant No. CAP 14733.

REFERENCES

1. **Marcus, P. I. and Puck, T. T.,** Host-cell interaction of animal viruses. I. Titration of cell-killing by viruses, *Virology*, 6, 405, 1958.
2. **Vogt, M. and Dulbecco, R.,** Properties of a HeLa cell culture with increased resistance to poliomyelitis virus, *Virology*, 5, 425, 1958.
3. **Marcus, P. I.,** Host-cell interaction of animal viruses. II. Cell-killing particle enumeration: survival curves at low multiplicities, *Virology*, 9, 546, 1959.
4. **Marcus, P. I.,** Symposium on the biology of cells modified by viruses or antigens. IV. Single-cell techniques in tracing virus-host interactions, *Bacteriol. Rev.*, 23, 232, 1959.
5. **Sekellick, M. J. and Marcus, P. I.,** Persistent infections by rhabdoviruses, in *Rhabdoviruses*, Vol. III, Bishop, D. H. L., Ed., CRC Press, Boca Raton, Fla., 1979, chap. 4.

6. **Pringle, C. R.,** Genetics of rhabdoviruses, in *Comprehensive Virology,* Vol. 9, Frankel- Conrat, H. and Wagner, R. R., Eds., Plenum Press, New York, 1977, 239.
7. **Wagner, R. R.,** Reproduction of rhabdoviruses, in *Comprehensive Virology,* Vol. 4, Frankel-Conrat, H. and Wagner, R. R., Eds., Plenum Press, New York, 1975, 1.
8. **Huang, A. S. and Baltimore, D.,** Defective interfering animal viruses, in *Comprehensive Virology,* Vol. 10, Frankel-Conrat, H. and Wagner, R. R., Eds., Plenum Press, New York, 1977, 73.
9. **Colonno, R. J., Lazzarini, R. A., Keene, J. D., and Banerjee, A. K.,** In vitro synthesis of messenger RNA by a defective interfering particle of vesicular stomatitis virus, *Proc. Natl. Acad. Sci. U.S.A.,* 74, 1884, 1977.
10. **Johnson, L. D. and Lazzarini, R. A.,** Replication of viral RNA by a defective interfering vesicular stomatitis virus particle in the absence of helper virus, *Proc. Natl. Acad. Sci. U.S.A.,* 74, 4387, 1977.
11. **Marcus, P. I. and Sekellick, M. J.,** Defective interfering particles with covalently linked (±)RNA induce interferon, *Nature (London),* 266, 815, 1977.
12. **Banerjee, A. K., Abraham, G., and Colonno, R. J.,** Vesicular stomatitis virus: mode of transcription, *J. Gen. Virol.,* 34, 1, 1977.
13. **Freeman, G. J., Rao, D. C., and Huang, A. S.,** Genome organization of vesicular stomatitis virus, in *Negative Strand Viruses,* Barry, R. D. and Mahy, B. W. J., Eds., Academic Press, London, 1978, 261—270.
14. **Marcus, P. I. and Sekellick, M. J.,** Cell killing by viruses. I. Comparison of cell-killing, plaque-forming, and defective-interfering particles of vesicular stomatitis virus, *Virology,* 57, 321, 1974.
15. **Marcus, P. I. and Sekellick, M. J.,** Cell killing by viruses. II. Cell killing by vesicular stomatitis virus: a requirement for virion-derived transcription, *Virology,* 63, 176, 1975.
16. **Marcus, P. I. and Sekellick, M. J.,** Cell killing by viruses. III. The interferon system and inhibition of cell killing by vesicular stomatitis virus, *Virology,* 69, 378, 1976.
17. **Marvaldi, J. L., Lucas-Lenard, J., Sekellick, M. J., and Marcus, P. I.,** Cell killing by viruses. IV. Cell killing and protein synthesis inhibition by vesicular stomatitis virus require the same gene functions, *Virology,* 79, 267, 1977.
18. **Marcus, P. I., Sekellick, M. J., Johnson, L. D., and Lazzarini, R. A.,** Cell killing by viruses. V. Transcribing defective interfering particles of vesicular stomatitis virus function as cell-killing particles, *Virology,* 82, 242, 1977.
19. **Marvaldi, J., Sekellick, M. J., Marcus, P. I., and Lucas-Lenard, J.,** Inhibition of mouse L cell protein synthesis by ultraviolet-irradiated vesicular stomatitis virus requires viral transcription, *Virology,* 84, 127, 1978.
20. **Marcus, P. I., Engelhardt, D. L., Hunt, J. M., and Sekellick, M. J.,** Interferon action: inhibition of vesicular stomatitis virus RNA synthesis induced by virion-bound polymerase, *Science,* 174, 593, 1971.
21. **Dubovi, E. J. and Youngner, J. S.,** Inhibition of pseudorabies virus replication by vesicular stomatitis virus. I. Activity of infectious and inactivated B particles, *J. Virol.,* 18, 526, 1976.
22. **Michalke, H. and Bremer, H.,** RNA synthesis in *Escherichia coli* after irradiation with ultraviolet light, *J. Mol. Biol.,* 41, 1, 1969.
23. **Hunt, D. M. and Wagner, R. R.,** Location of the transcription defect in Group I temperature-sensitive mutants of vesicular stomatitis virus, *J. Virol.,* 13, 28, 1974.
24. **Roy, P., Repik, P., Hefti, E., and Bishop, D. H. L.,** Complementary RNA species isolated from vesicular stomatitis virus (HR strain) defective virions, *J. Virol.,* 11, 915, 1973.
25. **Reichmann, M. E., Villarreal, L. P., Kohne, D., Lesnaw, J., and Holland, J. J.,** RNA polymerase activity and poly(A) synthesizing activity in defective T particles of vesicular stomatitis virus, *Virology,* 58, 240, 1974.
26. **Cooper, P. D. and Bellett, A. J. D.,** A transmissible interfering component of vesicular stomatitis virus preparations, *J. Gen. Microbiol.,* 21, 485, 1959.
27. **Doyle, M. and Holland, J. J.,** Prophylaxis and immunization in mice by virus-free defective T particles to protect against intracerebral infection by vesicular stomatitis virus, *Proc. Natl. Acad. Sci. U.S.A.,* 70, 2105, 1973.
28. **Baxt, B. and Bablanian, R.,** Mechanisms of vesicular stomatitis virus-induced cytopathic effects. II. Inhibition of macromolecular synthesis induced by infectious and defective-interfering particles, *Virology,* 72, 383, 1976.
29. **Dubovi, E. J. and Youngner, J. S.,** Inhibition of pseudorabies virus replication by vesicular stomatitis virus. II. Activity of defective-interfering particles, *J. Virol.,* 18, 534, 1976.
30. **Lazzarini, R. A., Weber, G. H., Johnson, L. D., and Stamminger, G. M.,** Covalently linked message and anti-message (genomic) RNA from a defective vesicular stomatitis virus particle, *J. Mol. Biol.,* 97, 289, 1975.
31. **Wagner, R. R., Schnaitman, T. A., and Snyder, R. M.,** Structural proteins of vesicular stomatitis viruses, *J. Virol.,* 3, 395, 1969.

32. **Kang, C. Y. and Previc, L.**, Proteins of vesicular stomatitis virus. I. Polyacrylamide gel analysis of viral antigens, *J. Virol.*, 3, 404, 1969.
33. **Marcus, P. I., Sekellick, M. J., and Fishman, M.**, Unpublished observations, 1974.
34. **Perrault, J. and Holland, J. J.**, Absence of transcriptase activity or transcription-inhibiting ability in defective interfering particles of vesicular stomatitis virus, *Virology*, 50, 159, 1972.
35. **Huang, A. S. and Manders, E. R.**, Ribonucleic acid synthesis of vesicular stomatitis virus. IV. Transcription by standard virus in the presence of defective interfering particles, *J. Virol.*, 9, 909, 1972.
36. **Kawai, A., Matsumoto, S., and Tanabe, K.**, Characterization of rabies viruses recovered from persistently infected BHK cells, *Virology*, 67, 520, 1975.
37. **McAllister, P. E. and Wagner, R. R.**, Differential inhibition of host protein synthesis in L cells infected with RNA⁻ temperature-sensitive mutants of vesicular stomatitis virus, *J. Virol.*, 18, 550, 1976.
38. **Repik, P., Flamand, A., and Bishop, D. H. L.**, Synthesis of RNA by mutants of vesicular stomatitis virus (Indiana serotype) and the ability of wildtype VSV New Jersey to complement the VSV Indiana tsG114 transcription defect, *J. Virol.*, 20, 157, 1976.
39. **Unger, J. T. and Reichmann, M. E.**, RNA synthesis in temperature-sensitive mutants of VSV, *J. Virol.*, 12, 570, 1973.
40. **Farmilo, A. J. and Stanners, C. P.**, Mutant of vesicular stomatitis virus which allows deoxyribonucleic acid synthesis and division in cells synthesizing viral ribonucleic acid, *J. Virol.*, 10, 605, 1972.
41. **Ball, L. A. and White, C. N.**, Order of transcription of genes of vesicular stomatitis virus, *Proc. Natl. Acad. Sci. U.S.A.*, 73, 442, 1976.
42. **Abraham, G. and Banerjee, A. K.**, Sequential transcription of the genes of vesicular stomatitis virus, *Proc. Natl. Acad. Sci. U.S.A.*, 73, 1504, 1976.
43. **Deutsch, V. and Berkaloff, A.**, Analyse d'un mutant thermolabile du virus de la stomatite vésiculaire (VSV), *Ann. Inst. Pasteur (Paris)*, 121, 101, 1971.
44. **Bablanian, R.**, Structural and functional alterations in cultured cells infected with cytocidal viruses, *Prog. Med. Virol.*, 19, 40, 1975.
45. **Mudd, J. A. and Summers, D. F.**, Protein synthesis in vesicular stomatitis virus infected HeLa cells, *Virology*, 42, 328, 1970.
46. **Yamazaki, S. and Wagner, R. R.**, Action of interferon: kinetics and differential effects on viral functions, *J. Virol.*, 6, 421, 1970.
47. **Wertz, G. W. and Youngner, J. S.**, Interferon production and inhibition of host synthesis in cells infected with vesicular stomatitis virus, *J. Virol.*, 6, 476, 1970.
48. **Wertz, G. W. and Youngner, J. S.**, Inhibition of protein synthesis in L cells infected with vesicular stomatitis virus, *J. Virol.*, 9, 85, 1972.
49. **Baxt, B. and Bablanian, R.**, Mechanism of VSV-induced cytopathic effects. I. Early morphologic changes induced by infectious and defective-interfering particles, *Virology*, 72, 370, 1976.
50. **Koschel, K., Aus, H. M., and ter Meulen, V.**, Lysosomal enzyme activity in poliovirus-infected HeLa cells and VSV-infected L cells: biochemical and histochemical comparative analysis with computer-aided techniques, *J. Gen. Virol.*, 25, 359, 1974.
51. **Norkin, L. C. and Ouellette, J.**, Cell killing by SV_{40}: variation in the pattern of lysosomal enzyme release, cellular enzyme release, and cell death during productive infection of normal and SV_{40}-transformed simian cell lines, *J. Virol.*, 18, 45, 1976.
52. **Emerson, S. U. and Wagner, R. R.**, Dissociation and reconstitution of the transcriptase and template activities of vesicular stomatitis B and T virions, *J. Virol.*, 10, 297, 1972.
53. **Colonno, R. J. and Banerjee, A. K.**, Mapping and initiation studies on the leader RNA of vesicular stomatitis virus, *Virology*, 77, 260, 1977.
54. **Schnitzlein, W. M. and Reichmann, M. E.**, The size and the cistronic origin of defective vesicular stomatitis virus particle RNAs in relation to homotypic and heterotypic interference, *J. Mol. Biol.*, 101, 307, 1976.
55. **Stamminger, G. M. and Lazzarini, R. A.**, Analysis of the RNA of defective VSV particles, *Cell*, 3, 85, 1974.
56. **Sekellick, M. J. and Marcus, P. I.**, Persistent infection. I. Interferon-inducing defective-interfering particles as mediators of cell sparing: possible role in persistent infection by vesicular stomatitis virus, *Virology*, 85, 175, 1978.
57. **Hunt, J. M. and Marcus, P. I.**, Mechanism of Sindbis virus-induced intrinsic interference with vesicular stomatitis virus replication, *J. Virol.*, 14, 99, 1974.
58. **Manders, E. K., Tilles, J. G., and Huang, A. S.**, Interferon-mediated inhibition of virion-directed transcription, *Virology*, 49, 573, 1972.
59. **Marcus, P. I. and Sekellick, M. J.**, Interferon action. III. The rate of primary transcription of vesicular stomatitis virus is inhibited by interferon action, *J. Gen. Virol.*, 38, 391, 1978.

60. **Stanners, C. P., Francoeur, A. M., and Lam, T.**, Analysis of VSV mutant with attenuated cytopathogenicity: mutation in viral function, P, for inhibition of protein synthesis, *Cell,* 11, 273, 1977.
61. **Abraham, G., Colonno, R. J., and Banerjee, A. K.**, Evidence for the synthesis of a "leader" RNA segment followed by the sequential transcription of the genes of vesicular stomatitis virus, in *Animal Virology,* Academic Press, New York, 1976, 439.
62. **Carver, D. H. and Marcus, P. I.**, Enhanced interferon production from chick embryo cells aged in vitro, *Virology,* 32, 247, 1967.
63. **Sekellick, M. J. and Marcus, P. I.**, Persistent infection by vesicular stomatitis virus: Interferon induction by *ts*-mutants, and cell sparing, *J. Supramol. Struct. Suppl.* 2, 246, 1978.
64. **Ehrenfeld, E. and Hunt, T.**, Double-stranded poliovirus RNA inhibits initiation of protein synthesis by reticulocyte lysates, *Proc. Natl. Acad. Sci. U.S.A.,* 68, 1075, 1971.
65. **Chao, J., Chao, L., and Speyer, J. F.**, The inhibition of f2 RNA directed *E. coli* protein synthesis by poly rI · rC in vitro, *Biochem. Biophys. Res. Commun.,* 45, 1096, 1971.
66. **Kerr, I. M., Brown, R. E., and Ball, L. A.**, Increased sensitivity of cell-free protein synthesis to double-stranded RNA after interferon treatment, *Nature (London),* 250, 57, 1974.
67. **Kerr, I. M., Brown, R. E., Clemens, M. J., and Gilbert, C. S.**, Interferon-mediated inhibition of cell-free protein synthesis in response to double-stranded RNA, *Eur. J. Biochem.,* 69, 551, 1976.
68. **Cordell-Stewart, B. and Taylor, M. W.**, Effect of viral double-stranded RNA on protein synthesis in intact cells, *J. Virol.,* 11, 232, 1973.
69. **Sen, G. C., Lebleu, B., Brown, G. E., Kawakita, M., Slattery, E., and Lengyel, P.**, Interferon, double-stranded RNA, and mRNA degradation, *Nature (London),* 264, 370, 1976.
70. **Cordell-Stewart, B. and Taylor, M. W.**, Effect of double-stranded RNA on mammalian cells in culture, *Proc. Natl. Acad. Sci. U.S.A.,* 68, 1326, 1971.
71. **Cordell-Stewart, B. and Taylor, M. W.**, Effect of viral stranded RNA on mammalian cells in culture: cytotoxicity under conditions preventing viral replication and protein synthesis, *J. Virol.,* 12, 360, 1973.
72. **Stewart, W. E., II, DeClercq, E., Billiau, A., Desmyter, J., and DeSomer, P.**, Increased susceptibility of cells treated with interferon to the toxicity of polyriboinosinic-polyribocytidylic acid, *Proc. Natl. Acad. Sci. U.S.A.,* 69, 1851, 1972.
73. **Carter, W. A. and DeClercq, E.**, Viral infection and host defense, *Science,* 186, 1177, 1972.
74. **Collins, F. D. and Roberts, W. K.**, Mechanism of mengo virus-induced cell injury in L cells: use of inhibitors of protein synthesis to dissociate virus-specific events, *J. Virol.,* 10, 969, 1972.
75. **Cantell, K., Skurska, Z., Paucker, K., and Henle, W.**, Quantitative studies on viral interference in suspended L-cells. II. Factors affecting interference by UV-irradiated Newcastle disease virus against vesicular stomatitis virus, *Virology,* 17, 312, 1962.
76. **Wagner, R. R., Levy, A. H., Synder, R. M., Ratcliff, G. A., Jr., and Hyatt, D. F.**, Biological properties of two plaque variants of vesicular stomatitis virus (Indiana serotype), *J. Immunol.,* 91, 112, 1963.
77. **Huang, A. S. and Wagner, R. R.**, Inhibition of cellular RNA synthesis by nonreplicating vesicular stomatitis virus, *Proc. Natl. Acad. Sci. U.S.A.,* 54, 1579, 1965.
78. **Wagner, R. R. and Huang, A. S.**, Inhibition of RNA and interferon synthesis in Krebs-2 cells infected with vesicular stomatitis virus, *Virology,* 28, 1, 1966.
79. **Huang, A. S., Greenawalt, J. W., and Wagner, R. R.**, Defective T particles of vesicular stomatitis virus. I. Preparation, morphology, and some biologic properties, *Virology,* 30, 161, 1966.
80. **Yaoi, Y., Mitsui, H., and Amano, M.**, Effect of UV-irradiated vesicular stomatitis virus on nucleic acid synthesis in chick embryo cells, *J. Gen. Virol.,* 8, 165, 1970.
81. **Baltimore, D., Huang, A. S., and Stampfer, M.**, Ribonucleic acid synthesis of vesicular stomatitis virus. II. An RNA polymerase in the virion, *Proc. Natl. Acad. Sci. U.S.A.,* 66, 572, 1970.
82. **Deutsch, V., Muel, B., and Brun, G.**, Action spectra for the rescue of temperature-sensitive mutants of vesicular stomatitis virus by ultraviolet-irradiated virions at nonpermissive temperature, *Virology,* 77, 294, 1977.
83. **Rubin, C. S. and Rosen, O. M.**, Protein phosphorylation, *Annu. Rev. Biochem.,* 44, 831, 1975.
84. **Sokol, F. and Clark, H.**, Phosphorylation of rhabdovirus proteins. in *Negative Strand Viruses,* Vol. 1, Mahy, B. W. J. and Barry, R. D., Eds., Academic Press, London, 1975, 25.
85. **Marvaldi, J. and Lucas-Lenard, J.**, Differences in the ribosomal protein gel profile after infection of L cells with wildtype or temperature-sensitive mutants of VSV, *Biochemistry,* 16, 4320, 1977.
86. **Allison, A. C.**, Lysosomes in virus-infected cells; in virus directed host response, *Perspect. Virol.,* 5, 29, 1967.
87. **Norkin, L. C.**, Cell killing by simian virus 40: impairment of membrane formation and function, *J. Virol.,* 21, 872, 1977.
88. **Carrasco, L.**, The inhibition of cell functions after viral infection. A proposed general mechanism, *FEBS Lett.,* 76, 11, 1977.

89. **Francoeur, A. M. and Stanners, C. P.**, Evidence against the role of K^+ in the shut off of protein synthesis by vesicular stomatitis virus, *J. Gen. Virol.*, 39, 551, 1978.
90. **Nishioka, Y. and Silverstein, S.**, Alterations in the protein synthetic apparatus of Friend erythroleukemia cells infected with vesicular stomatitis virus or herpes simplex virus, *J. Virol.*, 25, 422, 1978.
91. **Wertz, G. W.**, Method of examining viral RNA metabolism in cells in culture: metabolism of vesicular stomatitis virus RNA, *J. Virol.*, 16, 1340, 1975.
92. **Nuss, D. L. and Koch, G.**, Translation of individual host mRNAs in MPC-11 cells is differentially suppressed after infection by vesicular stomatitis virus, *J. Virol.*, 19, 572, 1976.
93. **Knipe, D. M., Baltimore, D., and Lodish, H. F.**, Separate pathways of maturation of the major structural proteins of vesicular stomatitis virus, *J. Virol.*, 21, 1128, 1977.
94. **Knipe, D. M., Baltimore, D., and Lodish, H. F.**, Maturation of viral proteins in cells infected with temperature-sensitive mutants of vesicular stomatitis virus, *J. Virol.*, 21, 149, 1977.
95. **Heine, J. W. and Schnaitman, C. A.**, Entry of vesicular stomatitis virus into L cells, *J. Virol.*, 8, 786, 1971.
96. **McSharry, J. J. and Choppin, P. W.**, Biological properties of the vesicular stomatitis virus glycoprotein. I. Effects of the isolated glycoprotein on host macromolecular synthesis, *Virology*, 84, 172, 1978.
97. **Lucas-Lenard, J.**, Personal communication, 1978.
98. **Knipe, D., Lodish, H. F., and Baltimore, D.**, Analysis of the defects of temperature-sensitive mutants of vesicular stomatitis virus: intracellular degradation of specific viral proteins, *J. Virol.*, 21, 1140, 1977.
99. **Holloway, A. F., Wong, P. K. Y., and Cormack, D. V.**, Isolation and characterization of temperature-sensitive mutants of vesicular stomatitis virus, *Virology*, 42, 917, 1970.
100. **Friedman, R. M. and Costa, J. R.**, Fate of interferon-treated cells, *Infect. Immun.*, 13, 487, 1976.
101. **Baxt, B., Sonnabend, J. A., and Bablanian, R.**, Effects of interferon on vesicular stomatitis virus transcription and translation, *J. Gen. Virol.*, 35, 325, 1977.
102. **Thacore, H.**, Effect of interferon on transcription and translation of vesicular stomatitis virus in human and simian cell cultures, *J. Gen. Virol.*, 41, 421, 1978.
103. **Greengard, P.**, Phosphorylated proteins as physiological effectors, *Science*, 199, 146, 1978.
104. **Kawai, A. and Matsumoto, S.**, Interfering and noninterfering defective particles generated by a rabies small plaque variant virus, *Virology*, 76, 60, 1977.
105. **Yang, Y. J., Stoltz, D. B., and Prevec, L.**, Growth of vesicular stomatitis virus in a continuous culture line of *Antheraea eucalypti* moth cells, *J. Gen. Virol.*, 5, 473, 1969.
106. **Mudd, J. A., Leavitt, R. W., Kingsbury, D. T., and Holland, J. J.**, Natural selection of mutants of vesicular stomatitis virus by cultured cells of *Drosophila melanogaster*, *J. Gen. Virol.*, 20, 341, 1973.
107. **Artsob, H. and Spence, L.**, Growth of vesicular stomatitis virus in mosquito cell lines, *Can. J. Microbiol.*, 20, 329, 1974.
108. **Baer, G. M., Shaddock, J. H., Moore, S. A., Yager, P. A., Baron, S. S., and Levy, H. B.**, Successful prophylaxis against rabies in mice and Rhesus monkeys: the interferon system and vaccine, *J. Infect. Dis.*, 136, 286, 1977.
109. **Wiktor, T. J., Koprowski, H., Mitchell, J. R., and Merigan, T. C.**, Role of interferon in prophylaxis of rabies after exposure, in *Antivirals with Clinical Potential*, Merigan, T. C., Ed., University of Chicago Press, 1976, 260.
110. **Ramseur, J. M. and Friedman, R. M.**, Prolonged infection of interferon-treated cells by vesicular stomatitis virus: possible role of temperature-sensitive mutants and interferon, *J. Gen. Virol.*, 37, 523, 1977.
111. **Wiktor, T. J. and Clark, H F.**, Chronic rabies virus infection of cell cultures, *Infect. Immun.*, 6, 988, 1972.
112. **Sulkin, S. E. and Allen, R.**, Interferon and rabies virus infection, in *The Natural History of Rabies*, Vol. 1, Baer, G. M., Ed., Academic Press, New York, 1975, 355.
113. **Bishop, D. H. L. and Flamand, A.**, Transcription processes of animal RNA viruses, in *Control Processes in Virus Multiplication*, Symp. 25, Soc. Gen. Microbiol., Burke, D. C. and Russell, W. C., Eds., Cambridge University Press, London, 1975, 117.
114. **Campbell, J. B., Maes, R. F., Wiktor, T. J., and Koprowski, H.**, The inhibition of rabies virus by arabinosyl cytosine. Studies on the mechanism and specificity of action, *Virology*, 34, 701, 1968.
115. **Wiktor, T. J. and Koprowski, H.**, Rhabdovirus replication in enucleated host cells, *J. Virol.*, 14, 300, 1974.
116. **Sarver, N., and Stollar, V.**, Sindbis virus-induced cytopathic effect in clones of *Aedes albopictus* (Singh) cells, *Virology*, 80, 390, 1977.
117. **Hsu, H. T.**, Cell fusion by a plant virus, *Virology*, 84, 9, 1978.
118. **Kascsak, R. J. and Lyons, M. J.**, Attempts to demonstrate the interferon defense mechanism in cultured mosquito cells, *Arch. Gesamte Virusforsch.*, 45, 149, 1974.

119. **Marcus, P. I., Sekellick, M. J., and Fuller, F. J.**, Double-stranded RNA: the interferon inducer of viruses, 4th Int. Virol. Cong., Abstr., 1978, 107.
120. **Imblum, R. L. and Wagner, R. R.**, Protein kinase and phosphoproteins of vesicular stomatitis virus, *J. Virol.*, 13, 113, 1974.
121. **Marcus, P. I. and Freiman, M. E.**, A novel method for cutting giant cells to study viral synthesis in anucleate cytoplasm, in *Methods in Cell Physiology*, Vol. II, Prescott, D. M., Ed., Academic Press, London, 1966, 93.
122. **Follett, E. A. C., Pringle, C. R., Wunner, W. H., and Skehel, J. J.**, Virus replication in enucleate cells: vesicular stomatitis virus and influenza virus, *J. Virol.*, 13, 394, 1974.
123. **Weck, P. K. and Wagner, R. R.**, Inhibition of RNA synthesis in mouse myeloma cells infected with vesicular stomatitis virus, *J. Virol.*, 25, 770, 1978.
124. **Clark, H F.**, Rabies viruses increase in virulence when propagated in neuroblastoma cell culture, *Science*, 199, 1072, 1978.
125. **Prescott, D. M., Myerson, D., and Wallace, J.**, Enucleation of mammalian cells with cytochalasin B, *Exp. Cell Res.*, 71, 480, 1972.
126. **Marcus, P. I.**, Cancer biology IV in *Advances in Pathobiology*, Vol. 6, Stratton Intercontinental, New York, 1977, 192.
127. **Sekellick, M. J. and Marcus, P. I.**, Persistent infection. II. Interferon-inducing temperature-sensitive mutants as mediators of cell sparing: possible role in persistent infection by vesicular stomatitis virus, *Virology*, 95, 36, 1979.
128. **Marcus, P. I.**, unpublished observations.
129. **Lucas-Lenard, J.**, personal communication.
130. **Marcus, P. I.**, unpublished observations.
131. **Lucas-Lenard, J.**, personal communication.
132. **Marcus, P. I.**, unpublished observations.
133. **Lucas-Lenard, J. and Wu, F. S.**, personal communication, 1978.
134. **Marcus, P. I. and Sekellick, M. J.**, unpublished observations.

Chapter 3

RHABDOVIRUS PSEUDOTYPES

Robin A. Weiss

TABLE OF CONTENTS

I. INTRODUCTION

Pseudotypes[1] are virus particles containing the genome of one virus and coat proteins of another. Pseudotypes represent an extreme form of phenotypic mixing, first recognized in bacteriophages,[2] whereby mixed assembly of coat proteins occurs on mixed infecton of structurally related viruses. Phenotypic mixing is an important phenomenon in the biology of retroviruses, where defective viruses are commonly rescued and propagated by pseudotype formation with helper viruses. Among rhabdoviruses, phenotypic mixing has not been analyzed in detail, perhaps because it is difficult to attain efficient mixed infection between different strains of highly lytic viruses. However, vesicular stomatitis virus (VSV), pseudotypes bearing the genome and nucleocapsid of the Indiana serotype and the glycoprotein of the New Jersey serotype, have been described by Deutsch[3] when UV-irradiated VSV_{NJ} was added to cells infected with *ts* 045(V), a glycoprotein mutant of VSV_{IND}. Pseudotype-like particles have also been prepared in vitro by Bishop et al.[4] in restoring infectivity to VSV_{IND} particles treated with bromelain or pronase by adding VSV_{NJ} envelope components. In both these cases, the infectivity of the rescued virions was neutralized specifically by anti-VSV_{NJ} serum, but propagation of the particles yielded VSV_{IND} progeny.

It is the capacity of VSV to assemble glycoproteins of other, quite unrelated, enveloped viruses (Figure 1) which has generated much interest in rhabdovirus pseudotypes in recent years, and which is the main topic of this chapter. The phenomenon was first described by Choppin and Compans,[5,6] who observed phenotypic mixing between VSV and the paramyxovirus, SV5, and it has since been extended to most, if not all, groups of enveloped animal viruses investigated (Table 1). In each case of phenotypic mixing between VSV and unrelated viruses, the mixed assembly has been restricted to the envelope glycoproteins, and no genetically recombinant viruses have been reported.

Zavada's discovery of pseudotype formation between VSV and avian or murine leukemia viruses[7] has stimulated the greatest interest because the pseudotypes have proved to be of great use in studying the biology of retroviruses. Moreover, because leukemia viruses are not lytic and are poor inducers of interferon, chronic infections can be readily established which are susceptible to superinfection with the rhabdovirus.

Previous reviews on pseudotypes have been written by Zavada.[8,9] Following Rubin's[1] designation, pseudotypes are denoted by the symbol of the genome of the particle followed by a bracketed symbol specifying the envelope; thus VSV(SV5) denotes a particle containing the VSV genome and nucleocapsid enveloped in a membrane containing SV5 glycoproteins. However, the distinction between pseudotypes and phenotypically mixed particles should not be defined too strictly. It is becoming increasingly clear that many particles which possess the envelope properties corresponding to the donor virus, as if complete genomic masking has occurred, nevertheless carry a minor proportion of glycoproteins coded by their own genomes, and may not in fact represent such extreme forms of phenotypic mixing as was formerly assumed.

II. DETECTION OF PSEUDOTYPES

A. Serological Detection

The detection of pseudotypes and phenotypically mixed particles depends on selective assays for genomes and envelopes.[9] VSV has proved to be a particularly useful tool in studying phenotypic mixing, not only for its propensity to become transvestite, but also on account of the ease of propagation to high titers in most cell types, the simplicity and speed of the quantitative plaque assay, and the lack of an infectious, persistent fraction on neutralization with hyperimmune antiserum. This means that even a small fraction of VSV pseudotypes resulting from mixed infections can be ti-

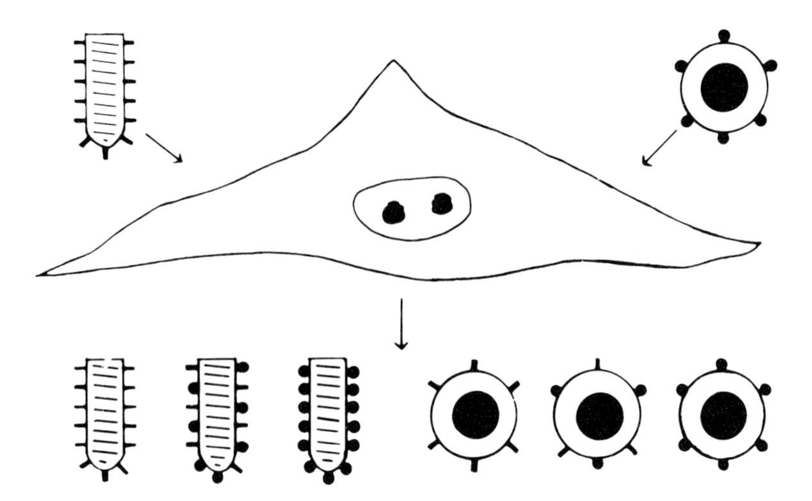

FIGURE 1. Production of phenotypically mixed particles following mixed infection of a rhabdovirus and an unrelated animal virus. When VSV is grown in cells chronically infected with retrovirus, reciprocal phenotypic mixing takes place, yielding, according to the conditions of culture and viral mutants employed, virions with mosaic envelopes or pure pseudotypes.

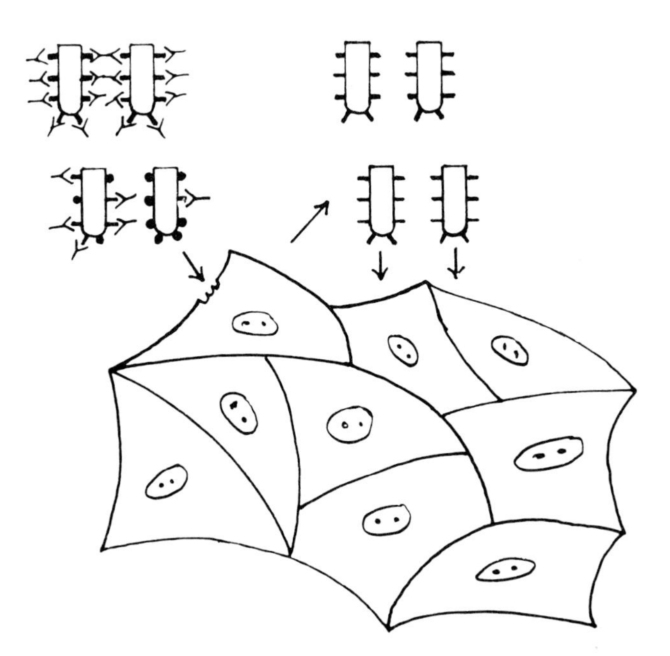

FIGURE 2. Assay of VSV pseudotype with envelope antigens of another virus. The VSV particles bearing a substantial proportion of spike (G) proteins are neutralized by specific anti-VSV antiserum, leaving pseudotypes as the only infectious particles. The pseudotype particles can only infect cells possessing appropriate receptors for the glycoproteins of the donor virus. Following virus adsorption the cells are washed to remove excess antibodies and are incubated at a temperature permissive for VSV replication. The progeny VSV particles released by a cell infected with a pseudotype carrying VSV spike proteins can infect neighboring cells. Thus, each cell infected by a pseudotype acts as an infectious center of pure VSV, which is assayed by cytopathic plaque formation. Heat inactivation of thermolabile G-protein mutants of VSV can be used in place of antiserum treatment provided that the donor glycoprotein is not also heat labile.

TABLE 1

Rhabdovirus Pseudotypes Obtained with Envelope Glycoproteins of Unrelated Viruses

Donor of envelope glycoprotein			
Family	Virus	Rhabdovirus	Ref.
Herpetoviridae	Herpes simplex Type I	VSV_{IND}	10
Orthomyxoviridae	Fowl plague	VSV_{IND}	11
Paramyxoviridae	SV5	VSV_{IND}	5, 6
	HVJ/Sendai	VSV_{NJ}	12
	Measles/SSPE	VSV_{IND}	13
Togaviridae	Sindbis	VSV_{IND}	14
	Langat	VSV_{IND}	14
Retroviridae	Avian leukosis/sarcoma	VSV_{IND}	7, 15—24
	Reticuloendotheliosis	VSV_{IND}	25
		VSV_{NJ}	25
	Murine leukemia (C-type)	VSV_{IND}	7, 26—35
	Mammary tumor (B-type)	VSV_{IND}	32
	Feline endogenous (C-type)	VSV_{IND}	31
	Bovine leukosis	VSV_{IND}	36, 37
		Chandipura	37
	Primate C-type	VSV_{IND}	31, 38,39
	D-type	VSV_{IND}	38—40
	Foamy	VSV_{IND}	41
	Human?	VSV_{IND}	42—45

trated as plaque-forming units (PFU) that resist neutralization by anti-VSV serum (Figure 2).

In mixed infections with other lytic viruses, PFU-bearing VSV genomes have been distinguished from the PFU of other viruses by the morphology or early appearance of the plaques,[5,10,11] by plating at temperatures or onto assay cells that yield only VSV plaques,[5,11,14] or in the case of the DNA virus, herpes simplex Type I (HSV), inhibiting HSV growth (but not the synthesis of herpes glycoproteins) by cytosine arabinoside.[10] The nonneutralized PFU fraction is presumed to be the pseudotype population of particles containing VSV genomes, and envelope glycoproteins composed mainly if not entirely of the second virus. The highest pseudotype fractions among VSV progeny have been obtained with HSV (30%)[10] and the avian leukosis virus, RAV-1 (20%)[17,21] but with a high total yield of VSV, pseudotype fractions as low as 10^{-6} can be readily discerned.[31,32]

The specificity of the pseudotypes can be ascertained by treatment with neutralizing antisera specific to the virus donating the envelope antigens. In addition, many enveloped viruses used for preparing VSV pseudotypes utilize cell surface receptors expressed on a restricted number of cell species or types in comparison to VSV. The restricted host range of the pseudotype may then be used to confirm its specificity.[7,10,17,18,23,29,35]

B. Heat Stabilization of Thermolabile G Protein Mutants

As an alternative to neutralization by anti-VSV serum, pseudotypes may be detected by heat inactivating phenotypically mixed stocks prepared with VSV possessing thermolabile (*tl*) G protein.[15] Two VSV_{IND} complementation group V mutants with a ther-

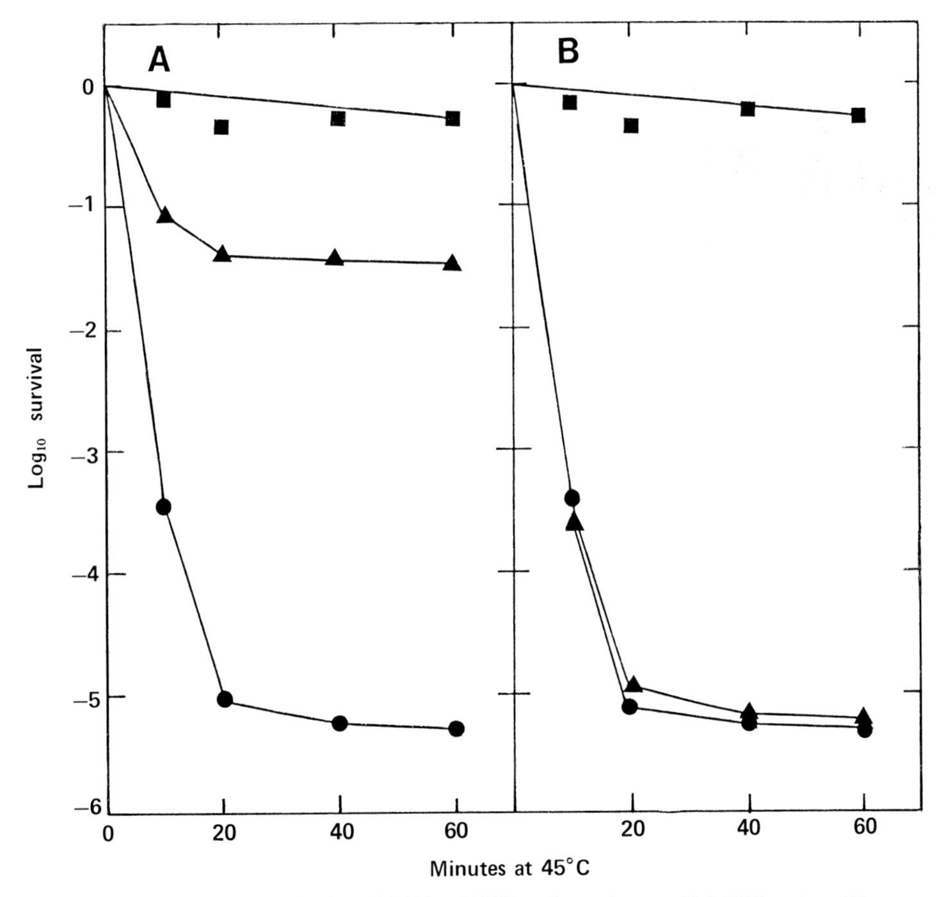

FIGURE 3. Thermal inactivation of VSV *ts* 045(V) and pseudotype. *Ts* 045(V) and a wild-type revertant were propagated at 32°C for 8 hr in normal chick cells or in chick cells preinfected with the avian leukosis virus, Carr-Zilber associated virus (CZAV). The harvested viruses were immersed in a 45°C water-bath for the times indicated and the surviving plaque-forming units titrated in chick cells susceptible to CZAV (panel A) and in duck cells resistant to CZAV (panel B).●, *ts* 045(V) grown in normal chick cells; ▲, *ts* 045(V) grown in chick cells preinfected with CZAV; ■, *wt* VSV grown in normal chick cells.

molabile function in the G protein have been used extensively, *ts* 045(V) and *tl* B17(I + V). These mutants are also temperature-sensitive *(ts)* for growth. When grown at the permissive temperature and heated at 45°C before plaque assay, there is a rapid loss of infectivity, giving a 10^{-5} heat resistant fraction for *ts* 045(V) and no discernible heat resistant fraction for *tl* B17(I + V). If these viruses are propagated in mixed infection with other enveloped viruses possessing heat stable glycoproteins, a heat-resistant pseudotype fraction is revealed (Figure 3). As is found with the anti-VSV resistant fraction, the heat-resistant pseudotypes possess the restricted host range and susceptibility to neutralization characteristic of the virus donating the envelope antigens.

Similar pseudotype titers are generally found in using either the neutralization or the heat-inactivation method (Figure 4). Zavada's *tl* B17(I + V) mutant[15] has been extensively used in pseudotype studies.[13,16-20,23,32,34,35,38-43] However, the envelope antigens of some retroviruses, such as bovine leukosis virus and baboon virus,[31,39] appear to be thermolabile at the temperature used for inactivation of *tl* B17(I + V), so that heat inactivation is not the method of choice for revealing all pseudotypes. In some mixed infections, *tl* B17(I + V) and *ts* 045(V) provide better pseudotype yields than *wt* (wild type) VSV when propagated at permissive temperature and treated with anti-VSV serum.

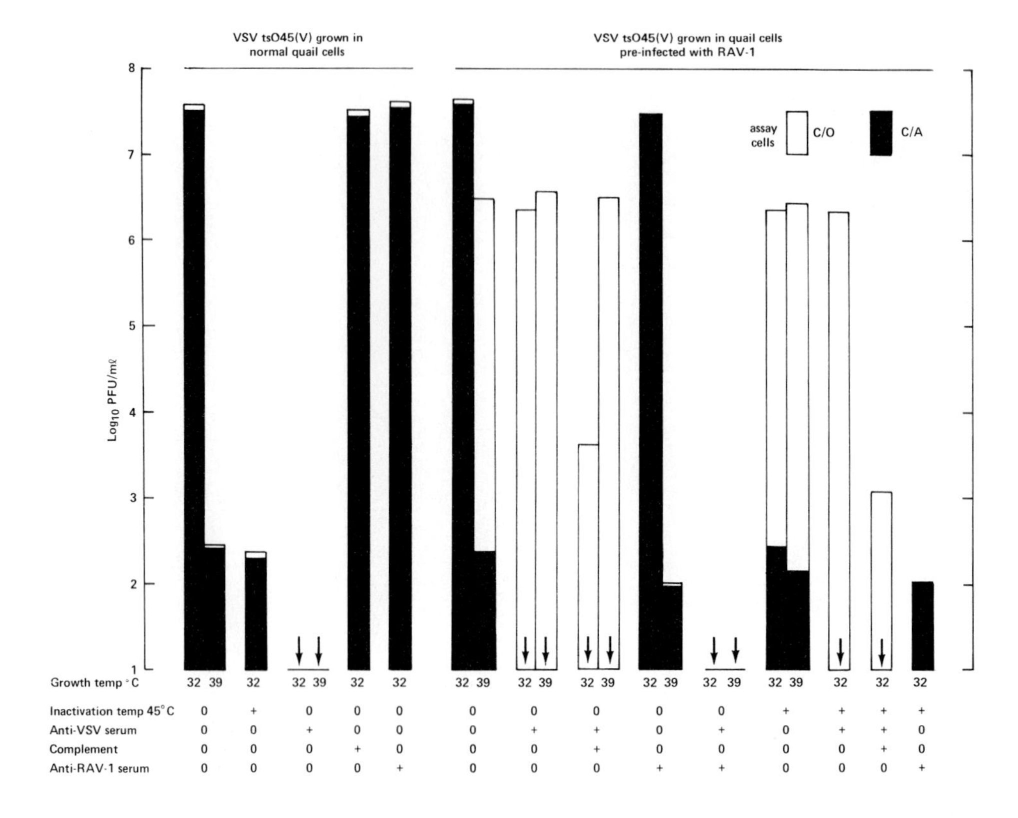

FIGURE 4. Properties of VSV *ts* 045(V) grown at permissive and restrictive temperatures in Japanese quail cells producing Rous associated virus 1 (RAV-1). Normal cells and cells preinfected with the avian leukosis virus RAV-1 were infected with VSV *ts* 045(V) (moi = 3) and incubated for 8 hr at 32°C or at 39°C. The harvested media were assayed at 32°C for PFU plating on chick C/O cells which bear cell surface receptors for RAV-1 (white columns) and on chick C/A cells, which lack receptors for RAV-1 (black columns superimposed on the white columns), both cell types being susceptible to VSV infection. The PFU were titrated before or after thermal inactivation at 45°C for 60 min, followed by incubation at room temperature with normal sheep serum, sheep anti-VSV serum, or chick anti-RAV-1 serum. Heat-inactivated complement or fresh complement were added 1 hr after mixing the antisera with virus and the reaction mixtures were incubated for a further hour before titrating the PFU on C/O and C/A cells.

The results with *ts* 045(V) grown in normal quail cells show that the yield of PFU at 39°C is 10^{-5} that obtained at 32°C, the fraction of PFU grown at 32°C and surviving thermal inactivation is also 10^{-5}, and anti-VSV serum completely neutralizes infectivity for both assay cell types. The yield of PFU obtained at 39°C from RAV-1-producing quail cells is 10^{4}-fold greater than that from normal quail cells at 39°C but the host range of the 'rescued' PFU is restricted to C/O cells bearing receptors to RAV-1. The rescued virus is strongly neutralized by anti-RAV-1 serum but is not significantly neutralized by anti-VSV serum even after addition of complement, indicating that VSV antigens are not present on the surface of these pseudotype particles.

The yield of PFU obtained at 32°C from RAV-1-producing cells is equivalent to that from normal cells. Following treatment of this stock with anti-VSV serum there is no surviving fraction plating on C/A cells, but 10^{-1} surviving fraction plating on C/O cells, thus revealing VSV(RAV-1) pseudotypes similar in yield and restriction of host range to the PFU rescued at 39°C. In contrast to the 39°C VSV(RAV-1) stock, however, the 32°C stock is reduced to 10^{-4} surviving PFU by the addition of complement to the neutralization reaction, indicating the presence of VSV antigens on the surface of the pseudotype virions despite their restricted host range. Treatment of the 32°C stock with anti-RAV-1 serum does not significantly reduce the infectivity for either assay cell type, whereas a combination of anti-VSV and anti-RAV-1 sera completely neutralizes all infectivity.

The heat-inactivation studies show that growth in RAV-1-producing cells at 32°C yields a 10^{-1} thermoresistant fraction, equivalent in titer to the pseudotype fraction resisting anti-VSV and to the total yield obtained at 39°C (which is not itself thermosensitive). The thermoresistant fraction of the 32°C stock is restricted to the host range of RAV-1, and is neutralized by anti-RAV-1 serum, but not by anti-VSV serum. Treatment with anti-VSV plus complement, however, causes a 10^{3} reduction in PFU, indicating the presence of VSV antigens on the surface of the thermoresistant pseudotype virions.

C. Rescue of Conditional G Protein Mutants

VSV_{IND} mutants in complementation Group V, such as *ts* 045(V) and *ts* M501(V) which are temperature sensitive for maturation of the G protein, can be rescued at the nonpermissive temperature by pseudotype formation with avian retroviruses.[7,16,23] Figure 4 depicts the rescue of infectious progeny of *ts* 045(V) from quail cells producing RAV-1 grown at 39°C. There is a 10^4-fold increase in yield of PFU compared with *ts* 045 propagated in uninfected chick cells. The rescued virus behaves as an envelope pseudotype in host range and neutralization tests. Approximately the same titer of *ts* 045(RAV-1) is obtained at 39°C as the pseudotype titer of the 32°C stock revealed by anti-VSV or heat treatment. In contrast to the pseudotype fraction obtained at 32°C, the *ts* 045(RAV-1) rescued at 39°C is not inactivated by anti-VSV serum plus complement. This finding indicates that no VSV antigens are exposed on the surface of the rescued virion, which will be discussed further in the section on pseudotype composition. Rescue of *ts* G protein mutants is thus an efficient way of preparing avian retrovirus pseudotypes, and it has also been used as a method of selecting new glycoprotein mutants.[24] However, it should be noted that pseudotype rescue at nonpermissive temperature by murine retrovirus is barely detectable.[23,30] Therefore, rescue of *ts* (V) mutants may not be applicable as a general method for preparing VSV pseudotypes.

III. COMPOSITION OF PSEUDOTYPES

A. Biochemical Studies

Very little biochemical work on pseudotypes has been published, owing to the difficulty in obtaining pseudotypes separated from the parental viruses. Whereas selective biological and immunological techniques can reveal small proportions of pseudotypes and phenotypically mixed particles, biochemical studies require larger amounts of pseudotypes free of other particles or proteins. Nevertheless McSharry et al.[6] have been successful in studying VSV(SV5) pseudotypes. The rhabdovirus particles can be separated from the paramyxovirus particles by isopycnic centrifugation in potassium tartrate, and the proteins of the separated particles were identified by gel electrophoresis. A study of phenotypically mixed stocks showed that both the SV5 spike proteins were assembled in the envelope of the VSV particles, in addition to all the VSV proteins, but that no other SV5 proteins, including the nonglycosylated membrane (matrix) protein were incorporated into the bullet-shaped VSV particles. The hemagglutinin and neuraminidase properties of SV5 glycoproteins were present on phenotypically mixed VSV, in that binding of bullet-shaped particles to erythrocytes at 0°C and subsequent elution at 37°C was observed.[5] Furthermore, neuraminidase activity was demonstrated in the VSV particles separated from SV5 by the tartrate gradient.[6]

B. Immunoelectron Microscopy

The incorporation of foreign viral glycoproteins into the envelope of VSV particles has also been visualized by immunoelectron microscopy of mixed infections of VSV and SV5,[5] VSV and MuLV,[34] and VSV and RAV-1 (Figure 5). Ferritin-labeled antibody specific to the donor virus antigens has revealed the frequent presence of antigens on the surface of the rhabdovirus particles, sometimes in clustered distribution, although it was not restricted to any particular region of the particles. In the double-labeled immunoelectron microscopy shown in Figure 5, hemocyanin-labeled anti-VSV antibody is observed on rhabdovirus particles which have also bound ferritin-labeled anti-RAV-1, indicating that individual particles can bear both VSV and RAV-1 envelope antigens.

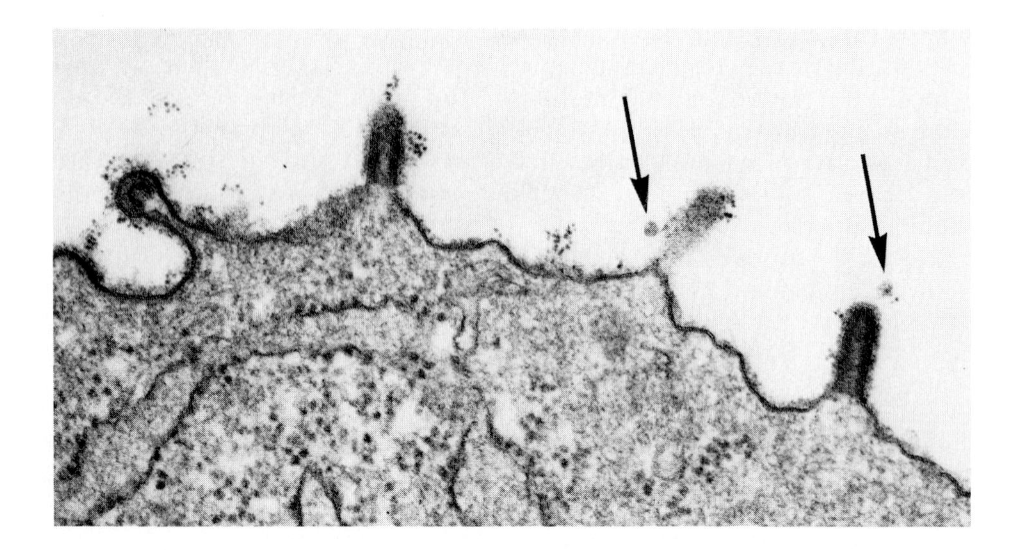

FIGURE 5. Electron micrograph of VSV and RAV-1 particles budding from a doubly infected chick embryo fibroblast in culture. The cell has been treated with ferritin-conjugated chick anti-RAV-1 serum and hemocyanin-conjugated sheep anti-VSV serum. The surface of the budding VSV is stained with clusters of ferritin particles in addition to the hemocyanin (arrows), suggesting a mosaic composition of antigens in the envelope.

C. Serological Analysis of Envelope Mosaics

Early experiments with phenotypic mixing indicated that VSV particles could be classified into three categories according to their response to neutralizing antisera specific to each parental virus.[5,7,11,44] Particles that were neutralized only by anti-VSV were not phenotypically mixed; particles that were neutralized by either anti-VSV or antiserum specific to the other virus (doubly neutralizable particles) were thought to have a mosaic composition; and particles that resisted neutralization by anti-VSV, but were neutralized by the alternative antiserum, were thought to represent pure pseudotypes in which all the VSV envelope spikes were substituted by those of the other virus. Recent studies indicate that particles of the third category, viz., the pseudotypes which resist neutralization by anti-VSV, in fact contain small amounts of VSV envelope antigen, presumably glycoproteins, detectable by their sensitivity to complement-mediated virolysis or immunoprecipitation.[18,23,30]

Figure 4 shows the effects of neutralizing antisera and complement on the infectivity of VSV *ts* 045(V) phenotypically mixed with RAV-1. Approximately 95% of the PFU grown in RAV-1-infected cells at 32°C are neutralized by anti-VSV serum and approximately 40% by anti-RAV-1 serum, indicating that at least 35% are doubly neutralizable particles containing significant amounts of both VSV and RAV-1 glycoproteins. Simultaneous treatment with both antisera neutralizes all the infectivity. Although 5% of the PFU survive neutralization by anti-VSV serum (at a concentration sufficient to neutralize pure-grown VSV to less than 10^{-7}), the addition of complement reduces the surviving fraction to 10^{-4}. A similar effect is seen with *ts* 045(V) grown at 32°C and heat-inactivated to reveal a 10^{-1} pseudotype fraction. This fraction is not significantly neutralized by anti-VSV serum alone, but is reduced to 10^{-4} by complement. These results indicate that >99% of the nonneutralizable pseudotype fraction nevertheless carries sufficient VSV envelope antigen to render the virions susceptible to virolysis.

The presence of small amounts of VSV G protein on the surface of *ts* M501 (MuLV) pseudotypes grown at 32°C, coupled with the failure to obtain *ts* 045 (MuLV) or *ts* M501 (MuLV) pseudotypes at 39°C, led Witte and Baltimore[30] to propose that VSV glycoprotein was necessary at least to initiate the budding of pseudotype virions. In cells infected with avian retroviruses, however, pseudotype formation takes place efficiently at 39°C.[7,16,23] Serological analysis of *ts* 45(RAV-1) obtained at 39°C shows that this pseudotype is not sensitive to virolysis by anti-VSV plus complement[23] (Figure 4). Apparently these particles represent pure pseudotypes devoid of VSV envelope antigens. It would be interesting to study such pseudotypes biochemically if they could be separated from the C-particles present in the same stocks.

IV. RECIPROCAL AND UNILATERAL PHENOTYPIC MIXING

As depicted in Figure 1, reciprocal phenotypic mixing may occur, yielding pseudotypes of the second virus enveloped partially or wholly in VSV glycoproteins. This has been demonstrated with RSV[19,22,23] and murine sarcoma virus (MSV).[46] In order to identify the sarcoma virus pseudotypes, *ts* VSV mutants must be employed so that selective focus assays for the sarcoma virus can be conducted at a temperature nonpermissive for *ts* VSV replication. The mutant *ts* T1026(I) is useful for studying reciprocal phenotypic mixing[22,46] because it is relatively inefficient in shutting off host cell and hence, retrovirus functions and *ts* 114G(I) has also been used to advantage[23] because it is nonleaky at 39°C and is thermolabile. Livingston et al.[46] revealed the MSV(VSV) pseudotypes as focus-forming units (FFU) that resisted neutralization by antiserum specific to murine leukemia virus, but were susceptible to anti-VSV. Sarcoma virus pseudotypes would be expected to possess a much wider host range for cell penetration than that of the parental sarcoma virus. Hence the RSV(VSV) pseudotypes were sought and detected as FFU plating on cells lacking receptors for the sarcoma virus glycoprotein.[19,22] The majority (>95%) of the FFU with the expanded host range were doubly neutralizable by either anti-VSV serum or anti-retrovirus serum, indicating that they were comprised of a mosaic of envelope antigens. This indicates that expansion of host range can be employed as a particularly sensitive method of detecting phenotypically mixed particles with relatively few donor envelope antigens. Unfortunately it is not applicable to the detection of VSV particles with phenotypically mixed envelopes as the host range of VSV itself is so wide.

VSV can also act as a helper virus[22,23] in rescuing the infectivity of envelope-defective strains of RSV which release RSV particles lacking the major envelope glycoprotein, gp85, but VSV does not complement nonenvelope defects of sarcoma viruses.[22,46] RSV(VSV) pseudotypes formed with defective RSV in the absence of a helper retrovirus are neutralized by anti-VSV serum only. In the presence of the helper virus, RAV-1, the titer of RSV(VSV) is enhanced rather than reduced, and is doubly neutralizable by either anti-VSV or anti-RAV-1 serum.[22] These results suggest that co-operative assembly of the unrelated VSV and RAV-1 glycoproteins occurs in the budding virions of RSV.

On account of the broadened host range conferred on retroviruses by VSV, including the rescue of infectivity for envelope-defective sarcoma viruses, suitable safety precautions should be taken in conducting phenotypic mixing experiments so that tumor virus pseudotypes do not infect hosts which are nonsusceptible to the parental tumor virus. Indeed, since many cells lines (e.g., CHO and L cells) commonly used in rhabdovirus research frequently release C-type particles, care should be taken that virologists and their assistants are not unwittingly exposed to C-type (VSV) pseudotypes.

In contrast to the reciprocal pseudotype formation by VSV and retroviruses, phenotypic mixing of VSV with fowl plaque virus (FPV)[11] and Sindbis virus[14] was found to be unilateral. VSV(FPV) and VSV(Sindbis) pseudotypes were readily demonstrable, but the reverse pseudotypes FPV(VSV) and Sindbis(VSV) were not found even under sensitive selective conditions for their detection. Similarly, in mixed infections of Sindbis and RSV, RSV(Sindbis) pseudotypes were detected, but not Sindbis(RSV) pseudotypes.[14] The reasons for unilateral phenotypic mixing are not understood. The envelope of Sindbis virus appears to be closely bound to the nucleocapsid; it is possible that the transmembrane termini of the envelope glycoproteins interact directly with the core proteins, as togaviruses do not have matrix proteins, and this might explain the apparent stringency of glycoprotein assembly.

The possibility that reciprocal phenotypic mixing occurs between the other virus groups known to form rhabdovirus pseudotypes has not been studied. It would be particularly interesting to search for HSV(VSV) pseudotypes in mixed infections of VSV and HSV in addition to the VSV(HSV) already reported,[10] because the VSV envelope is derived by budding from the plasma membrane of the host cell whereas the HSV envelope is derived from the nuclear membrane.

V. APPLICATIONS OF RHABDOVIRUS PSEUDOTYPES

A. Genetic and Structural Studies

The application of phenotypic mixing techniques to the study of conditional mutants has been fruitful for learning more about the assembly of rhabdoviruses. The rescue of *ts* VSV_{IND} mutants at nonpermissive temperature and the stabilization of thermolabile defects helped to identify complementation Group V with the G protein.[3,16] Studies of reciprocal pseudotype formation with Group III mutants indicate that the G protein is assembled into retrovirus particles under conditions not permissive for M protein interactions or for the assembly of VSV virions.[23] The rescue of *ts* mutants by glycoproteins of unrelated viruses has recently been applied to select new G protein mutants of VSV_{IND} and in principle can be applied to all classes of viruses able to form pseudotypes.[24]

Rhabdovirus pseudotypes have also been employed to aid the characterization of retrovirus *ts* mutants. Using *wt* VSV and *ts* MuLV, Breitman and Prevec[28] found that VSV(MuLV) pseudotypes could not be obtained at the nonpermissive temperature for the MuLV mutant, *ts* 3, which has a temperature-sensitive block in the release of budding virions from the cell membrane. However, negative evidence of phenotypic mixing must be interpreted with caution, lest it be a special rather than general case. Tato et al.[22] have investigated an RSV mutant, LA30A, which has an early thermolabile defect. The observation that *wt* VSV(LA30A) pseudotypes are thermolabile helped to pinpoint the lesion to the RSV envelope glycoproteins.

The demonstration of the presence of VSV envelope antigens on the surface of nonneutralizable pseudotypes[19,23,30] by complement-mediated lysis (see Section IIIC) indicates that antibody can bind to virus particles without neutralizing them, supporting a multi-hit model for neutralization of enveloped viruses.[47] A detailed analysis of the kinetics of pseudotype neutralization, however, awaits the development of monoclonal neutralizing antibodies, because the antisera used to date may also contain secondary specificities which might be nonneutralizing in the absence of complement. Indeed, the hyperimmune anti-VSV sera are not specific to the G protein, reacting with the four nonglycosylated structural proteins too; yet the resistance of *ts* 045 pseudotypes prepared at 39°C to complement-mediated antibody inactivation indicates that the other virus-coded antigens are not exposed on the surface of intact pseudotype virions in a manner that causes virolysis. A recent study employing neutralizing, monoclonal anti-G antibody confirms the multi-hit model for inactivation of pseudotype virions.[23]

B. Detection of Glycoproteins in Latent Infections

The full replication of the virus donating antigens to rhabdovirus pseudotypes is not essential, provided that glycoproteins are expressed at the cell surface. Inhibition of HSV replication by cytosine arabinoside did not diminish the production of VSV(HSV) pseudotypes.[10] Neither is the release of retrovirus particles necessary for efficient formation of pseudotypes. This was demonstrated in a study of the assembly of glycoproteins coded by the endogenous avian C-type viral genome (chick helper factor) into VSV envelopes.[17] Only the cells of those embryos known to express chick helper factor formed pseudotypes of VSV, which possessed the host range and neutralization properties specific to the endogenous virus.

It has been of considerable interest, therefore, to investigate whether pseudotypes might result from propagation of VSV in human tumor cells. The early promise of phenotypic mixing in human tumor cells has been somewhat disappointing because, while pseudotypes were unquestionably demonstrated,[42,44] the origin or passage history of the cells has been obscure. The human mammary carcinoma and human sarcoma cells used by Zavada et al.,[42,43] may have been contaminated with variants of HeLa cells or with the D-type retrovirus present in many European HeLa cell sublines. This retrovirus does phenotypically mix with VSV, although in some chronically infected cell lines only doubly neutralizable VSV particles can be detected.[38-40] However, the human tumor cells first studied[42,43] did not release retrovirus particles, but yielded a substantial VSV pseudotype fraction as well as doubly neutralizable particles. Recently, using more sensitive techniques for preserving phenotypically mixed VSV,[32] new pseudotypes have been detected in a human melanoma line and in some cultures of normal human embryo cells.[45] So far as is known, these cultures have not been subject to laboratory contamination with D-type or other retroviruses, and it is hoped that the characterization of the pseudotypes will eventually throw light on new antigens of presumed viral origin.

The rhabdovirus pseudotype method has obvious potential application to the search for possible membrane glycoproteins of latent viruses that might underlie some human autoimmune and neurological disorders. Wild et al.[13] have demonstrated the formation of VSV pseudotypes in Vero cells infected productively or latently with measles virus strains isolated from patients suffering subacute sclerosing panencephalitis. To date, no pseudotypes have been reported from VSV grown directly in human pathological tissues, other than the tumor cells already discussed.

C. Seroepidemiological Studies

Rhabdovirus pseudotypes are frequently more convenient than the parental virus for quantitative neutralization tests. For instance, VSV(Sindbis) is more efficiently neutralized in a direct antibody test than Sindbis virus itself.[14] The pseudotypes represent particularly useful tools for serological studies of the virus donating the envelope antigens when neutralization of the virus in question is tedious or not sufficiently quantitative. For example, the titration of neutralizing antibodies in mice specific to mammary tumor virus (MMTV) is simply accomplished in a 48 hr plaque assay.[32] This replaces a test which relied on the inoculation of MMTV into large numbers of susceptible newborn mice, with an incubation period of several months in order to quantitate a reduction in tumor incidence. Pseudotype neutralization assays of antibodies to avian,[18] bovine,[36,37] and primate[38-42] retroviruses have been similarly adopted. The technique has been applied to the search for antibodies specific to retroviruses in human sera,[39,42] and is proving to be of practical application in the discovery and elimination of bovine leukosis virus in cattle.[37] Furthermore, any pseudotype of VSV formed by the envelope antigen of an unknown virus, as might occur in chronically infected tissues or tumors in man (see Section VB), could be exploited in a seroepidemiological survey without the necessity of isolating or identifying the latent virus concerned.

D. Receptor Studies

VSV has an extraordinarily broad host range, replicating in cells of almost all species of vertebrate and insect tested. The viruses used to form VSV pseudotypes usually have a more restricted host range, and the pseudotypes can be used to test whether the restrictions apply at the stage of viral reception and penetration of the cell surface. VSV (retrovirus) pseudotypes in particular have been exploited profitably for such studies.[48]

The avian leukosis and sarcoma group of viruses is classified into subgroups according to the specificity of the envelope antigens. The subgroup classification is based on three criteria; cross-reacting neutralization antigens (assumed to be the major glycoprotein, gp85), host range patterns (based on genetic determination of cell surface receptors), and viral cross-interference patterns (due to competitive blocking of the receptors). As might be expected from previous studies of RSV pseudotypes, the VSV pseudotypes fall precisely into the envelope classification of avian tumor viruses.[7,15,17,18] Since avian tumor viruses adsorb to genetically resistant cells, the receptors being of crucial importance for penetration or uncoating rather than for adsorption,[48] the specific host range of the pseudotypes suggests that similar restrictions on penetration apply to rhabdovirus pseudotype virions as to the retrovirus particles.

Certain strains of avian sarcoma virus are oncogenic in mammals; these strains, but not the nononcogenic strains, transform mammalian cells in culture, albeit at a much lower frequency than the transformation of susceptible avian cells. It was not clear to what extent the mammalian-tropic host range was determined by the sarcoma virus envelope. VSV pseudotype studies demonstrated that the envelope specificity was indeed of crucial importance for the infection of mammalian cells, but revealed that a postpenetration restriction on transformation also operates.[18]

Rhabdovirus pseudotypes have also been of great use in clarifying host range phenomena among mammalian retroviruses. For instance, there is at present no convenient culture system for *de novo* infection and replication of MMTV, perhaps because an appropriate target cell type is not available for this tissue-specific and hormone-dependent virus. Pseudotype experiments indicate that many cell types of mice and other species are susceptible to VSV(MMTV), though NIH-3T3 and BALB-3T3 cells appear to lack MMTV receptors.

Murine C-type viruses are broadly classified into three groups: ecotropic (infectious for mouse cells), xenotropic (noninfectious for mouse cells but infectious for several other mammalian and avian species), and amphotropic (infectious for murine and foreign cells). Studies with VSV(MuLV) pseudotypes have demonstrated that a major block to infection of murine cells with xenotropic viruses, and of nonmurine cells with ecotropic viruses, is the lack of appropriate cell surface receptors.[29,31,32] Ecotropic viruses can be further classified into N-tropic, B-tropic, and NB-tropic viruses according to whether their replication is restricted by the *n* or *b* alleles of the host *Fv-1* locus. By contrast with the interspecies tropisms of MuLV strains, the restriction of N-tropic substrains of ecotropic viruses in B-type murine cells was found to be a postpenetration phenomenon, and this was first indicated by VSV pseudotype experiments.[26,27]

The differentiated state also affects the susceptibility of cells to retrovirus replication. For instance, the undifferentiated embryonal carcinoma cells of teratocarcinomas derived from strain 129 mice are completely resistant to replication of ecotropic MuLV strains, whereas many of the differentiated cells derived in culture from these embryonal carcinoma cells become permissive to MuLV.[33] As VSV replicates well in the undifferentiated cells they could be challenged with pseudotypes and were found to be highly susceptible to VSV(MuLV) infection. Thus the embryonal carcinoma cells express cell surface receptors for ecotropic MuLV, although MuLV will not replicate in these cells.[33] VSV pseudotypes should prove to be of use in analyzing further examples of cell or tissue tropism among retroviruses, particularly those that correlate with recombinant glycoprotein specificities.

Pseudotypes can be further exploited to analyze the genetic basis of receptor determination. The host genetics of avian tumor virus infection has been previously established with the use of RSV pseudotypes. For most avian subgroups host susceptibility is determined by single loci with dominant alleles coding for cell surface receptors.[48] Susceptibility of chickens to reinfection by RAV-O, the endogenous C-type virus, is more restricted. Experiments with VSV(RAV-O) pseudotypes confirm that at least two genetic loci affect early steps of RAV-O infection, a receptor gene, and an unlinked, dominant 'inhibitor' gene coding for a factor that blocks the receptors.[19,48] Receptor genes may also be mapped to linkage groups by examining susceptibility to VSV pseudotypes in interspecies somatic hybrids between resistant and susceptible species. By this method the determinant for the ecotropic MuLV receptor has been mapped to chromosome 5 in mouse-hamster hybrids,[35] and similar studies of human-mouse hybrids allocates the receptor gene for C-type viruses belonging to the RD114-Baboon subgroup to human chromosome 19.[49]

VI. CONCLUSIONS AND PROSPECTS

Apart from their intrinsic interest, rhabdovirus pseudotypes have proved useful for studies of virion assembly, the selection and characterization of mutants, the detection of latent infections, and serological, host range and receptor studies. Phenotypic mixing techniques will continue to be useful as an adjunct of the studies, and it is hoped that the formation of rhabdovirus pseudotypes will be exploited further as a probe for glycoprotein expression of latent and perhaps hitherto unknown human infections.

The facility with which VSV and Chandipura virus assemble glycoproteins of other viruses into their envelopes is, perhaps, just an indicator of a wider phenomenon in membrane glycoprotein assembly, as VSV also appears to incorporate host antigen.[38,50] The detection of infectious pseudotypes has so far been confined to those generated by other viruses, but this may be a reflection of the sensitivity of the assays for functional viral glycoproteins, rather than a restriction of phenotypic mixing to viral antigens. It will be of interest in the future, therefore, to investigate to what extent glycoproteins of host origin are incorporated into rhabdovirus envelopes, and whether they play a functional role in the biology of the virus.

REFERENCES

1. **Rubin, H.**, Genetic control of cellular susceptibility to pseudotypes of Rous sarcoma virus, *Virology*, 26, 270, 1965.
2. **Novick, A. and Szilard, L.**, Virus strains of identical phenotype but different genotype, *Science*, 113, 34, 1951.
3. **Deutsch, V.**, Parental G protein reincorporation by a vesicular stomatitis virus temperature-sensitive mutant of complementation group V at nonpermissive temperature, *Virology*, 69, 607, 1976.
4. **Bishop, D. H. L., Repik, P., Obijeski, J. F., Moore, N. F., and Wagner, R. R.**, Restitution of infectivity to spikeless vesicular stomatitis virus by solubilized viral compnents, *J. Virol.*, 16, 75, 1975.
5. **Choppin, P. W. and Compans, R. W.**, Phenotypic mixing of envelope proteins of the parainfluenza virus SV5 and vesicular stomatitis virus, *J. Virol.*, 5, 609, 1970.
6. **McSharry, J. J., Compans, R. W., and Choppin, P. W.**, Proteins of vesicular stomatitis virus and of phenotypically mixed vesicular stomatitis virus-simian virus 5 virions, *J. Virol.*, 8, 722, 1971.

7. **Zavada, J.**, Pseudotypes of vesicular stomatitis virus with the coat of murine leukaemia and of avian myeloblastosis viruses, *J. Gen. Virol.*, 15, 183, 1972.
8. **Zavada, J.**, Viral pseudotypes and phenotypic mixing, *Arch. Virol.*, 50, 1, 1976.
9. **Zavada, J.**, Assay methods for viral pseudotypes, in *Methods in Virology*, 6, Maramorosch, K. and Koprowski, H., Eds., Academic Press, New York, 1977, 109.
10. **Huang, A. S., Palma, E. L., Hewlett, N., and Roizman, B.**, Pseudotype formation between enveloped RNA and DNA viruses, *Nature (London)*, 252, 743, 1974.
11. **Zavada, J. and Rosenbergova, M.**, Phenotypic mixing of vesicular stomatitis virus with fowl plague virus, *Acta Virol. (Engl. Ed.)*, 16, 103, 1972.
12. **Kimura, Y.**, Phenotypic mixing of vesicular stomatitis virus with HVJ (Sendai virus), *Jpn. J. Microbiol.*, 17, 373, 1973.
13. **Wild, F., Cathala, F., and Huppert, J.**, Vesicular stomatitis virus (measles) pseudotypes: tool for demonstrating defective measles infections, *Intervirology*, 6, 185, 1975.
14. **Zavadova, Z., Zavada, J., and Weiss, R. A.**, Unilateral phenotypic mixing of envelope antigens between togaviruses and vesicular stomatitis virus or avian RNA tumour virus, *J. Gen. Virol.*, 37, 557,1977.
15. **Zavada, J.**, VSV pseudotype particles with the coat of avian myeloblastosis virus, *Nature New Biol.*, 240, 122, 1972.
16. **Zavada, J. and Zavodska, E.**, Complementation and phenotypic stabilization of vesicular stomatitis virus temperature-sensitive and thermolabile mutants by avian myeloblastosis virus, *Intervirology*, 2, 25, 1973.
17. **Love, D. N. and Weiss, R. A.**, Pseudotypes of vesicular stomatitis virus determined by exogenous and endogenous avian RNA tumor viruses, *Virology*, 57, 271, 1974.
18. **Boettiger, D., Love, D. N., and Weiss, R. A.**, Virus envelope markers in mammalian tropism of avian RNA tumor viruses, *J. Virol.*, 15, 108, 1975.
19. **Weiss, R. A., Boettiger, D., and Love, D. N.**, Phenotypic mixing between vesicular stomatitis virus and avian RNA tumor viruses, *Cold Spring Harbor Symp. Quant. Biol.*, 39, 913, 1975.
20. **Ogura, H. and Friis, R.**, Further evidence for the existence of a viral envelope protein defect in the Bryan high-titer strain of Rous sarcoma virus, *J. Virol.*, 16, 443, 1975.
21. **Weiss, R. A., Boettiger, D., and Murphy, H. M.**, Pseudotypes of avian sarcoma viruses with the envelope properties of vesicular stomatitis virus, *Virology*, 76, 808, 1977.
22. **Tato, F., Beamand, J. A., and Wyke, J. A.**, A mutant of Rous sarcoma virus with a thermolabile defect in the virus envelope, *Virology*, 88, 71, 1978.
23. **Weiss, R. A. and Bennett, P.**, Assembly of glycoproteins in mutants of vesicular stomatitis virus studied by phenotypic mixing with retroviruses, *Virology*, in press.
24. **Lodish, H. F. and Weiss, R. A.**, Selective isolation of mutants of vesicular stomatitis virus defective in production of the glycoprotein, *J. Virol.*, 30, 177, 1979.
25. **Kang, C.-Y. and Lambright, P.**, Pseudotypes of vesicular stomatitis virus with the mixed coat of reticuloendotheliosis virus and vesicular stomatitis virus, *J. Virol.*, 21, 1252, 1977.
26. **Huang, A., Besmer, P., Chu, L., and Baltimore, D.**, Growth of pseudotypes of vesicular stomatitis virus with N-tropic murine leukemia virus coats in cells resistant to N-tropic viruses, *J. Virol.*, 12, 659, 1973.
27. **Krontiris, T. G., Soeiro, R., and Fields, B. N.**, Host restriction of Friend leukemia virus. Role of the viral outer coat, *Proc. Natl. Acad. Sci. U.S.A.*, 70, 2549, 1973.
28. **Breitman, M. and Prevec, L.**, The use of vesicular stomatitis virus pseudotype production in the study of a temperature-sensitive murine leukemia virus, *Virology*, 76, 643, 1977.
29. **Besmer, P. and Baltimore, D.**, Mechanism of restriction of ecotropic and xenotropic murine leukemia viruses and formation of pseudotypes between the two viruses, *J. Virol.*, 21, 965, 1977.
30. **Witte, O. N. and Baltimore, D.**, Mechanism of formation of pseudotypes between vesicular stomatitis virus and murine leukemia virus, *Cell*, 11, 505, 1977.
31. **Schnitzer, T. J., Weiss, R. A., and Zavada, J.**, Pseudotypes of vesicular stomatitis virus with the envelope properties of mammalian and primate retroviruses, *J. Virol.*, 23, 449, 1977.
32. **Zavada, J., Dickson, C., and Weiss, R. A.**, Pseudotypes of vesicular stomatitis virus with envelope antigens provided by murine mammary tumor virus, *Virology*, 82, 221, 1977.
33. **Teich, N. M., Weiss, R. A., Martin, G. R., and Lowy, D. R.**, Virus infection of murine teratocarcinoma stem cell lines, *Cell*, 12, 973, 1977.
34. **Chan, J. C., Hixson, D. C., and Bowen, J. M.**, Detection of vesicular stomatitis virus (murine leukemia virus) pseudotypes by immunoelectron microscopy, *Virology*, 88, 171, 1978.
35. **Ruddle, N. H., Conta, B. S., Leinwand, L., Kozak, C., Ruddle, F., Besmer, P., and Baltimore, D.**, Assignment of the receptor for ecotropic murine leukemia virus to mouse chromosome 5, *J. Exp. Med.*, in press.

36. **Zavada, J., Černy, L., Altstein, A. D., and Zavadova, Z.**, Pseudotype particles of vesicular stomatitis virus with surface antigens of bovine leukaemia virus — VSV (BLV) — as a sensitive probe for detecting antibodies in the sera of spontaneously infected cattle, *Acta Virol. (Engl. Ed.),* 22, 91, 1978.
37. **Zavada, J., Černy, L., Zavadova, Z., Bozonova, J., and Altstein, A. D.**, A rapid neutralization test for antibodies to bovine leukemia virus employing rhabdovirus pseudotypes, *J. Natl. Cancer Inst.*, in press, 1978.
38. **Thiry, L., Cogniauz-le-Clerc, J., Content, J., and Tack, L.**, Factors which influence inactivation of vesicular stomatitis virus by fresh human serum, *Virology,* 87, 384, 1978.
39. **Thiry, L., Sprecher-Goldberger, S., Bossens, M., Cogniaux-le-Clerc, J., and Vereerstraeten, P.**, Neutralization of Mason-Pfizer virus by sera from patients treated for renal diseases, *J. Gen. Virol.*, 41, 587, 1978.
40. **Altstein, A. D., Zhdanov, V. M., Omelchenko, T. N., Dzagurov, S. G., Miller, G. G., and Zavada, J.**, Phenotypic mixing of vesicular stomatitis virus and D-type oncornavirus, *Int. J. Cancer,* 17, 780, 1976.
41. **Schnitzer, T. J., Bharakhda, J., and Chamove, A.**, A spontaneous primate lymphoma: isolation and characterization of a simian foamy virus, *Infect. Immun.*, submitted.
42. **Zavada, J., Zavadova, Z., Malir, A., and Kocent, A.**, VSV pseudotype produced in cell line derived from human mammary carcinoma, *Nature New Biol.*, 240, 124, 1972.
43. **Zavada, J., Zavadova, Z., Widmaier, R., Bubenik, J., Indrova, M., and Altaner, Č.**, A transmissible antigen detected in two cell lines derived from human tumours, *J. Gen. Virol.*, 24, 327, 1974.
44. **Zavada, J., Bubenik, J., Widmaier, R., and Zavadova, Z.**, Phenotypically mixed vesicular stomatitis virus particles produced in human tumor cell lines, *Cold Spring Harbor Symp. Quant. Biol.*, 39, 907, 1975.
45. **Zavada, J., Bozonova, J., Zavadova, Z., and Svec, J.**, Pseudotypes of vesicular stomatitis virus as a probe for detecting viral antigens in chronic infections, in *Proceedings of the International Symposium on Chronic Virus Infections,* Slovak Academy of Sciences, Tratislava, in press.
46. **Livingston, D. M., Howard, T., and Spence, C.**, Identification of infectious virions which are vesicular stomatitis virus pseudotypes of murine type C virus, *Virology,* 70, 432, 1976.
47. **Della-Porta, A. J. and Westaway, E. G.**, A multi-hit model for the neutralization of animal viruses, *J. Gen. Virol.*, 38, 1, 1977.
48. **Weiss, R. A.**, Receptors for RNA tumor viruses, in *Cell Membrane Receptors for Viruses, Antigens and Antibodies, Polypeptide Hormones, and Small Molecules,* Beers, R. F., Jr. and Bassett, E. G., Eds., Raven Press, New York, 1976, 237.
49. **Schnitzer, T. J., Weiss, R. A., Juricek, D., and Ruddle, F.**, unpublished data, 1979.
50. **Hecht, T. T. and Summers, D. F.**, Interactions of vesicular stomatitis virus with murine cell surface antigens, *J. Virol.*, 19, 833, 1976.

Chapter 4

PERSISTENT INFECTIONS BY RHABDOVIRUSES

Margaret J. Sekellick and Philip I. Marcus

TABLE OF CONTENTS

I. INTRODUCTION

A *sine qua non* of persistent virus infection is survival of both host and virus, i.e., normally "susceptible" host cells in the presence of "lethal" virus. Persistent infections by animal viruses under natural conditions or in experimental systems such as cells in culture have been well documented.[1] Animal and host cell systems have been described for many kinds of viruses with responses ranging from slow or latent infection to chronic infections with a constant shedding of virus without cytopathic effects.[1,2]

The mechanism(s) underlying the initiation and maintenance of such infections is not understood, especially for normally cytopathic RNA viruses. Their definition represents a challenging enigma for the virologist. Defective interfering virus parti cles,[3-15] the interferon system,[10,15-23] temperature-sensitive viral mutants,[10,23-26] and endogenous RNA-dependent DNA polymerase activity[27-30] (of either cellular or latent oncornavirus origin) have all been implicated in the establishment and maintenance of persistent infection. Whatever the mechanism(s), persistent infection of susceptible cells by normally cytopathic viruses must be associated with reduced cytopathicity, at least in a population sense, in order that both virus and host cell persist.[15,31,32] However, at the single-cell level survival of a "persistently" infected cell is "all-or-none" — a cell either survives infection by a virus or it is killed.

Throughout the text we use the term "carrier" (culture or cells) to describe a virus-cell relationship dependent on an extracellular or horizontal transmission phase for perpetuating the virus. Implicit in this definition is the potential for isolating from such carrier cultures "cured" cells, which by all criteria give rise to populations that resemble normal uninfected cells. Cloning,[8] treatment with antiviral antibody,[8,10] treatment with interferon,[10] and shift to nonpermissive temperature[10,25] all are means which have been used successfully to eliminate (cure) virus from the carrier cultures. All rhabdovirus persistent infections examined to date appear to be carrier cultures requiring an extracellular transmission of the virus. This virus-carrier state contrasts with an "integrated" state of the virus, a condition in which the virus-cell relationship depends on an intracellular or vertical transmission for perpetuating the virus, and some measure of control over the expression of cell killing. Persistent infections by some paramyxoviruses best exemplify this situation.[28,30]

In this review we will examine persistent infections induced by rhabdoviruses, in particular in in vitro systems developed with vesicular stomatitis virus (VSV) and rabies virus. We will attempt to identify the essential components in these systems and define the mechanism(s) which may be operating to permit coexistence of "lethal" virus and "susceptible" host cells. We will show how seemingly unrelated reactants in persistent infection can be accommodated by a single mode of action, and specifically identify two major components in the rhabdovirus-carrier state: the interferon system and CPE-suppressing particles.

II. REACTANTS IN PERSISTENT VIRUS INFECTION

A. Defective Interfering Virus Particles

Defective interfering (DI) virus particles have been postulated to play a key role in persistent infection of cells by vesicular stomatitis virus[4-6,8,10,13-15] and rabies virus[7,11] — model systems for studying this phenomenon. Huang and Baltimore[4] were the first to suggest a role for defective interfering virus particles in viral disease processes and persistent virus infections, proposing that DI particles acted to modulate the yield of infectious virus produced in a cyclical manner. Palma and Huang[6] provided experimental evidence to support such a hypothesis, having demonstrated in vitro the cyclic overlapping production of infectious (B particles) and defective interfering particles of VSV yielded from a culture of doubly infected Chinese hamster ovary (CHO) cells to which uninfected susceptible cells were freshly added at each passage. A similar observation was made by Kawai et al.[7] (see below) who have reported the cyclic overlapping production of infectious and defective virus in Baby hamster kidney (BHK) cells persistently infected with rabies virus.

Holland and co-workers[5,8] extensively investigated the persistence of VSV in BHK cells, monitoring the changes that occurred in these carrier cells over a period of several years. They found that persistent infection could be easily and regularly established by challenging BHK cells with a particular mutant of VSV (*ts* G31 [III]), at semipermissive temperature (37°C); but only if the infecting virus population contained homologous DI (T) particles — in this case a short T particle. Clonally purified infectious VSV-*ts* G31(III) was not capable of initiating persistent infection in these cells in the absence of added homologous DI particles since infected cells were destroyed totally — a result consistent with single cell survival experiments[33] showing that a single particle of VSV-*ts* G31(III) is lethal for cells at temperatures as high as 40.0°C. Clearly, in the absence of some mechanism to spare the cell, VSV-*ts* G31(III) alone would not be capable of initiating persistent infection even at a totally restrictive temperature. However, Holland and Villarreal[5] reported that in the presence of added homologous short DI particles, some cells survived the initial infection and were able to replicate, although periodic "crises" occurred involving destruction of some cells before the carrier culture finally stabilized. The stabilized persistently infected cell population elaborated a new defective interfering particle which was morphologically distinguishable as a long DI (DI-LT) particle. This DI-LT was capable of supporting persistent infection in BHK cells even when wild-type VSV (B-particles) was used to initiate the infection.

More recently, Holland et al.[8] reported the isolation of subviral ribonucleoprotein complexes (T particle RNP) from the cytoplasmic extracts of the VSV-BHK carrier cells at a time when they were producing few, if any, mature DI particles. Upon coinfection with complete B particles this T particle RNP interfered with the yield of homologous virus and permitted establishment of persistent infection in susceptible cells.

The mechanism by which virus is maintained in this system in the absence of a cytopathic effect has not been defined precisely. Interferon mediated-interference is not a likely candidate since heterologous influenza virus formed normal plaques on monolayers of the persistently infected cells, and attempts by others[7,15] to produce interferon from BHK cells treated with conventional and very efficient interferon inducers have failed. However, cured cells can be obtained by plating carrier cells under conditions where cell to cell spread of infectious virus is prevented, such as at very low density (two or three cells per dish), or in the presence of viral antibody.[8] These results firmly implicate an extracellular mechanism for maintenance of the virus, permitting us to define the persistent infection as a virus-carrier state. Although Holland et al.[8] reported that temperature sensitive and small plaque mutants were selected for in their carrier cultures, they did not define the role of these mutants in persistent infection. However, they did report that B particles cloned from these cultures could not establish persistent infection in the absence of added DI particles. Clearly then, DI particles appear to play a critical role in the establishment, if not maintenance, of the carrier state. In this context, Villarreal and Holland[34] also showed that in carrier BHK cells the amount of full-sized virion RNA is greatly suppressed and usually not detectable — characteristics expected from DI particle-induced homotypic interference with B particle replication at the level of RNA synthesis.[35,36]

Nishiyama[10] reports detecting DI particles in a culture of mouse L cells persistently infected with VSV, along with other factors frequently found associated with carrier cultures — small plaque temperature-sensitive mutants and interferon. He was not able to separate the individual contribution of DI particles in maintaining the carrier state from these other factors present in the culture. The fact that all of these factors are present in the carrier suggests that a delicate balance of these reactants may be operating in order to maintain both the virus and the cell.

Holland et al.[8] sought to determine whether DI particles were essential for the initiation of persistent infection by rabies virus. They found that cloned B virions could readily establish persistent infection in BHK cells over a broad range of multiplicities (0.01 to 100). They proposed that DI particles were present in their clonal pools of virus, or were generated very rapidly, since small T particles were found by the 8th day after initiation of infection. A similar experiment with VSV would not usually lead to the establishment of persistent infection. Presumably, the high potential capacity of generating DI particles from rabies virus makes it a more efficient effector of persistent infection than VSV.

It seems reasonable to conclude that the presence of DI particles per se is not sufficient to establish a persistent infection. Apparently, a special kind of DI particle must be present, or selected for, to permit survival of the host cell. For example, Holland and Villarreal[5] have shown that most preparations of DI particles do not contribute to the establishment of persistent infection. Furthermore, the experiments of Marcus and Sekellick[15,37] established that at least two kinds of DI particles do not prevent the lethal action of VSV when cells are simultaneously infected with defective and infectious particles. In this context, the DI-LT particles isolated by Holland from the VSV-BHK carrier culture may represent a special kind of particle. Holland and Villarreal[5] have shown that the B particles obtained from cells simultaneously infected with B and DI-LT particles accumulated very few primary transcripts when tested in vitro and compared with an equivalent number of B particles grown separately. Conceivably, this DI-LT particle may act by limiting the synthesis and/or action of the putative cell-killing factor of VSV,[91] thereby permitting the doubly infected cell to survive an otherwise lethal infection. Mechanistically, the DI-LT subvirions seem related to the CPE-suppressing particles produced by a rabies virus-BHK carrier culture studied by Kawai et al.[7,11] and the defective particles of lymphocytic choriomeningitis virus which block cell killing.[38] We also note, and will discuss below, that preparations of DI-LT particles function as good inducers of interferon.[15]

Kawai et al.[7] characterized the virus produced by BHK cells persistently infected with the HEP-Flury strain of rabies virus. They found that by the 22nd transfer of the persistently infected cells the original large plaque virus (LPV) used to initiate infection had been replaced totally by a small plaque virus (SPV). They also demonstrated an overlapping cyclic production of defective and infectious virus particles. In these studies Kawai and colleagues made an important observation: one class of DI particles associated with the carrier culture, when added to fresh cells, were able to suppress the cytopathic effect of standard rabies virus! In a follow-up study of these "CPE-suppressing defective particles", Kawai and Matsumoto[11] found that the LPV produced persistent infection in BHK cells more quickly than the SPV strain of rabies virus and that this was correlated with the earlier appearance of CPE-suppressing defective particle activity. The CPE-suppressing defective particles appeared to be important in the maintenance of the persistent infection: cloned SPV (probably free of CPE-suppressing particles) produced a more extensive cytopathic effect in BHK cells than that produced by the original LPV virus (probably containing, or generating, CPE-suppressing particles). They also found evidence for the production of large amounts of noninterfering, non-CPE suppressing defective particles, the significance of which is unknown. No interferon or interferon-like substance was found in the medium bathing the persistently infected culture, and no antiviral state in the cells themselves was demonstrable using VSV as a heterologous challenge virus.

1. Influence of cell type on DI particle generation

In terms of generating DI particles, the cell is an important component. Experiments of Huang and Baltimore,[4] Holland et al.,[8,39] and most recently those of Kang and Allen[40] demonstrated that the host cell can influence the type of DI particle generated and, in the extreme, control whether they are made at all. Kang and Allen[40] have examined the problem of host cell influence on DI particle production and find that extensive pretreatment of cells with actinomycin D blocks production of VSV DI particles, even under conditions of high multiplicity passage. Clearly, the cell is more than just a passive carrier of virus in persistent infection, and may play a dynamic role in determining which reactants in viral persistence are brought into operation. The use of cells like HeLa[39] or Madin-Darby bovine kidney,[4] deficient in their capacity to generate DI particles, might prove useful for assessing the role of these particles in persistent infection. In one such study with HeLa cells, Holland et al.[8] noted the eventual appearance of DI particles, possibly reflecting a selection of cells able to replicate defective particles. Conceivably, the DI particles were unique.

B. Viral Mutants

1. Temperature-sensitive mutants

There are numerous observations which show that persistent infection of cells by normally cytopathic virus is accompanied by a shift from a "wild-type" to a "temperature-sensitive" phenotype.[24] Youngner and colleagues[25] have examined this shift to temperature sensitivity in detail in mouse L cells persistently infected with VSV. Persistent infections were initiated with wild-type VSV, or a *ts* RNA^+ mutant,[113] by infecting the cells at very low multiplicities along with a large excess of defective virus particles. With continued passage of the culture, there was a dramatic shift in the nature of the virus produced by the persistently infected cells. By 63 days after initiation of infection, all of the virus released by the persistently infected cells was temperature-sensitive (*ts*) at 39.5°C and produced small plaques at either 32 or 37°C in primary chick-embryo cell cultures. Furthermore, they have shown that persistent infection of mouse L cells by temperature-sensitive mutants could be accomplished in the apparent absence of added defective interfering particles if the cells were infected at very low multiplicities (that is $m = 10^{-5}$ or 10^{-4}) with a *ts*-mutant obtained from the original persistently infected culture. However, in order for the *ts*-mutant to establish persistent infection at higher multiplicities of infection, there was an absolute requirement for coinfection with defective interfering particles.

Youngner et al.[25] reported that they could not detect interferon in the culture medium of their persistently infected mouse L cells, either at early or late passage levels. In contrast, Nishiyama[10] was able to detect low levels of interferon in a mouse L-cell culture carrying VSV, but was not able to separate its contribution to persistent infection from other reactants in the system (small plaque temperature-sensitive mutants and defective-interfering particles). The use of antiserum to mouse interferon in these systems to determine whether interferon is playing an essential regulatory role seems warranted.[15]

The detection of interferon in mouse L cells persistently infected with VSV, and the isolation of *ts*-mutants therefrom,[10] tempted us to consider whether the temperature-sensitive small-plaque mutants may take on another role in persistent infection similar to one we have proposed for DI particles of the [±]RNA class, namely, as inducers of the interferon system.[15,41] In Table 1 we provide evidence that indeed certain *ts*-mutants of VSV are extremely efficient inducers of interferon — both at a nonpermissive temperature (40.5°C) and one used to establish persistent infection (37°C) (data not shown). Only mutants which produce primary transcripts *(tra⁺)* and do not inhibit cell-

TABLE 1

Characteristics of VSV *ts*-Mutants

Complementation group	*ts*-Mutant designation	Phenotype (40°C)[a]				Interferon yield[b]
		tra	*psi*	*ckp*	*ifp*	
	Wild-type[c]	+	+	+	−	<10
I	*G*114	−	−	−	−	420[d]
	*G*11	+	−	−	+	3,400
	*T*1026	+	−[e]	−	+	24,300
	*O*5	+[f]	+	+	−	30
II	*G*22	+	−	−	+	11,400
	*O*52	+[f]	n.t.[g]	+	−	360
III	*T*54	+	n.t.	+	−	360
	*O*23	+[f]	n.t.	+	−	150
IV	*G*41	+	−	−	+	12,600
	*W*10	+	+	+	−	<10
V	*O*110	n.t.	+	+	−	<10
	*O*44	n.t.	n.t.	+	−	<10

[a] The "+" and "−" designations denote the presence or absence, respectively, of measurable activity for each virus at 40°C, a nonpermissive temperature.[26,33] Abbreviations: *tra* = primary transcription; *psi* = protein synthesis inhibition; *ckp* = cell killing particle activity; *ifp* = interferon inducing particle activity.

[b] Interferon yields are expressed as the total number of PR_{50} (VSV) units of interferon produced from 1×10^7 primary chick embryo cells[42] in 24 hr at 40.5°C following infection at optimal multiplicity (see Figure 1) for each mutant.

[c] Three strains of wildtype VSV-Indiana have been tested for interferon inducing particle activity as described in the legend to Figure 1 (HR, MS, and Balt.). All three strains failed to induce detectable levels of interferon on primary chick emoryo cells at 40.5°C.

[d] Interferon yields of ≤ 10 per cent of the maximum obtainable from mutants designated *ifp*$^+$ (≤ 500 units in these experiments) are considered of marginal significance due to the possibility they may represent induction by low levels of [±]RNA DI particles (capable of inducing 20,000 to 40,000 units of interferon under optimal conditions[41]) contaminating the *ts*-mutant stocks, or leak of genetic expression combined with a delay in the manifestation of the *psi* function so as to permit some accumulation and activity of the interferon-inducer moiety. (We think the latter activity normally is masked by the *psi*$^+$ function of VSV.[91]) For these reasons mutants yielding interferon at a level ≤ 10 per cent of the maximum have been designated *ifp*$^-$.

[e] Stanners, Francoeur and Lam.[48]

[f] Reference 112

[g] n.t. = not tested.

protein synthesis (*psi*$^-$) or kill cells (*ckp*$^-$) are capable of inducing high levels of interferon (*ifp*$^+$) — phenotypically these mutants are (*tra*$^+$, *psi*$^-$, *ckp*$^-$, *ifp*$^+$). Even clonally purified, gradient banded *ifp*$^+$ mutants tested at low multiplicities yield high levels of interferon, ruling out contamination by [±]RNA DI particles. Wild-type virus and mutants which inhibit cell-protein synthesis and kill cells, phenotypically (*tra*$^+$, *psi*$^+$, *ckp*$^+$, *ifp*$^-$) yield little or no detectable interferon.

VSV-G11(I) is a prototype of the mutants showing the *tra*$^+$, *psi*$^-$, *ckp*$^-$, *ifp*$^+$ phenotype. We have examined in more detail its ability to induce interferon in aged, primary chick-embryo cells[42] at a restrictive temperature (40.5°C). As illustrated in Figure 1, clonally purified and gradient-banded VSV-*ts* G11(I) induced a maximal yield of inter-

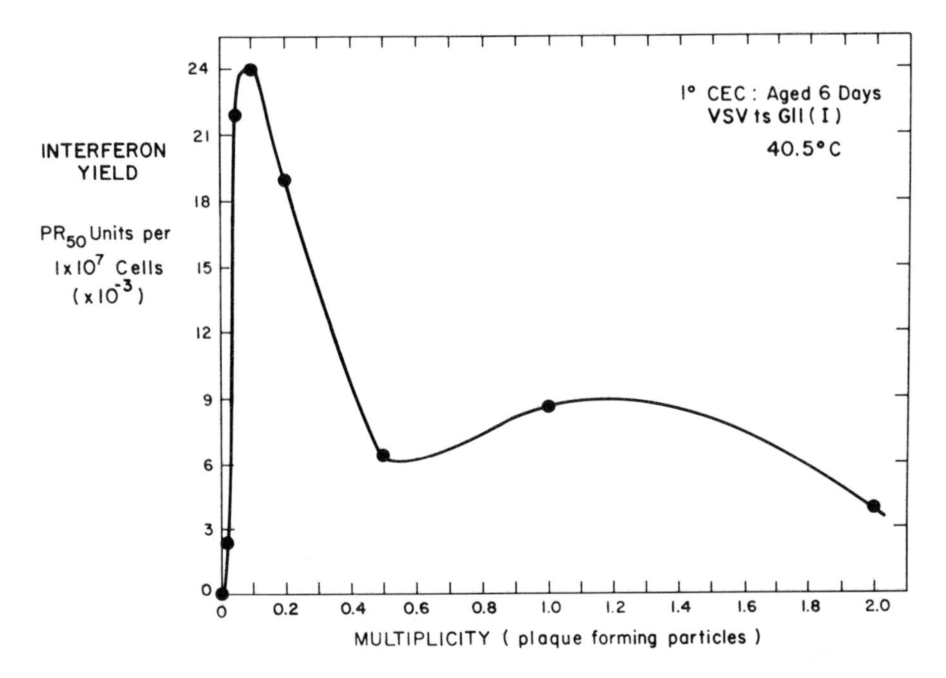

FIGURE 1. Interferon inducing capacity of VSV- *ts* G11(I). Primary chick embryo cells, aged in vitro for 6 days,[90] were infected with varying multiplicities of VSV- *ts* G11(I) obtained as clonally purified, once gradient-banded stocks. DEAE-dextran at 10μg/mℓ was present during virus attachment. Following virus attachment at 37.5°C for 30 min, cell monolayers were washed three times with warm medium and incubated at 40.5°C with 3 mℓ of attachment solution per 60 mm dish. After 24 hr the supernatant fluid was collected and assayed for interferon. Activity was recorded as a 50% VSV plaque reduction unit (PR_{50} unit/mℓ). Interferon yield is reported as the total amount of interferon released into the medium from a monolayer of 1×10^7 cells in 24 hr.

feron at $m_{pfp} \simeq 0.1$, indicating that there may be up to ten times more particles in a standard preparation of *ts* G11(I) that are capable of inducing interferon than can be scored as infectious virus. To learn whether the inducing capacity of *ts* G11 depends on formation of an inducer moiety we subjected the virus to inactivation by heat at 50°C. Any preexisting inducer should be relatively stable to heat — as was the case for DI-011 particles.[41] However, the data in Figure 2 show that the induction of interferon by this mutant was as sensitive to heat (50°C) inactivation as was plaque-forming particle and virion-associated transcriptase (data not shown) activity, indicating that the interferon inducer did not exist preformed but had to be synthesized by the infecting virion. The interferon-inducing capacity of *ts* G11(I) was relatively sensitive to UV radiation (data not shown) suggesting that transcription of a significant portion of the genome was required to produce the interferon-inducer moiety. In contrast, the *ifp*$^+$ capacity of a [±]DI-011 particle, already containing the inducer molecule, is extremely resistant to inactivation by UV radiation.[41] (The section on Interferon considers in greater detail the significance of the interferon inducing capacity of *ts*-mutants).

The selection of *ts*-mutants of VSV in insect cells[43,44] is remarkably similar to that observed in vertebrate cells[25] in spite of the virtual lack of CPE in the infected invertebrate host cell.[45,46] Thus, Printz[43] demonstrated a selection for temperature-sensitive mutants of VSV when the virus was passaged in fruit flies (*Drosophila melanogaster*), a result confirmed by Mudd et al.[44] using *Drosophila* cells in culture. In the latter system, mutants were selected for that were both temperature-sensitive and small plaque-formers in chick-embryo cells. Several mutants isolated from the culture fluid of the persistently infected cells were shown to be RNA^- at 37°C, a nonpermissive temperature for fruit-fly cells. In general, insect cells are not sensitive to the cytopathic

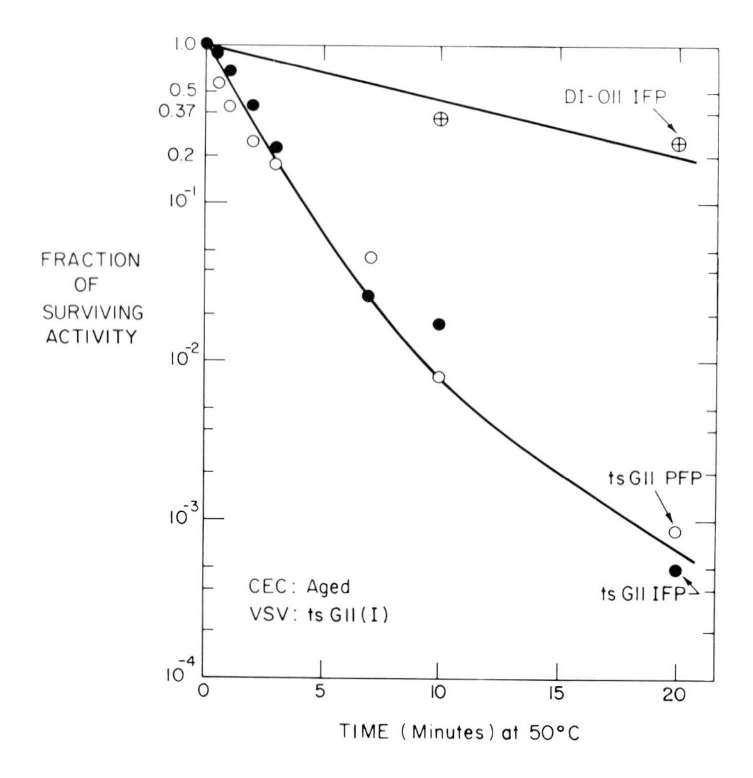

FIGURE 2. Surviving activity curves for interferon-inducing particles of VSV DI-011 and mutant *ts* G11(I) subjected to heat (50°C). IFP activity was assayed on primary chick embryo cells aged in vitro as described previously,[90] and at an optimal multiplicity for induction.[26,41] The experimental conditions for induction of interferon and its assay are described in the legend of Figure 1. The plaque-forming (PFP) activity for *ts* G11(I) was assayed on GMK-Vero cells at 30°C.

effect of wild-type VSV, even after high multiplicity infection (m = 100) where all of the cells in the population can be shown to be positive for viral antigens.[45] However, Sarver and Stollar[46] have recently reported the isolation of clones of mosquito (*Aedes albopictus*) cells which undergo an extensive cytopathic effect (CPE) after infection with Sindbis virus. Both the yield and the cytopathic effect of VSV is enhanced considerably on the clone of cells sensitive to Sindbis virus-induced CPE. A comparison of mechanisms for initiating and maintaining VSV in these two populations of cells, should persistent infection of the sensitive clone prove possible, should prove interesting.

2. Small plaque mutants

Many investigators have observed the emergence of small-plaque virus mutants in the course of the evolution of persistent infection with both VSV[8,10,20,25,44] and rabies virus[7] and have suggested that the small-plaque phenotype somehow contributes to the stabilization of the virus-cell relationship. An explanation has only recently been proposed.

Stanners and colleagues, while studying the properties of VSV *ts*T1026(I), discovered it to be a double mutant.[47-51] The mutant contained a temperature-sensitive lesion in the virion-bound RNA polymerase and a non-*ts* mutation expressed as a delay before host-cell protein synthesis is inhibited, a function they termed P.[48,49] Although this mutant was not derived from a persistent infection, it has several properties commonly associated with virus derived from such cultures, that is, (1) it forms small plaques relative to the wild-type virus and (2) it has reduced cytopathicity in vitro at elevated temperatures[33,47] and displays decreased virulence in hamsters.[50] However, unlike many *ts*-mutants derived directly from persistent infections, it seems ideally suited to establish a virus-carrier state. Mutant *ts*T1026(I) can be used routinely to initiate a persistent infection in the apparent absence of added defective interfering virus particles or their generation,[114] and even when infection is initiated at a high multiplicity (m_{pfp} = 2 to 20).[51] Independently of us[26] (Table 1), Francoeur et al.[51] observed that *ts*T1026(I) is a potent inducer of interferon. They also obtained evidence that the small-plaque phenotype of *ts*T1026 revertant is due to the induction of interferon and its subsequent action. If *ts*T1026 revertants are plaqued on GMK-Vero cells, a line genotypically incapable of responding to inducers of the interferon system,[15,52] normal-sized plaques are obtained.[51]

The role of P function in persistent infection by *ts*T1026(I) is unclear; however, it is known that revertants which no longer have a temperature-sensitive transcriptase but are still P^- for inhibition of host-cell-protein synthesis, can induce interferon, but are no longer able to establish persistent infection since infected cells are killed.[49,51] Also, their results indicated that the *ts*-mutation in the virion transcriptase was insufficient by itself to permit persistent infection since infected cells were either killed or cured, depending on the temperature.[48,49,51] They concluded that both the P^- and *ts*-mutations are needed in order to establish persistent infection (at least in cells with an inducible interferon system). These results leave unanswered the question of why small-plaque viral mutants are selected for in persistent infections of cells (BHK)[8] which are insensitive to inducers of the interferon system. Possibly they possess the attributes described by Youngner and Quagliana[53] for *ts*-mutants of VSV, and can outcompete wild-type (large-plaque) virus.

We note that *ts*T1026(I) possesses the phenotype predictive of an *ifp*$^+$ mutant, based on our analysis of *ts*-mutants from all five complementation groups of the VSV-Indiana serotype (see Table 1), namely, *tra*$^+$, *psi*$^-$, *ckp*$^-$. The P function of Stanners appears to delay host protein-synthesis inhibition long enough to permit expression of the *ifp*$^+$ character which we think is normally masked by the *psi*$^+$ nature of wild-type VSV.[26] Wertz and Youngner[54] astutely reached a similar conclusion while studying a small-plaque mutant of VSV which induced significant levels of interferon.

3. Host-range mutants

Host-range mutants of VSV have been described but, as yet, not implicated in persistent infections.[55-57] One intriguing explanation for the host-range temperature-dependent mutants described by Szilágyi and Pringle[57] occurs to us, namely that the defects reflect a delay in the expression of protein-synthesis inhibition and a subsequent unmasking of the *ifp*$^+$ phenotype. Cells with an inducible interferon system (like those from the chick embryo: nonpermissive host cell) could limit spread of the virus through interferon action, whereas cells deficient in the interferon system (like BHK: permissive host cell) could not.[57] In a quantitative sense Stanner's *ts*T1026(I) resembles a host-range mutant, producing plaques of limited size on one host, and unlimited size on another host lacking the interferon system.[51] On this basis these host-range temperature-dependent mutants[55-57] might not be able to initiate persistent infection in a cell lacking an inducible interferon system unless defective-interfering virus particles were added.

C. Interferon

One impressive feature of interferon action is its capacity to induce in cells total resistance to the cytopathic effects of otherwise lethal viruses. Cells exposed to high enough concentrations of interferon are not destroyed by viruses,[58-60] though there are exceptions.[61,62] Some investigators have implicated the interferon system in persistent infections of a wide range of viruses,[10,15-23] including rhabdoviruses.[10,15,20,23] Indeed, there is clear evidence that in contrast to homotypic interference induced by conventional [−]RNA DI-particles, interferon-mediated interference can spare cells from the lethal effects of VSV.[15,37,60] The first of these important points has been established using single-cell survival-curve analysis to demonstrate that the cell-killing capacity of a thrice-gradient purified stock of VSV-HR (CKP-3) is unaffected by the presence of large amounts of thrice-gradient purified, conventional-[−]RNA-defective interfering particles (DIP-3) since the slope of the survival curve is the same in the presence or absence of a multiplicity of 70 defective interfering virus particles.[37] Figure 3 illustrates the second point, namely, when the concentration of interferon was increased (the amount of interferon in PR_{50} units is shown to the right of its corresponding curve) there was a progressive decrease in the slope of the survival curve, demonstrating that the higher the concentration of interferon the greater its protective effect against cell killing by VSV. Note that 54 units of interferon suppresses cell-killing-particle activity almost totally. Thus, while homotypic interference by conventional [−]RNA, non-CPE-suppressing DI particles does not appear to spare cells from the lethal action of VSV, the interferon system does.[37,60]

Wagner et al.[20] described the isolation and properties of two plaque-size variants (large and small) from a stock culture of VSV. Both types maintained their characteristic plaque size upon subsequent passage at 37°C. The small plaque variant (SPV) represented about 5% of the total virus in the original pool and gave high yields of virus on primary chick-embryo cells, but did not induce detectable levels of interferon. The small-plaque mutant replicated poorly in mouse L cells, induced low levels of interferon, and did not produce any cytopathic effect in these cells at $m_{PFP} \leqslant 4$. In contrast, the large-plaque variant (LPV) produced high yields of virus in mouse L cells, no interferon could be detected and all of the cells were killed. Mouse L cells which survived infection with the small-plaque variant of VSV could repopulate the culture. However, these cells now underwent crisis whenever dispersed by trypsin — a treatment which might have destroyed cell-associated interferon and tipped the balance of the virus-carrier state in favor of the virus (see next section). Cell destruction, when it appeared periodically, was accompanied by an increase in virus yield. Wagner et al. went on to show that the small-plaque variant was four to six times more sensitive to interferon action than the large-plaque variant, offering this difference as the major factor contributing to the longevity of the carrier culture.

In light of the results of more recent studies by other investigators,[63,64] it seems likely that the SPV was also a temperature-sensitive (*ts*) mutant of VSV, especially since *ts* mutants can represent about 5% of the population of wild-type VSV stocks. In this regard, Wagner et al.[20] did not report testing the SPV at temperatures above or below 37°C. The failure to detect interferon in primary, chick-embryo cell cultures infected with SPV may have been due to the relative inefficiency of young chick cell monolayers to produce interferon.[42] Finally, whether interferon-inducing DI particles of the [±]RNA type were present in the SPV virus population remains unanswered.

Nishiyama[10] has reported the presence of low levels of interferon in the culture fluid of mouse L cells persistently infected with VSV. In keeping with this observation, these VSV-carrier cells were also resistant to superinfection by a heterologous virus (Mengo). However, the role of other reactants present in the carrier culture (small-plaque temperature-sensitive mutants and defective-interfering particles) was not separated or evaluated with respect to the interferon system and the maintenance of persistent infection.

Wiktor and Clark[65] demonstrated that a line of Syrian hamster cells (Nil-2) capable of inducing and reacting to interferon, and persistently infected with HEP-Flury strain of rabies virus, displayed a cyclical resistance to challenge with a heterologous virus, VSV. They detected low levels of interferon in the medium bathing a BHK-HEP Flury rabies persistently infected culture at some of the cell passages, but could not correlate its presence with other parameters of the persistent infection being measured. For example, resistance of the cultures to VSV replication was high even when fewer than ten units of interferon were demonstrable. However, these observations do not eliminate the interferon system from consideration because high levels of an interferon-mediated antiviral state can be established in the absence of comparably high levels of free interferon.[21] Defective interfering particle production was not assayed in either of their persistently infected cell lines,[65] since the first identification of biologically active DI particles of rabies virus appears to have occurred some 2 years later,[66] although reference to an autointerference phenomenon dates back some 20 years.[67]

III. THE INTERFERON SYSTEM: A COMMON PATHWAY TO PERSISTENT INFECTION

Interferon action would provide a simple explanation for the survival of cells in a virus-carrier state if inducers of interferon constituted a common reactant in persistent infection of normal cells (i.e., cells capable of producing, and responding to, interferon) by rhabdoviruses. We think they do, and provide in this section the evidence and arguments to support this thesis.

Marcus and Sekellick[37] reported that standard [−]RNA stranded defective interfering particles such as DI-HR, derived from the HR strain of VSV-Indiana, do not block the lethal effect of this virus upon simultaneous infection of cells with both defective and infectious (B) virus particles, even though a yield of infectious B particles is totally prevented in doubly infected cells. On that basis we concluded[15] that if defective interfering particles are playing a temporizing role in persistent infection of cells by VSV, then some mechanism other than conventional homotypic interference by DI particles must exist to prevent the demise of the host cell. The discovery by Lazzarini and colleagues[68] of a DI particle of VSV which contained covalently bonded self-complementary RNA provided a possible link between DI particle-interference and the interferon system. The reports by Perrault[69,70] indicate that the [±]RNA class of DI particles may be distributed widely throughout the numerous collections of VSV. We tested the [±]RNA DI-011 particles of Lazzarini et al.[68] and discovered that they were extremely efficient inducers of interferon.[41] The interferon-inducing capacity of the DI-011 particle is illustrated in Figure 4, and shows that maximal amounts of interferon were found in the medium when "aged" primary chick embryo cells (CEC)[42] were inoculated with a multiplicity of 0.3 DI particles, the equivalent of about one physical particle.[115] Furthermore, analysis of the dose-response curve shows that under appropriate conditions a single physical particle suffices to induce a quantum yield of interferon or generate an antiviral state. Infection with additional particles may actually act to reduce the yield of interferon — an effect presently under study. Under these same conditions, preparations of conventional minus-stranded DI particles from wild-type VSV ([−]DI-HR) failed to induce detectable levels of interferon. This unique capacity of [±]RNA DI particles to induce interferon, coupled with the demonstration that interferon action prevents cell killing by VSV (see Figure 3),[60] led us to postulate that DI particles of the [±]RNA class, initially present, or arising during infection, activate the interferon system and prevent cell-killing (CK) particle activity — providing a means whereby normally susceptible cells may survive in the presence of lethal virus.[15]

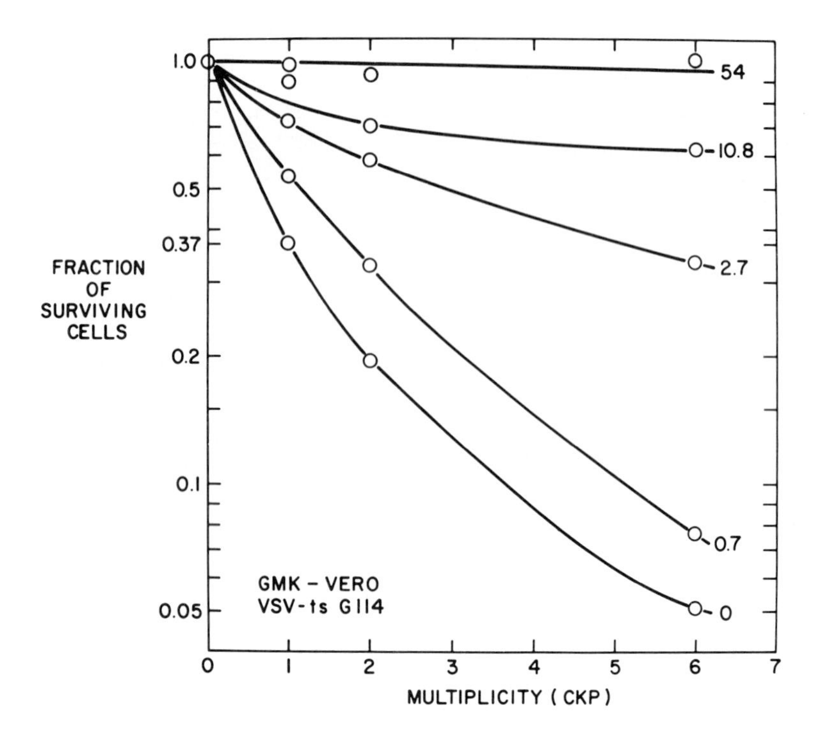

FIGURE 3. Survival curves of GMK-Vero cells exposed to different concentrations of human interferon for 24 hr at 37.5°C and challenged with varying multiplicities of VSV *ts* G-114(I). Following virus attachment and entry, infected and control cell monolayers were dispersed with trypsin, and the cells were counted and plated in growth medium containing VSV antiserum. All plates were held at 30.0°C for 24 hr (pulse-infection) before shifting to 40.0°C for colony formation from surviving cells. The abscissa represents the CKP multiplicities calculated from the 37% survival level in mock-treated cells ("O" interferon). The concentrations of interferon, in PR_{50} (VSV) units/mℓ, are indicated to the right of each survival curve. Individual points represent the averages of duplicate plates and agreed within ±20% of the mean value. (From Marcus, P. I. and Sekellick, M. J., *Virology*, 69, 378, 1976. With permission.)

In support of this hypothesis we have shown that GMK-BSC cells are spared from killing if they are pretreated with interferon-inducing [±]DI-011 particles 24 hr prior to challenge with VSV.[15] Figure 5 documents the dramatic effect that DI-011 particle pretreatment has on the survival of VSV-infected cells. Under the conditions of this experiment, simultaneous infection of GMK-BSC cells did not protect against the expression of CK particle activity by VSV. Analyzing single-cell survival curves, we found that GMK-BSC cells retained the capacity to form colonies, i.e., were spared from killing, if they were pretreated with interferon-inducing [±]DI-011 particles 24 hr prior to challenge with homologous virus (VSV) (Figure 6). The bottom curve in Figure 6 labeled CKP, demonstrates the lethal action of VSV in untreated cells and is identical to the curve generated by infecting cells *simultaneously* with cell killing and DI-011 particles (data not shown). The upper curve demonstrates that *pretreatment* of GMK-BSC cells with DI-011 particles renders virtually all of the cells refractory to cell killing by VSV over the range of cell-killing particle multiplicities tested. The multiplicity range used in these experiments includes infection by a single virion, representing a condition very similar to that existing in cells persistently infected with VSV.[8,10,25]

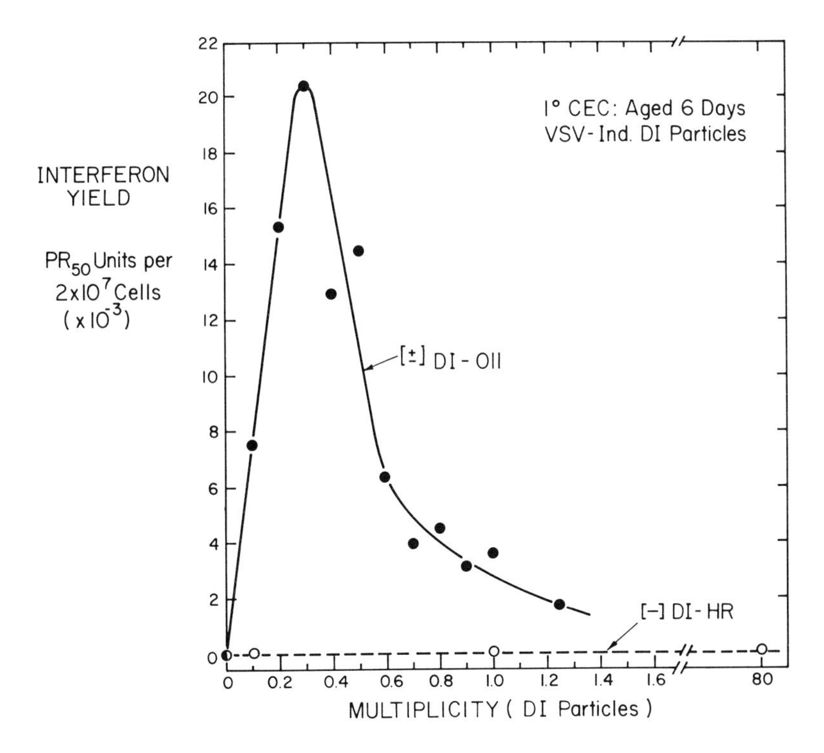

FIGURE 4. Interferon inducing capacity of [±]RNA DI-011 (●) and [−]RNA DI-HR (●) particles. Primary chick embryo cells aged in vitro for 6 days were infected with varying multiplicities of DI particles obtained as thrice gradient-purified stocks. DEAE-dextran at 10 μg/mℓ was present during virus attachment. Following virus attachment at 37.5°C for 30 min, cell monolayers were washed three times with warm medium and incubated at 40.5°C with 3 mℓ of attachment solution per 60-mm dish. After 24 hr the supernatant fluid was collected and assayed for interferon. Activity was recorded as a 50% VSV plaque reduction unit (PR_{50} unit/mℓ). Interferon yield is reported as the total amount of interferon released into the medium from a monolayer of 2×10^7 cells in 24 hr. This period sufficed to produce maximal yields of interferon over the entire range of multiplicities tested. DI particle titers were determined by measuring the reduction in yield of thrice gradient purified homologous VSV-HR B particles. (From Marcus, P. I. and Sekellick, M. J., *Nature (London)*, 266, 815, 1977. With permission.)

As already noted (Figures 5 and 6), simultaneous infection with DI-011 and infectious particles did not spare cells from the lethal action of VSV: a period of pretreatment is required to develop interference to cell killing. Maximal interference was achieved by pretreating the cells for 24 hr with DI-011 particles. Shorter periods of pretreatment produced proportionally fewer cells surviving challenge with VSV as measured by their capacity to form macroscopically visible colonies in 9 to 11 days. Furthermore, if cells were pretreated with [±]DI-011 particles in the presence of cycloheximide (25 μg/mℓ) and the drug was washed out at the time of challenge with VSV, the cells were not spared from the lethal action of the virus. We concluded that cellular protein synthesis was required during the pretreatment phase to develop cell resistance to the lethal action of VSV.[15] We also observed that [±]DI-011 particles which had sustained the equivalent of 25 lethal hits of UV radiation (one lethal hit to infectivity = 52.3 ergs/mm^2) were still fully capable of inducing cell sparing in mouse L cells against the lethal action of wild-type VSV. DI-011 particles exposed to this dose of UV radiation lose their capacity to function as defective-interfering agents for homotypic interference but are undiminished in their capacity as interferon-inducing particles.[41]

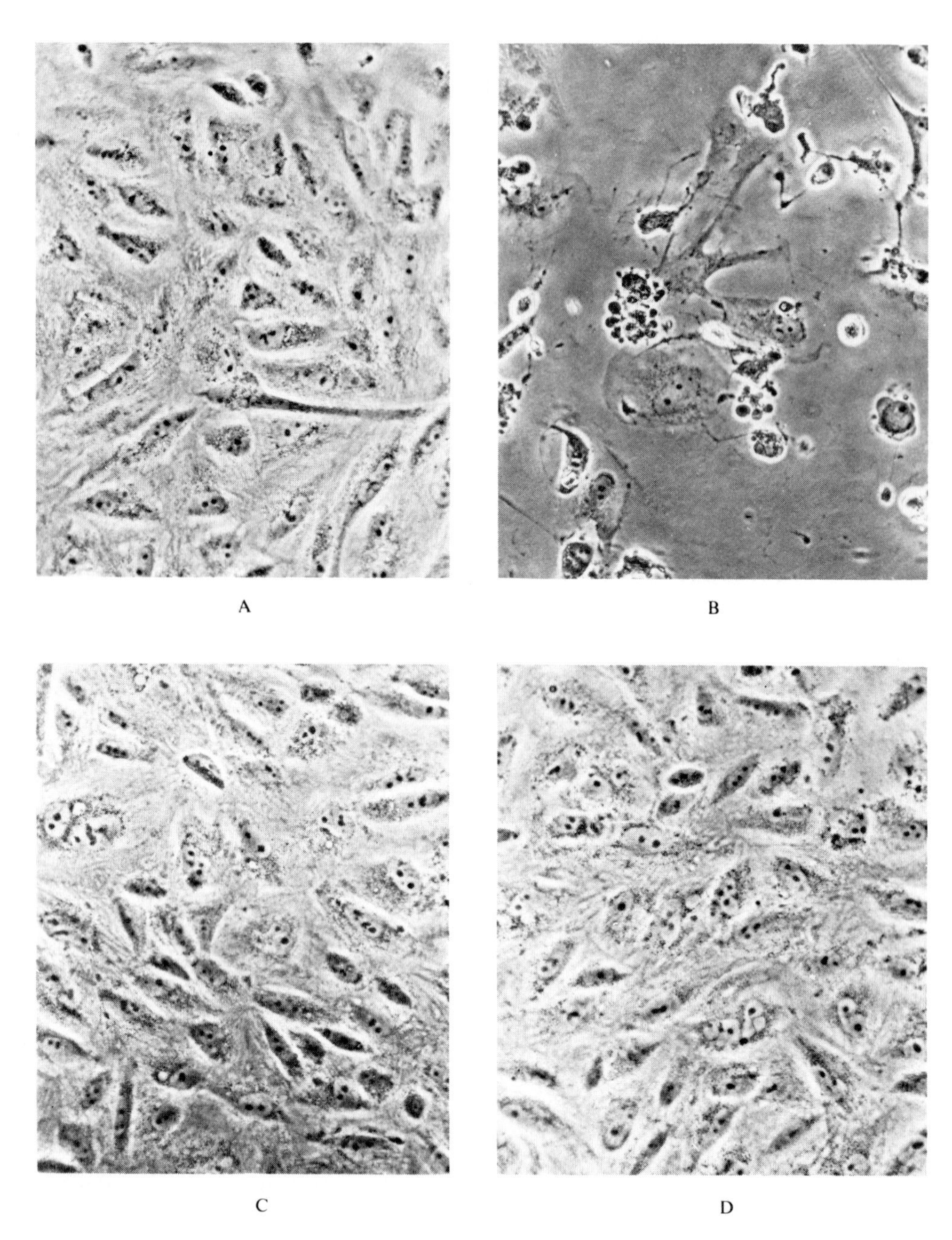

FIGURE 5. Photomicrographs of GMK-BSC cells illustrating cell sparing against VSV induced by pretreatment with [±]DI-011 particles. Confluent monolayers of cells in 60-mm dishes were infected with DI-011 particles at m_{DIP} = 3 in the presence of 10 μg/mℓ of DEAE-dextran. After adsorption at 37.5°C for 60 min, the unattached virus was removed and the cell monolayers were washed three times prior to adding 3 mℓ of medium containing VS virus antiserum (1:400 dilution) to each plate. Following incubation at 37.5°C for 24 hr, the medium was removed by aspiration and the cell monolayers were washed three times and then infected with thrice gradient-purified VSV (B particles) at a multiplicity of 10 in the presence of 10 μg/mℓ of DEAE-dextran. This challenge virus was allowed to adsorb for 30 min at 37.5°C. Unadsorbed virus was removed by aspiration, and the cell monolayers were washed three times prior to adding 3 mℓ of medium containing VSV antiserum (1:400 dilution) to each plate. Control plates were mock-infected but otherwise manipulated similarly. Cell monolayers were photographed following incubation at 37.5°C for 18 hr. Conditions: A , uninfected cells; B, VSV at m_{PFP} = 10; C, pretreatment with [±]DI-011 particles at m_{DIP} = 3; D, pretreatment with [±]DI-011 particles at m_{DIP} = 3 and challenge with VSV at m_{PFP} = 10. (From Sekellick, M. J., and Marcus, P. I., *Virology,* 85, 175, 1978. With permission.)

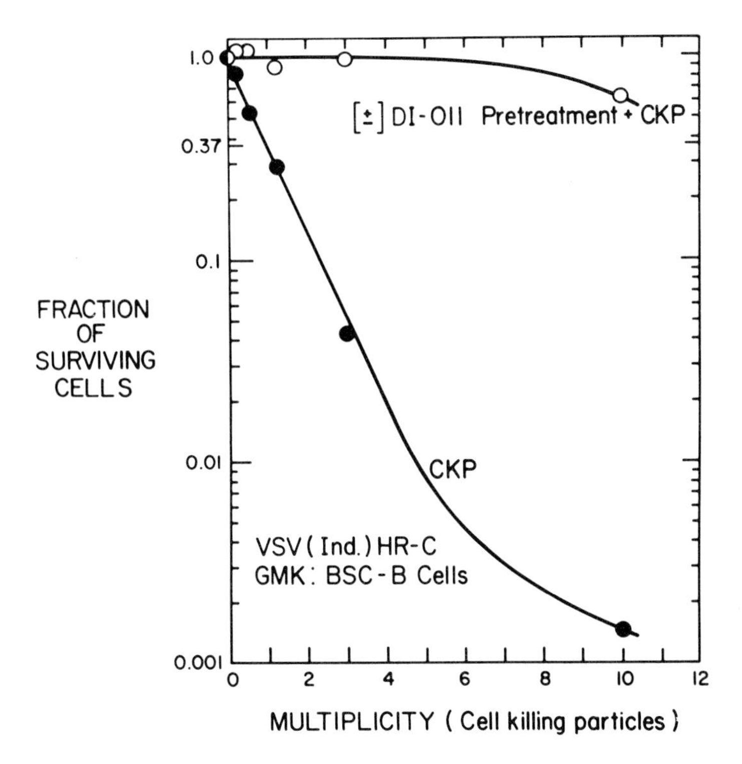

FIGURE 6. Survival curves of GMK-BSC cells infected with varying multiplicities of thrice gradient purified VS virus CK particles; after 24 hr pretreatment (○), or mock-treatment (●) with DI-011 particles. Following virus attachment and entry, infected and control cell monolayers were dispersed with trypsin, the cells counted in a hemocytometer, and plates inoculated with 200, 2000, or 20,000 cells per 60 mm dish in growth medium containing VSV antiserum (1:400 dilution). The plates were incubated at 37.5°C for 9 days to allow development of macroscopically visible colonies from surviving cells. The abscissa represents the CK particle (CKP) multiplicities calculated from the 37% survival level in cells infected with CK particles only. Individual points represent the average of duplicate plates and agreed within ±15% of the mean value. (From Sekellick, M. J. and Marcus, P. I., *Virology*, 85, 175, 1978. With permission.)

To evaluate further the role of interferon in cell sparing induced by [±]DI-011 particle pretreatment, we carried out single-cell survival experiments with GMK-Vero cells, a line genotypically incapable of responding to interferon inducers.[52] Figure 7 shows that the slope of the survival curve is unaffected by preinfection with DI-011 particles or their presence as coinfecting particles with infectious VSV. Under the latter condition, homotypic interference develops and viral replication is blocked.[41] Nonetheless, these cells do not become refractory to cell killing by VSV, presumably due to the inherent inability of GMK-Vero cells to respond to inducers of interferon.[52]

We sought to determine whether the cell sparing induced by DI-011 particles was heterologous in nature, as would be expected were the interferon system responsible. We repeated the experiment illustrated in Figure 6 but used Newcastle disease virus (NDV) as the challenge virus. The results are shown in Figure 8. These single-cell survival curves established that a 24-hr pretreatment of GMK-BSC cells with DI-011 particles prevented the expression of almost all cell-killing particle activity by the otherwise lethal NDV. The antiviral state induced by DI-011 particle pretreatment was also effective in preventing cell killing by Sindbis virus, another heterologous virus (data not

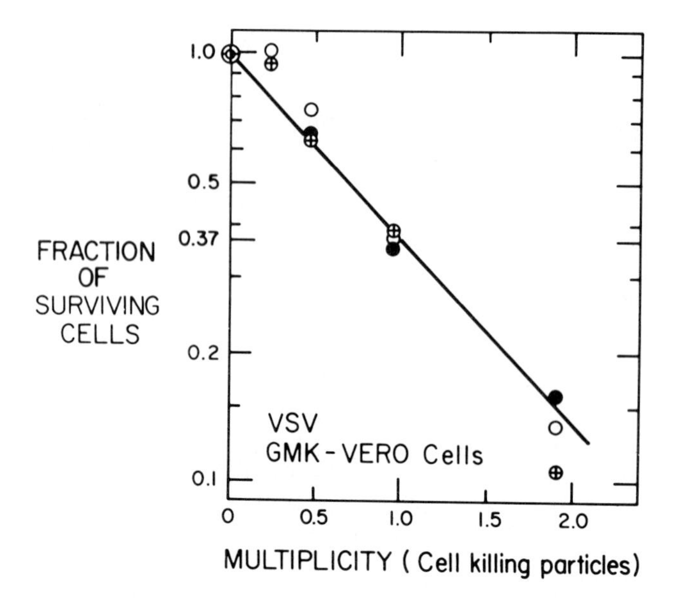

FIGURE 7. Survival curves of GMK-Vero cells infected with varying multiplicities of thrice gradient purified VSV CK particles only (○), after pretreatment with DI-011 particles (●) or simultaneously with DI-011 particles (⊕). The solid line represents a best fit of the experimental points obtained under all conditions tested. The experimental details are described in the legend to Figure 6. (From Sekellick, M. J. and Marcus, P. I., *Virology*, 85, 175, 1978. With permission.)

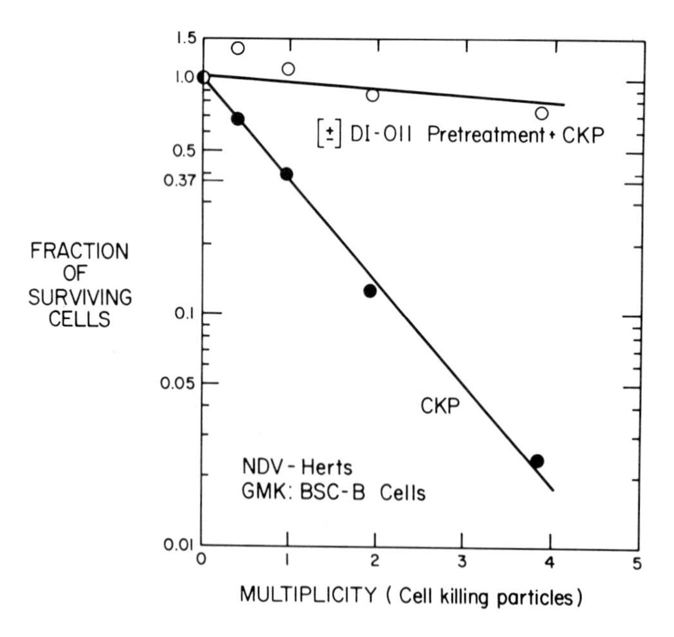

FIGURE 8. Survival curves of GMK-BSC cells infected with various multiplicities of Newcastle disease virus (NDV-Herts) only (●), or after a 24 hr pretreatment with DI-011 particles (○). Surviving cells were incubated in medium containing NDV-antiserum (diluted 1:400) to prevent spurious cell killing by any newly released virus. The experimental details are described in the legend to Figure 6. (From Sekellick, M. J. and Marcus, P. I., *Virology*, 85, 175, 1978. With permission.)

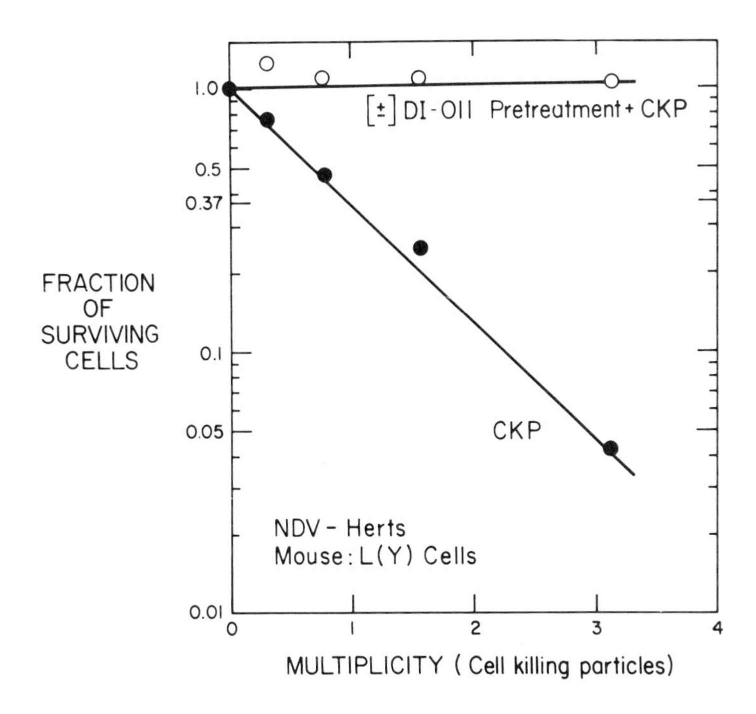

FIGURE 9. Survival curves of mouse L(Y) cells infected with various multiplicities of Newcastle disease virus (NDV-Herts) CK particles only (●), or after pretreatment for 24 hr with DI-011 particles (○). Surviving cells were incubated in medium containing NDV-antiserum as described in the legend of Figure 6. (From Sekellick, M. J. and Marcus, P. I., *Virology*, 85, 175, 1978. With permission.)

shown). Clearly, with respect to challenge viruses, the state of cell sparing induced by DI-011 particles is heterologous in nature — a necessary condition for interferon-mediated interference.

We examined mouse L cells, another line used frequently in studies of persistent infection[10,20,23,25] and capable of producing and responding to interferon,[71] for its cell-sparing response following pretreatment with [±]DI-011 particles. Data shown in Figure 9 demonstrate that these cells become totally refractory to the cell-killing particle activity of NDV once they have been pretreated with the interferon-inducing DI-011 particles. In contrast, the pretreatment of BHK cells with [±]DI-011 particles did not prevent cell killing by VSV or NDV, nor did it induce an antiviral state.[116]

We reasoned that if interferon induced by the [±]RNA class of DI particles and its subsequent action constitute the basis of cell sparing in GMK-BSC and mouse L cells, then its neutralization with specific antibody should prevent its action[21,72] and hence the development of cell sparing.[60] As a final test of our hypothesis we added mouse-interferon antiserum to medium bathing the [±]DI-011 particle-infected mouse L cells during the 24-hr period of pretreatment prior to challenge with VSV. The results are illustrated graphically in Figure 10. Following pretreatment of mouse L cells with [±]DI-011 particles, a cytopathic effect failed to develop in these cells when challenged with wild-type VSV (Figure 10C). However, the protection against cell killing by VSV is totally obliterated when antiserum to mouse interferon is present during the pretreatment period (Figure 10D). To confirm the action of the mouse-interferon antiserum, we measured the yield of infectious VSV under the same experimental conditions described above. The results are presented in Table 2 and show that DI-011 particle pretreatment results in a 400-fold reduction in the yield of VSV. This interference with virus replication is prevented completely when interferon antiserum is included in the

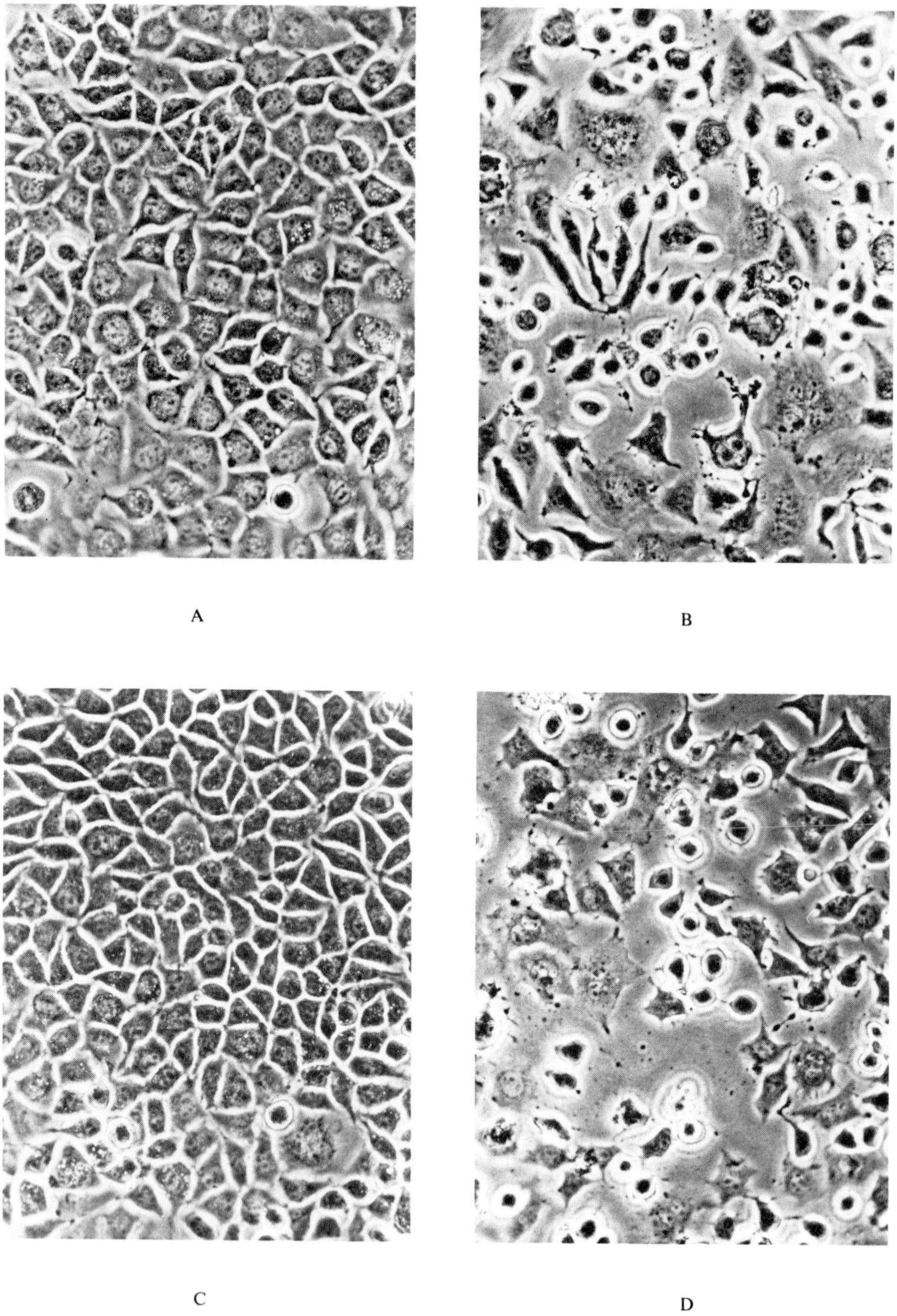

A

B

C

D

FIGURE 10. Photomicrographs of mouse L(Y) cells: A, uninfected cells; B, VSV at m_{PFP} = 10; C, pretreatment with DI-011 particles at m_{DIP} = 3 and then challenge with VSV at m_{PFP} = 10; D, pretreatment with DI-011 particles at m_{DIP} = 3 in the presence of mouse interferon antiserum and then challenge with VSV at m_{PFP} = 10. The experimental details are described in the legend to Figure 5. (From Sekellick, M. J. and Marcus, P. I., *Virology*, 85, 175, 1978. With permission.)

TABLE 2

Yield of Vesicular Stomatitis Virus from Mouse L(Y) Cells Following Pretreatment With DI-011 Particles and Interferon-antiserum

Treatment[a]	Yield of VS virus (PFP/2 × 10^6 cells)
VS virus	1.4×10^9
Interferon-antiserum and VS virus	1.8×10^9
DI-011 particle pretreatment and VS virus	3.6×10^6
DI-011 particle pretreatment and interferon-antiserum and VS virus	1.3×10^9

[a] Pretreatment consisted of infecting confluent cell monolayers with thrice-gradient purified [±]DI-011 particles at $m_{DIP} = 3$. When interferon-antiserum was used it was added to the medium immediately after virus adsorption was completed. Cell monolayers were challenged with VSV 24 hr after infection with DI-011 particles, and harvested and plaque assayed 18 hr later. (From Sekellick, M. J. and Marcus, P. I., *Virology*, 85, 175, 1978. With permission.)

medium during the pretreatment phase. These results also demonstrate that after 24 hr in the cell, the homotypic interference capacity of the DI-011 particles is nil — a finding in accord with recent observations.[73]

A number of investigators[5,7,16,25] have reported a failure to detect interferon in the culture medium from persistently infected cells and have, on that basis, ruled out a role for the interferon system. However, the absence of detectable interferon should not be taken as absolute proof of the absence of interferon-mediated interference. Indeed, Glasgow[74] has described some extreme steps that may be necessary to detect low levels of interferon. Detection of an antiviral state or the use of interferon antiserum may prove to be more reliable indicators for the presence and action of interferon as a number of workers have recently demonstrated.[15,21,23,75,76] Indeed, if the interferon system is playing a pivotal role in allowing cells to survive, then treatment with interferon antiserum should prevent the establishment or maintenance of a persistent infection and result in cell death, a prediction borne out by the experiment presented in Figure 10 and the results of Ramseur and Friedman.[23] These investigators reported that a culture of mouse L cells infected for prolonged periods with VSV occasionally contained low levels of interferon in the spent growth medium. When mouse interferon antiserum was included in fresh growth medium they observed an increased yield of infectious virus, and the development of a cytopathic effect (crisis).[23]

Three other observations lend further credence to our view that in interferon-competent cells the interferon system represents a common pathway to persistent infection. Youngner and co-workers[25] reported that (1) actinomycin D treatment of VSV-carrier L cells enhanced the yield of infectious virus almost 100-fold 48 hr after exposure to the drug, (2) shift-down to 32°C, a permissive temperature for the *ts*-mutants in the carrier culture and a suboptimal one for interferon production,[106] resulted in complete destruction of the cells (the ultimate crisis) by 3 days, and (3) addition of mouse interferon antiserum resulted in an increased yield of infectious virus and development of a cytopathic effect (crisis).[117] We think these results reflect the loss of interferon-mediated viral interference and cell sparing, respectively.

We have reported[41] that a DI-LT particle isolated from a persistent infection of BHK-21 cells by Holland and Villarreal[5] had the capacity of inducing interferon. On the other hand we have not been able to detect cell sparing or an antiviral state in BHK-21 cells pretreated with [±]DI-011 particles,[15] an observation consistent with the report by Kawai et al.[7] that UV-irradiated Newcastle disease virus did not induce interferon in these cells — suggesting that they are phenotypically like Vero-GMK in their inability to respond to inducers of the interferon system. Thus, even though DI particles which have the capacity to induce the interferon system are isolable from BHK cells persistently infected with VSV, the failure of these cells to respond to interferon inducers appears to rule out the interferon system as a mediator of cell sparing in some,[7,8] but possibly not all,[65] systems using BHK as the host cell.

We conclude that in cells with a fully functional interferon system interferon-inducing defective particles and/or interferon-inducing viral mutants (see above), may be important in determining the ease (frequency) and/or stability (longevity) with which persistent infection may be established and maintained, respectively. In other cells, like GMK-Vero or BHK, deficient in some element of the interferon system yet capable of supporting persistent infection, not only of rhabdoviruses,[7,8] but many other kinds of viruses as well,[9,12,77,78] other mechanisms would appear to be operative for cell survival. As discussed below, the CPE-suppressing defective particles of Kawai et al.[7,11] provide one such mechanism.

In the in vivo situation, which we will consider only briefly, persistent infection of animals probably does not occur in the absence of the interferon system. Gresser and colleagues[79] have demonstrated convincingly that interferon plays a significant role in the disease process induced in VSV-infected mice using injections of interferon-antiserum to negate interferon action in these animals. As with animals infected with other types of viruses, treatment with interferon-antiserum accelerated the disease process and dramatically shortened the survival time of the VSV-infected animals.[79] Although it remains to be determined whether treatment with interferon antiserum will affect the course of neurovirulence seen in mice[80,81] or hamsters[50] infected with various mutants of VSV, the appearance of interferon in animals inoculated with rabies virus,[83,84] or VSV,[85] and the dramatic enhancement of viral virulence in animals given interferon-antibody,[79] portend a significant effect. The interferon system might also be brought into play in instances where large numbers of DI particles are inoculated into animals along with the wild-type virus.[82,86] For example, Crick and Brown[86] were able to inactivate the homotypic interference capacity of VSV DI particles (with acetyl ethyleneimine [AEI]), inoculate them intracerebrally into mice, and soon after demonstrate interference to challenge by heterologous viruses. Not withstanding the lack of demonstrable circulating interferon in those mice,[86] we can conceive of the AEI-inactivated DI particles acting functionally like heat or UV-inactivated DI-011 particles to induce interferon and a heterologous antiviral state[15,41] which spares the animals from the lethal effect of challenge viruses.[86] The use of mouse interferon-antiserum[79] in this kind of experiment should help to resolve this important point, as would a search for interferon-inducing [±]RNA particles[41,68-70] within the defective particle population.

IV. CPE-SUPPRESSING DEFECTIVE PARTICLES: CELL SPARING IN THE ABSENCE OF THE INTERFERON SYSTEM

Not all cell types persistently infected with rhabdoviruses appear to possess a functional interferon system,[7,8,87-89] making it rather difficult to attribute cell sparing to the action of interferon.[60] An important observation by Kawai et al.[7,11] provides an explanation for cell sparing in the absence of the interferon system. Kawai and colleagues[7,11] discovered that BHK cells persistently infected with a large-plaque variant (LPV) of the HEP-Flury strain of rabies virus contained a defective particle with a special property — it was capable of suppressing the CPE of complete virus, even when added simultaneously with infectious virions. Kawai and Matsumoto[11] found that the LPV strain of rabies produced persistent infection in BHK cells more quickly than a small-plaque variant (SPV), and that this occurrence was correlated with the earlier appearance of a so-called, CPE-suppressing, defective-particle activity in the LPV-infected cultures. CPE-suppressing defective particles appeared to play a decisive role in the initiation and maintenance of the persistent infection: cloned SPV (presumably free of CPE-suppressing particles) produced a more extensive cytopathic effect in BHK cells than the original LPV virus (presumably containing or generating CPE-suppressing particles). They also found evidence for the production of large amounts of noninterfering, non-CPE suppressing defective particles whose significance in persistent infection is unknown. No interferon or interferon-like substance was found in the medium bathing the persistently infected culture, and no heterologous antiviral state in the cells themselves was demonstrable using VSV as the challenge virus.[7]

These results contrast sharply with our own in which we demonstrated that conventional [−]RNA DI particles of VSV conferred no protection against the lethal action of that virus, whether added before or simultaneously with the challenge infection.[15,37,41] However, the data of Kawai et al.[7,11] provide evidence for a class of defective particles capable of preventing cell killing by complete virus. Apparently, the CPE-suppressing particles act early enough in virus replication to block the formation of a putative cell-killing factor,[90,91] or, in some way, its expression. The relatively slow rate at which rabies virus CPE progresses,[92,93] may permit large numbers of CPE-suppressing particles to accumulate and block cell killing — contributing in turn to the slow development of CPE. Perhaps the accumulation of large numbers of viral RNP cores in persistently infected cells lacking the interferon system[8] contributes in a like manner to the suppression of cell killing by the carried infectious virus.

From studies still in progress we have obtained evidence for a CPE-suppressing particle activity in high multiplicity passages of VSV *ts* G11(I) — as tested on the interferon-incompetent GMK-Vero cells. In this context we note a recent report by Dutko and Pfau[38] which demonstrates that a single defective particle of an arenavirus, lymphocytic choriomeningitis virus can prevent the action of a cell-killing particle of that virus.

Defective particles capable of suppressing cell killing by viruses, in the absence of the interferon system, may emerge as a new and important class of subvirions — capable of regulating viral functions and certainly deemed worthy of further study.

V. POTENTIAL REACTANTS IN PERSISTENT INFECTION

The interferon system, DI particles, *ts* -, and small-plaque (host range?)-mutants and CPE-suppressing particles all are reactants that have been associated with persistent infection and may be required in various combinations to establish a virus-carrier state with rhabdoviruses. Potential reactants include reverse transcriptase, antiviral antibody, and genetically resistant host cells. These will be discussed briefly below.

A. Reverse Transcriptase

The available evidence does not favor involvement of reverse transcriptase (derived either from the carried virus or a latent oncornavirus) in persistence of rhabdoviruses in in vitro systems. Holland et al.[8] have tested their long-term VSV-BHK carrier culture for the presence of stable DNA provirus and could not detect any at the level of one DNA copy of the VSV RNA genome per 40 cells. Similarly, evidence for DNA provirus was not obtained when they assayed for transfection of virus infectivity or viral antigen-synthesizing capacity. Comparable experiments with a rabies-BHK carrier also failed to detect viral DNA. A study of short-term persistent infections of mouse L cells — a line of cells known to contain large numbers of an endogenous retrovirus, also failed to provide evidence for proviral DNA.

Testing this point further, Youngner et al.[25] attempted to increase the yields of virus from mouse L cells persistently infected with VSV, hoping to activate an integrated viral genome with inhibitors which affect DNA synthesis or functioning. Bromodeoxyuridine did not increase infectious virus yield as would be expected if the culture harbored a latent intact VSV-DNA provirus.[94] Only actinomycin D treatment of the carrier cells was able to increase virus yield over control values — a result consistent with suppression of a cellular dependent viral interference mechanism (interferon or DI particle generation).

The aforementioned results are perhaps not unexpected in light of the relative ease with which persistent infections involving rhabdoviruses can be cured by cloning in the presence, or absence, of specific viral antiserum[8,10] — supporting the view that the persistent state is maintained by a mechanism involving an extracellular sojourn of virus. In contrast to these results, Sato, et al.[29] have presented evidence for the induction of a latent reverse-transcriptase activity in rubella virus-infected BHK cells (a host cell used in several model, rhabdovirus-persistent infection systems), leaving open the possibility that a rhabdovirus-persistent infection may yet be found that depends on reverse transcriptase for its maintenance.

B. Resistant Host Cells

Selection for a host cell inherently resistant to the lethal action of rhabdoviruses, yet capable of sustaining viral replication, would represent an obvious means for establishing persistent infection. A host cell displaying these attributes does not appear to have been isolated, in spite of many cell generations exposed to virus in carrier cultures. In this context we note that many invertebrate cells appear genetically endowed with a means for preventing cell killing by VSV.[43-45] However, resistance to CPE seems subject to genetic modification since Sarver and Stollar[46] have isolated clones of mosquito cells which do display CPE upon infection with VSV (and other viruses). Furthermore, leafhopper cells succumb to a plant rhabdovirus, potato yellow dwarf virus, with characteristic CPE,[95] and even undergo cell fusion[95] similar to that induced by VSV in vertebrate cells.[96,97] We conclude that host cells resistant to the cell-killing potential of rhabdoviruses do not constitute a reactant of major involvement in persisent infection.

The lymphocyte may function as a unique host cell for infection by rhabdoviruses and conceivably contribute to the persistence of these viruses in vivo. Appropriately stimulated lymphocytes may serve as hosts for VSV,[98] and in the resting state may possibly harbor the virus for many days before treatment with mitogens triggers virus replication and the appearance of infectious virions. The factors controlling the apparent persistence of VSV in the resting lymphocyte and its restricted replication in some lines of human lymphoblastoid cells are ill-defined[98,99] yet intriguing enough to warrant further investigation.

C. Antiviral Antibody

Although viral antibodies are not usually present in in vitro systems of persistent infection, their appearance in animals in response to viral infection make them a reactant of potential importance in vivo.[50,80-82,86] With a view toward simulating more closely the in vivo situation, Sinarachatanant and Huang[100] infected Chinese hamster ovary cells (probably deficient in the interferon system[89]) with both standard wild-type VSV and DI particles, and then added viral antiserum to the medium. Successive passages received fresh cells (presumably to replace killed cells) and viral antiserum at a concentration designed to neutralize about 90% of the virus. (Higher levels of antibody essentially cured the cultures.) They were able to produce a cyclic production of infectious virus and of cells which mimicked the fluctuations of these two reactants often observed in persistent infection.[7,8,65] Although these experiments did not involve a persistent infection in the usual sense (since fresh cells were overtly added at each passage), they do serve to point out again that the production of virus can be aborted with specific viral antibody — akin to the curing of persistent infection — and that the surviving cells would go on to repopulate the culture. Similar patterns of cycling can take place in the absence of specific viral antiserum, and at least one system with rabies virus appears to involve fluctuations in the level of interferon,[65] while another appears under the control of CPE-suppressing particles.[7]

VI. INITIATION, MAINTENANCE, CRISIS AND CURE

No integrated genetically stable relationship between virus and cell has been reported for persistent infections with rhabdoviruses. All virus-cell systems in which rhabdoviruses have established persistent infection — some for years — appear to represent virus-carrier cultures, i.e., the persistence of the virus depends upon cycling infection that includes a mandatory extracellular phase.[7,8,10,25] The carrier state is most easily demonstrated by the relative ease with which specific viral antibody can cure the persistent infection and provide genetically stable cells, differing in no discernable way from the susceptibility characteristics of the original cell population.[8,10] Simple cloning procedures may also lead to the loss of persistent infection and the isolation of susceptible wild-type cells.[8] Cloning in the presence of virus antiserum provides the most reliable means of aborting the virus-carrier state and obtaining cells permanently free of virus. In a comparable sense, virus-carrier systems kept in balance by interferon action may be made to produce virus "permanently free of the cell" through the addition of antibody specific for interferon: the ensuing loss of cells (crisis) being accompanied by the production of virus.[15,21,23,59,76,79]

Since the carrier state appears to represent adequately, and perhaps exclusively, all cases of persistent infection in rhabdoviruses, we will examine the data discussed above in terms of four components of persistent infection in vitro: (1) initiation, (2) maintenance, (3) crisis, and (4) cure, with the particular bias that we are analyzing a virus-cell carrier state.

A. Initiation

Since a single particle of VSV (and presumably of rabies virus) can kill a cell,[37] the initiation of a carrier culture (persistent infection) in a population of otherwise susceptible cells by a lethal virus must entail some mechanism(s) to ensure survival of the host. This axiom is often relegated to a subordinate role in view of the usual emphasis on "persistence" of the virus. Certainly, any satisfactory explanation of the virus-carrier state must account for the "persistence" of both virus and cell.[15]

Based on the reactants thus far associated with the rhabdovirus carrier state we envision successful initiation of persistent infection when a significant portion of the cell population is protected from the lethal action of the parental virus — in the extreme, one cell. We postulate that in host cells where interferon action can provide this protection an inducer of interferon must be present initially, or emerge soon, in the virus population. The reactants for this event have been described above and may take the form of (1) [±]RNA DI particles, (2) *ts*-mutants with a *tra*$^+$, *psi*$^-$, *ckp*$^-$, *ifp*$^+$phenotype, or (3) small plaque (or similar) mutants which can express the *ifp*$^+$ character because they are slowed or deficient in the rate at which they inhibit cellular protein synthesis. In the absence of these IFP$^+$ reactants initially, or their failure to emerge rapidly, the overt addition of interferon may serve to ameliorate cell killing and allow the establishment (initiation) of a carrier state.[10,23] The discovery of [±]DI particles[41] and *ts*-mutants[26] (and probably small-plaque mutants[20,51,85]) of VSV which are extremely efficient inducers of interferon (a single particle may suffice to produce a quantum yield of interferon[26,41,101]) provides reactants which are singularly qualified to promote cell sparing and hence the initiation of persistent infection by that virus. The appearance of interferon in many rabies virus-infected cell cultures[83] and in animals[84] suggests that further exploration may reveal interferon inducers of comparably high efficiency.

In host cells lacking some component of the interferon system, cell survival and hence initiation of the carrier state may well depend on the initial presence or rapid emergence of CPE-suppressing defective particles of the type discovered by Kawai and Matsumoto,[11] and which also appear to be present in high multiplicity-generated stocks of an arenavirus,[38] and a reovirus of fish.[102]

Thus, reactants have been described that will afford the host cell protection against the lethal action of rhabdoviruses: when the interferon system is operative the reactants are presumably [±]DI particles, temperature-sensitive or small-plaque mutants; and when inoperative, CPE-suppressing defective particles.

Some host cells, those of invertebrates, appear to have evolved a system capable of controlling the formation and/or expression of cell-killing factor by several families of viruses — including rhabdoviruses — and do not appear to require any special attributes (reactants) of the virus to ensure survival. In cells of that type, i.e., fly,[44] moth,[103] or mosquito,[45,88] persistent infection follows as a usual consequence of the initial infection. There are exceptions which deserve further exploration,[46] including the cytopathic effect of a plant rhabdovirus in leafhopper cells.[95]

We propose that initiation of persistent infection in cell cultures competent for the interferon system takes place when the virus population used to initiate infection contains interferon-inducing (IF) particles, or when they emerge rapidly enough for interferon action to produce cell sparing.[15,60]In support of this thesis we note that all reactants thus far identified in persistent infection have been shown to be capable of inducing interferon at 37°C.[15,20,23,26,41,51,76] In cell cultures deficient in some component of the interferon system we envision CPE-suppressing defective particles as the reactant responsible for cell sparing.[11,38,102] In most invertebrate cells (apparently lacking the interferon system[87,88]) the host may regulate the expression of cell killing by the virus by some yet unknown mechanism. However, in all cases a *sine qua non* of persistent infection is fulfilled, namely, the cells are spared the lethal action of the carried virus.

B. Maintenance

Once initiated, maintenance of a virus-carrier state must depend upon a balance between (1) cell killing and sparing, and (2) virus replication and inhibition. We propose that an interferon-induced antiviral state — often present in the absence of readily demonstrable levels of interferon,[74] and/or the presence of CPE-suppressing defective particles constitute reactants in persistent infection which function to maintain an equilibrium between cell survival and virus production. The emergence and continuous production of [±]DI particles and/or *ts*- or small-plaque mutants which are interferon inducers would provide the carrier culture with a stable, albeit cyclic, source and reservoir of interferon, while concomitantly providing a means of regulating virus replication — even in the possible absence of DI particles.[25] The studies of Youngner and colleagues[25,53,104] clearly demonstrated that *ts*/small-plaque mutants displayed a selective advantage and outcompeted wild-type virus during the establishment and maintenance of persistent infection. The replication advantage of these mutants may prevail over their enhanced sensitivity to interferon if the studies of Wagner et al.[20] are representative of most small-plaque mutants.

In addition to sparing cells from the lethal action of the carrier virus, interferon action on rhabdoviruses[105] may also serve to maintain levels of infectious and DI particles compatible with continued survival (persistence) of both cell and virus. In this regard homotypic interference with infectious virus replication may also contribute to the maintenance of cell-killing particles at acceptable levels. However, the possible absence of DI particles in some virus-carrier lines[25] suggests that the interferon system alone may suffice to regulate virus replication. In general, the action of interferon and homotypic interference (when operative), would act to maintain the levels of infectious or defective cell-killing particles below the thresholds needed to overcome the cell-sparing action of interferon.[58]

Perturbations in the balance of reactant ([±]DI particles, *ts* - and small-plaque mutants, and CPE-suppressing particles) production and their expression through product action (cell sparing and viral interference) would tend to throw the virus-carrier cell population into crisis — as discussed next.

C. Crisis

Crisis in persistent infection is seen as a waning of cell sparing, by whatever mechanism, and the onset of a cytopathic effect, often accompanied by enhanced virus production. Total cell destruction and loss of the virus-carrier culture represents the ultimate crisis. Crisis situations brought about by experimental manipulation provide insight into the factors contributing to maintenance of the "steady state" of a carrier culture.

The central role we propose for the interferon system in persistent infection of interferon-competent cells predicts that crises will ensue when experimental conditions are created which tend to suppress the induction or action of interferon. In a well-established persistent infection of L cells by VSV in which *ts*-mutants had replaced virtually all of the wild-type virus, a shift-down to permissive temperature brought about complete cell destruction within 2 to 3 days, accompanied by a high yield of virus.[25] We interpret these results as reflecting the inefficient induction of interferon at low temperatures,[106] even by *ts*-mutants which are highly efficient inducers at high nonpermissive temperatures,[26,118] and a waning of interferon-mediated cell sparing and antiviral state. Consistent with this interpretation is the enhanced yield of VSV observed when the VSV-L cell carrier culture is treated with actinomycin D[25] or, as tested more directly, with antibody specific for interferon.[117] This treatment would stop forthwith any further development of an antiviral state[72] and allow for loss of the existing interference.[107] If the interferon system does command a crucial role in persistent infection then predictably this treatment would precipitate a crisis. In interferon-treated, wild-type VSV-infected mouse L cell cultures which "persisted" for weeks, Ramseur and Friedman[23] observed that the addition of antimouse interferon globulin brought on a crisis, with eventual total cell destruction and a marked increase in virus yield. In this system of "prolonged" VSV infection, Ramseur and Friedman suggested that the *ts*-mutants which emerged over the weeks were probably responsible for interferon production.[23] We now know that many *ts*-mutants of VSV are highly efficient inducers of interferon.[15,26,51,118]

The replication advantage afforded the *ts*-mutants under the conditions of temperature shift-down, coupled with the inefficient induction of interferon by *ts*-mutants at permissive temperatures[26,118] and the half-life of the antiviral state, might easily tip the balance in favor of virus replication and cell killing, i.e., a crisis. In cells presumably lacking the interferon system a temperature shift-down also resulted in a crisis and increased virus production.[5] Again, this treatment would presumably confer a replicative advantage on any *ts*-mutants, and possibly on small-plaque mutants. This advantage may allow critical amounts of putative cell-killing factor[91] to be synthesized and/or lead to a dimunition of CPE-suppressing particle activity. Further studies on this new class of defective particles seem warranted.

D. Cure

Cure of a virus-carrier culture reflects the total elimination of virus, and the return of the host cells to a spectrum of viral sensitivities characteristic of the parent population. No viral antigens, nor homotypic or heterotypic interference characteristic of the carrier cells is demonstrable in a cured population of cells. As with the onset of crises, the experimental means used to cure the virus-carrier state may provide some insight into the mechanisms underlying its maintenance. Rhabdovirus-carrier cultures have been cured of virus by (1) cloning the cells,[8] (2) including viral specific antiserum in the growth medium,[10] (3) cloning in the presence of viral antibody,[8] (4) shifting the culture to a temperature nonpermissive for *ts*-mutants,[10,25] and (5) adding high levels of interferon.[10] To our knowledge no cures have been effected (attempted?) through homotypic interference, i.e., the addition of large numbers of DI particles — even though the purified particles are not intrinsically toxic[15,37,41,82,108] and might, if added periodically, be expected to suppress all helper virus replication.

All of the methods used to cure VSV-carrier cultures emphasize that persistent infection is maintaind by overt infection of cells by virus through an obligatory extracellular phase. Evidence for integration of the viral genome is all negative. While these conclusions seem firm for the well-studied VSV-carrier system, and may well account for the self-limiting nature of VSV infection in its natural host,[109] further studies are required before similar statements can be made about rabies virus.

VII. PERSPECTIVES ON PERSISTENT INFECTION AND VIRAL DISEASE

We have reviewed persistent infection in rhabdoviruses and conclude that the VSV-cell systems (and with less certainty those of rabies virus) examined to date qualify as virus-carrier states, meaning that continued persistence of the virus depends upon an obligatory extracellular phase, while total integrity of the virus-cell populations requires a concinnity of balance between cell sparing and killing, and virus replication and inhibition. We can identify two mechanisms capable of sparing otherwise sensitive host cells from the lethal action of the carried virus: (1) *the interferon system* — its activation dependent upon interferon-competent cells and the action of interferon inducers, and (2) *the CPE-suppressing particle* — its generation and action required in cells incompetent in some aspect of the interferon system. Considering the first of these two mechanisms and emphasizing survival of the host cell as the critical element in the establishment of a virus-carrier state, we have identified three different reactants and linked them through a common function — interferon induction. We now propose that these reactants, (1) [±]RNA DI particles, (2) *ts*-mutants, and (3) small-plaque mutants constitute the paramount determinants in the establishment (initiation and maintenance) of persistent infection of rhabdoviruses in interferon-competent cells. Each of these reactants has been shown to be an efficient inducer of interferon. Furthermore, interferon action is singularly effective in sparing cells from cell-killing particle activity. We contend that the production of interferon by these reactants, and the subsequent development of an antiviral-cell sparing state through interferon action represents the primary mechanism for persistent infection in vitro with rhabdoviruses. We think that conventional [−]RNA DI particles which fail to induce interferon and cell sparing lack the attributes required for the establishment of successful persistent infection, and suggest that their role in initiating a virus-carrier state is palliative in that they reduce the rate of spread of infectious and cell-killing particles through the population until an interferon-inducing reactant is generated and becomes dominant. In that context we would modify the original role proposed for defective particles in viral disease[4] and point out that the suppression of virus replication is but one facet of their biologic potential. The capacity of some kinds of defective particles to induce interferon and/or suppress the lethal action of viruses may emerge as even more important attributes.

The dramatic effects of interferon[110,111] and of interferon-antiserum[79] on the course of virus disease lead us to favor the interferon-inducing (IF) particle as an element of major importance in the outcome of a virus disease. The extraordinarily efficient interferon-inducing capacity[41] of the newly recognized class of [±]RNA DI particles[68] and their apparent ubiquitous distribution[69,70] portend an important role for them in viral diseases. Temperature-sensitive and small-plaque mutants with interferon-inducing capacity exceeding that of parental virus may assume this same role — especially in viral diseases of the nervous system.

CPE-suppressing particles[7,11,38,102] are also emerging as another special type of defective particle, important perhaps in sparing interferon-incompetent cells from the lethal action of viruses.

Defective virus particles with capabilities beyond those of a homotypic interfering agent are emerging with potential importance as controlling elements in virus disease. We anticipate their zenith with great expectations and zetetic zeal.

ACKNOWLEDGMENTS

The research from our laboratory referred to in this chapter was aided in part by National Science Foundation grant No. PCM-76-00467, National Cancer Institute grant No. CA20882, and benefited from use of the Cell Culture Facility supported by National Cancer Institute Grant No. CAP 14733.

REFERENCES

1. **Schlessinger, D.,** Persistent viral infections, in *Microbiology,* Am. Soc. Microbiology, Washington, D.C., 1977, 433.
2. **Fenner, F., McAuslan, B. R., Mims, C. A., Sambrook, J., and White, D. O.,** Persistent infections, in *The Biology of Animal Viruses,* Academic Press, New York, 1974, chap. 12.
3. **Chambers, V. C.,** The prolonged persistence of western equine encephalomyelitis virus in cultures of strain L cells, *Virology,* 3, 62, 1957.
4. **Huang, A. S. and Baltimore, D.,** Defective viral particles and viral disease processes, *Nature (London),* 226, 325, 1970.
5. **Holland, J. J. and Villarreal, L. P.,** Persistent noncytocidal vesicular stomatitis virus infections mediated by defective particles that suppress virion transcriptase, *Proc. Natl. Acad. Sci. U.S.A.,* 71, 2956, 1974.
6. **Palma, E. L., and Huang, A. S.,** Cyclic production of vesicular stomatitis caused by defective interfering particles, *J. Infect. Dis.,* 129, 402, 1974.
7. **Kawai, A., Matsumoto, S., and Tanabe, K.,** Characterization of rabies viruses recovered from persistently infected BHK cells, *Virology,* 67, 520, 1975.
8. **Holland, J. J., Villarreal, L. P., Welsh, R. M., Oldstone, M. B. A., Kohne, D., Lazzarini, R., and Scolnick, E.,** Long-term persistent vesicular stomatitis virus and rabies virus infection of cells *in vitro, J. Gen. Virol.,* 33, 193, 1976.
9. **Rima, B. K., Davidson, W. B., and Martin, S. J.,** The role of defective interfering particles in persistent infection of Vero cells by measles virus, *J. Gen. Virol.,* 35, 89, 1977.

10. **Nishiyama, Y.**, Studies of L cells persistently infected with VSV: factors involved in the regulation of persistent infection, *J. Gen. Virol.*, 35, 265, 1977.
11. **Kawai, A. and Matsumoto, S.**, Interfering and noninterfering defective particles generated by a rabies small plaque variant virus, *Virology*, 76, 60, 1977.
12. **Ahmed, R. and Graham, A. F.**, Persistent infections in L cells with temperature-sensitive mutants of reovirus, *J. Virol.*, 23, 250, 1977.
13. **Huang, A. S. and Baltimore, D.**, Defective interfering animal viruses, in *Comprehensive Virology*, Vol. 10, Fraenkel-Conrat, H. and Wagner, R. R., Eds., Plenum Press, New York, 1977, 73.
14. **Huang, A. S.**, Viral pathogenesis and molecular biology, *Bacteriol. Rev.*, 41, 811, 1977.
15. **Sekellick, M. J. and Marcus, P. I.**, Persistent infection. I. Interferon-inducing defective-interfering particles as mediators of cell sparing: possible role in persistent infection by vesicular stomatitis virus, *Virology*, 85, 175, 1978.
16. **Baron, S.**, The defensive and biological roles of the interferon system, in *Interferons and Interferon Inducers*, Finter, N. B., Ed., North-Holland, Amsterdam, 1973, chap. 13.
17. **Henle, W., Henle, G., Deinhardt, F., and Bergs, V. V.**, Studies on persistent infections in tissue cultures. IV. Evidence for the production of an interferon in MCN cells by myxoviruses, *J. Exp. Med.*, 110, 525, 1959.
18. **Ho, M. and Enders, J. F.**, Further studies on an inhibitor of viral activity appearing in infected cell cultures and its role in chronic viral infections, *Virology*, 9, 446, 1959.
19. **Glasgow, L. A. and Habel, K.**, The role of interferon in vaccinia virus infection of mouse embryo tissue culture, *J. Exp. Med.*, 115, 503, 1962.
20. **Wagner, R. R., Levy, A. H., Snyder, R. M., Ratcliff, G. A., Jr., and Hyatt, D. F.**, Biologic properties of two plaque variants of vesicular stomatitis virus (Indiana serotype), *J. Immunol.*, 91, 112, 1963.
21. **Inglot, A. D., Albin M., and Chudzio, T.**, Persistent infection of mouse cells with Sindbis virus: role of virulence of strains, auto-interfering particles and interferon, *J. Gen. Virol.*, 20, 105, 1973.
22. **Enzmann, P. J.**, Induction of an interferon-like substance in persistently infected *Aedes albopictus* cells, *Arch. Gesamte Virusforsch.*, 40, 382, 1973.
23. **Ramseur, J. and Friedman, R. M.**, Prolonged infection of interferon-treated cells by vesicular stomatitis virus: possible role of temperature-sensitive mutants and interferon, *J. Gen. Virol.*, 37, 523, 1977.
24. **Preble, O. T. and Youngner, J. S.**, Temperature-sensitive viruses and the etiology of chronic and inapparent infections, *J. Infect. Dis.*, 131, 467, 1975.
25. **Youngner, J. S., Dubovi, E. J., Quagliana, D. O., Kelly, M., and Preble, O. T.**, Role of temperature-sensitive mutants in persistent infections initiated with vesicular stomatitis virus, *J. Virol.*, 19, 90, 1976.
26. **Sekellick, M. J. and Marcus, P. I.**, Persistent infection by vesicular stomatitis virus: interferon induction by ts-mutants, and cell sparing, *J. Supramol. Struct.*, 2 (Suppl.), 246, 1978.
27. **Furman, P. A. and Hallum, J. V.**, RNA-dependent DNA polymerase activity in preparations of a mutant of Newcastle disease virus arising from persistently infected L cells, *J. Virol.*, 12, 548, 1973.
28. **Simpson, R. W. and Iinuma, M.**, Recovery of infectious proviral DNA from mammalian cells infected with respiratory syncytial virus, *Proc. Natl. Acad. Sci. U.S.A.*, 72, 3230, 1975.
29. **Sato, M., Yamada, T., Yamamoto, K., and Yamamoto, N.**, Evidence for hybrid formation between Rubella virus and a latent virus of BHK21/W1-2 cells, *Virology*, 69, 691, 1976.
30. **Zhdanov, V. M.**, Integration of viral genomes, *Nature (London)*, 256, 471, 1975.
31. **Walker, D. L.**, The viral carrier state in animal cell cultures, in *Progress in Medical Virology*, Vol. 6, Melnick, J. L., Ed., S. Karger, Basel, 1964, 111.
32. **Walker, D. L.**, Persistent viral infection in cell cultures, in *Medical and Applied Virology*, Sanders, M. and Lennette, E. H., Eds., Warren Green, St. Louis, 1968, 99.
33. **Marvaldi, J. L., Lucas-Lenard, J., Sekellick, M. J., and Marcus, P. I.**, Cell killing by viruses. IV. Cell killing and protein synthesis inhibition by vesicular stomatitis virus require the same gene functions, *Virology*, 79, 267, 1977.
34. **Villarreal, L. P. and Holland, J. J.**, RNA synthesis in BHK21 cells persistently infected with vesicular stomatitis virus and rabies virus, *J. Gen. Virol.*, 33, 213, 1976.
35. **Huang, A. S. and Manders, E. R.**, RNA synthesis of vesicular stomatitis virus. IV. Transcription by standard virus in the presence of defective interfering particles, *J. Virol.*, 9, 909, 1972.
36. **Perrault, J. and Holland, J. J.**, Absence of transcriptase activity and transcription-inhibiting ability in defective interfering particles of vesicular stomatitis virus, *Virology*, 50, 159, 1972.
37. **Marcus, P. I. and Sekellick, M. J.**, Cell killing by viruses. I. Comparison of cell killing, plaque-forming and defective-interfering particles of vesicular stomatitis virus, *Virology*, 57, 321, 1974.
38. **Dutko, F. J. and Pfau, C. J.**, Arenavirus defective interfering particles mask the cell-killing potential of standard virus, *J. Gen. Virol.*, 38, 195, 1978.

39. **Holland, J. J., Villarreal, L. P., and Breindl, M.,** Factors involved in the generation and replication of rhabdovirus defective T particles, *J. Virol.*, 17, 805, 1976.
40. **Kang, C. Y. and Allen, R.,** Host function-dependent induction of defective interfering particles of vesicular stomatitis virus, *J. Virol.*, 25, 202, 1978.
41. **Marcus, P. I. and Sekellick, M. J.,** Defective interfering particles with covalently linked [±]RNA induce interferon, *Nature (London)*, 266, 815, 1977.
42. **Carver, D. H. and Marcus, P. I.,** Enhanced interferon production from chick embryo cells aged in vitro, *Virology*, 32, 247, 1967.
43. **Printz, P.,** Adaption du virus de la stomatite vesiculaire a *Drosophila melanogaster*, *Ann. Inst. Pasteur Paris*, 119, 520, 1970.
44. **Mudd, J. A., Leavitt, R. W., Kingsburg, D. T., and Holland, J. J.,** Natural selection of mutants of vesicular stomatitis virus by cultured cells of *Drosophila melanogaster*, *J. Gen. Virol.*, 20, 341, 1973.
45. **Artsob, H. and Spence, L.,** Growth of vesicular stomatitis virus in mosquito cell lines, *Can. J. Microbiol.*, 20, 329, 1974.
46. **Sarver, N. and Stollar, V.,** Sindbis virus-induced cytopathic effect in clones of *Aedes albopictus* (Singh) cells, *Virology*, 80, 390, 1977.
47. **Farmilo, A. J. and Stanners, C. P.,** Mutant of vesicular stomatitis virus which allows deoxyribonucleic acid synthesis and division in cells synthesizing viral ribonucleic acid, *J. Virol.*, 10, 605, 1972.
48. **Stanners, C. P., Francoeur, A. M., and Lam, T.,** Analysis of VSV mutant with attenuated cytopathogenicity: mutation in viral function, P, for inhibition of protein synthesis, *Cell*, 11, 273, 1977.
49. **Stanners, C. P. and Lam, T.,** The role of the *ts* L and P⁻ mutations of VSV T1026 in persistent infection, in *Negative Strand Viruses and the Host Cell*, Barry, R. D., and Mahy, B. W. J., Eds., Academic Press, New York, 1978, 577—582.
50. **Stanners, C. P. and Goldberg, V. J.,** On the mechanism of neurotropism of vesicular stomatitis virus in newborn hamsters. Studies with temperature-sensitive mutants, *J. Gen. Virol.*, 29, 281, 1975.
51. **Francoeur, A. M., Lam, T., and Stanners, C. P.,** The role of interferon in persistent infection with T1026, *J. Supramol. Struct.*, 2 (Suppl.), 245, 1978.
52. **Desmyter, J., Melnick, J. L., and Rawls, W. E.,** Defectiveness of interferon production and of rubella virus interference in a line of African green monkey kidney cells (Vero), *J. Virol.*, 2, 955, 1968.
53. **Youngner, J. S. and Quagliana, D. O.,**Temperature-sensitive mutants of VSV are conditionally defective particles that interfere with and are rescued by wildtype virus, *J. Virol.*, 19, 102, 1976.
54. **Wertz, G. W. and Youngner, J. S.,** Interferon production and inhibition of host synthesis in cells infected with vesicular stomatitis virus, *J. Virol.*, 6, 476, 1970.
55. **Simpson, R. W. and Obijeski, J. F.,** Conditional lethal mutants of VSV. I. Phenotypic characterization of single and double mutants exhibiting host restriction and temperature sensitivity, *Virology*, 57, 357, 1974.
56. **Obijeski, J. F. and Simpson, R. W.,** Conditional lethal mutants of VSV. II. Synthesis of virus-specific polypeptides in nonpermissive cells infected with "RNA⁻" host-restricted mutants, *Virology*, 57, 369, 1974.
57. **Szilágyi, J. F. and Pringle, C. R.,** Virion transcriptase activity differences in host range mutants of vesicular stomatitis virus, *J. Virol.*, 16, 927, 1975.
58. **Yamazaki, S. and Wagner, R. R.,** Action of interferon: kinetics and differential effects on viral functions, *J. Virol.*, 6, 421, 1970.
59. **Friedman, R. M. and Costa, J. R.,** Fate of interferon-treated cells, *Infect. Immun.*, 13, 487, 1976.
60. **Marcus, P. I. and Sekellick, M. J.,** Cell killing by viruses. III. The interferon system and inhibition of cell killing by vesicular stomatitis virus, *Virology*, 69, 378, 1976.
61. **Joklik, W. K. and Merigan, T. C.,** Concerning the mechanism of action of interferon, *Proc. Natl. Acad. Sci. U.S.A.*, 56, 558, 1966.
62. **Horak, I., Jungwirth, C., and Bodo, G.,** Poxvirus specific cytopathic effect in interferon-treated L-cells, *Virology*, 45, 456, 1971.
63. **Pringle, C. R.,** The induction and genetic characterization of conditional lethal mutants of vesicular stomatitis virus, in *The Biology of Large RNA Viruses*, Barry, R. D., and Mahy, B. W. J., Eds., Academic Press, New York, 1970, 567.
64. **Flamand, A.,** Etude génétique du virus de la stomatite vésiculaire: Classement de mutants thermosensibles spontanés en groupes de complémentation, *J. Gen. Virol.*, 8, 187, 1970.
65. **Wiktor, T. J. and Clark, H F.,** Chronic rabies virus infection of cell cultures, *Infect. Immun.*, 6, 988, 1972.
66. **Crick, J. and Brown, F.,** An interfering component of rabies virus which contains RNA, *J. Gen. Virol.*, 22, 147, 1974.
67. **Koprowski, H.,** Biological modification of rabies virus as a result of its adaptation to chicks and developing chick embryos, *Bull. W.H.O.*, 10, 709, 1954.

68. **Lazzarini, R. A., Weber, G. H., Johnson, L. D., and Stamminger, G. M.,** Covalently linked message and anti-message (genomic) RNA from a defective vesicular stomatitis virus particle, *J. Mol. Biol.,* 97, 289, 1975.
69. **Perrault, J.,** Cross-linked double-stranded RNA from a defective vesicular stomatitis virus particle, *Virology,* 70, 360, 1976.
70. **Perrault, J. and Leavitt, R. W.,** Characterization of snap-back RNAs in vesicular stomatitis defective interfering virus particles, *J. Gen. Virol.,* 38, 21, 1978.
71. **Youngner, J. S., Scott, A. W., Hallum, J. V., and Stinebring, W. R.,** Interferon production by inactivated Newcastle disease virus in cell cultures and in mice, *J. Bacteriol.,* 92, 862, 1966.
72. **Vengris, V. E. Stollar, B. D., and Pitha, P.M.** Interferon externalization by producing cell before induction of antiviral state, *Virology,* 65, 410, 1975.
73. **Sekellick, M. J. and Marcus, P. I.,** The half-life of defective particle interference, in preparation.
74. **Glasgow, L. A.,** Cytomegalovirus interference in vitro, *Infect. Immun.,* 9, 702, 1974.
75. **Fauconnier, B.,** Viral auto-inhibition studied by the effect of anti-interferon serum on plaque formation, *Arch. Gesamte Virusforsch.,* 31, 266, 1970.
76. **Vileček, J., Yamazaki, S., and Havell, E. A.,** Interferon induction by vesicular stomatitis virus and its role in virus replication, *Infect. Immun.,* 18, 863, 1977.
77. **Stanwick, T. L. and Hallum, J. V.,** Role of interferon in six cell lines persistently infected with rubella virus, *Infect. Immun.,* 10, 810, 1974.
78. **ter Meulen, V. and Martin, S. J.,** Genesis and maintenance of a persistent infection by canine distemper virus, *J. Gen. Virol.,* 32, 431, 1976.
79. **Gresser, I., Tovey, M. G., Maury, C., and Bandu, M. T.,** Role of interferon in the pathogenesis of virus disease in mice as demonstrated by the use of anti-interferon serum. II. Studies with herpes simplex, Maloney sarcoma, vesicular stomatitis, Newcastle disease, and influenza viruses, *J. Exp. Med.,* 144, 1316, 1976.
80. **Rabinowitz, S. G., DalCanto, M. C., and Johnson, T. C.,** Comparison of central nervous system disease produced by wildtype and temperature-sensitive mutants of vesicular stomatitis virus, *Infect. Immun.,* 13, 1242, 1976.
81. **Rabinowitz, S. G., Johnson, T. C., and DalCanto, M. C.,** The uncoupled relationship between the temperature-sensitivity and neurovirulence in mice of mutants of vesicular stomatitis virus, *J. Gen. Virol.,* 35, 237, 1977.
82. **Doyle, M. and Holland, J. J.,** Prophylaxis and immunization in mice by use of virus-free defective T particles to protect against intracerebral infection by vesicular stomatitis virus, *Proc. Natl. Acad. Sci. U.S.A.,* 70, 2105, 1973.
83. **Wiktor, T. J. and Clark, H F.,** Growth of rabies virus in cell culture, in *The Natural History of Rabies,* Vol. 1, Baer, G., Ed., Academic Press, New York, 1975, 170.
84. **Sulkin, S. E. and Allen, R.,** Interferon and rabies virus infection, in *The Natural History of Rabies,* Vol. 1, Baer, G. M., Ed., Academic Press, New York, 1975, 355.
85. **Youngner, J. S. and Wertz, G.,** Interferon production in mice by vesicular stomatitis virus, *J. Virol.,* 2, 1360, 1968.
86. **Crick, J. and Brown, F.,** *In vivo* interference in vesicular stomatitis virus infection, *Infect. Immun.,* 15, 354, 1977.
87. **Murray, A. M. and Morahan, P. S.,** Studies on interferon production in *Aedes albopictus* mosquito cells, *Proc. Soc. Exp. Biol. Med.,* 142, 11, 1973.
88. **Kascsak, R. J. and Lyons, M. J.,** Attempts to demonstrate the interferon defense mechanism in cultured mosquito cells, *Arch. Gesamte Virusforsch.,* 45, 149, 1974.
89. **Morgan, M. J.,** The production and action of interferon in Chinese hamster cells, *J. Gen. Virol.,* 33, 351, 1976.
90. **Marcus, P. I. and Sekellick, M. J.,** Cell killing by viruses. II. Cell killing by vesicular stomatitis virus: a requirement for virion-derived transcription, *Virology,* 63, 176, 1975.
91. **Marcus, P. I. and Sekellick, M. J.,** Cell killing by rhabdoviruses, in *Rhabdoviruses,* Vol. III, Bishop, D. H. L., Ed., CRC Press, Boca Raton, Fla., 1979, Chap. 2.
92. **Kaplan, M. M., Wiktor, T. J., Maes, R. F., Campbell, J. B., and Koprowski, H.,** Effect of polyions on the infectivity of rabies virus in tissue culture: construction of a single-cycle growth curve, *J. Virol.,* 1, 145, 1967.
93. **Hummeler, K., Koprowski, H., and Wiktor, T. J.,** Structure and development of rabies virus in tissue culture, *J. Virol.,* 1, 152, 1967.
94. **Lowy, D. R., Rowe, W. P., Teich, N., and Hartley, J. W.,** Murine leukemia virus: high-frequency activation *in vitro* by 5-iododeoxyuridine and 5-bromodeoxyuridine, *Science,* 174, 155, 1971.
95. **Hsu, H. T.,** Cell fusion by a plant virus, *Virology,* 84, 9, 1978.
96. **Takehara, T.,** Polykaryocytosis induced by VSV infection in BHK-21 cells, *Arch. Virol.,* 49, 297, 1976.

97. **Nishiyama, Y., Ito, Y., Shimokata, K., Kimura, Y., and Nagata, I.**, Polykaryocyte formation induced by VSV in mouse L cells, *J. Gen. Virol.*, 32, 85, 1976.
98. **Nowakowski, M., Bloom, B. R., Ehrenfeld, E., and Summers, D. F.**, Restricted replication of VSV in human lymphoblastoid cells, *J. Virol.*, 12, 1272, 1973.
99. **Bloom, B. R., Senik, A., Stoner, G., Ju, G., Nowakowski, M., Kano, S., and Jimenez, L.**, Studies on the interactions between viruses and lymphocytes, *Cold Spring Harbor Symp. Quant. Biol.*, 41, 73, 1977.
100. **Sinarachatanant, P. and Huang, A. S.**, Effects of temperature and antibody on the cyclic growth of VSV, *J. Virol.*, 21, 161, 1977.
101. **Marcus, P. I., Sekellick, M. J., and Fuller, F. J**, Double-stranded RNA: the interferon inducer of viruses, *Int. Virol.*, 4(Abstr.), 107, 1978.
102. **Macdonald, R. D. and Yamamoto, T.**, Quantitative analysis of defective interfering particles in infectious pancreatic necrosis virus preparations, *Arch. Virol.*, 57, 77, 1978.
103. **Yang, Y. J., Stoltz, D. B., and Prevec, L.**, Growth of VSV in a continuous culture line of *Antheraea eucalypti* moth cells, *J. Gen. Virol.*, 5, 473, 1969.
104. **Thacore, H. and Youngner, J. S.**, Cells persistently infected with Newcastle disease virus. I. Properties of mutants isolated from persistently infected L cells., *J. Virol.*, 4, 244, 1969.
105. **Marcus, P. I. and Sekellick, M. J.**, Interferon action. III. The rate of primary transcription of VSV is inhibited by interferon action, *J. Gen. Virol.*, 38, 391, 1978.
106. **Isaacs, A.**, Production and action of interferon, *Cold Spring Harbor Symp. Quant. Biol.*, 27, 343, 1962.
107. **Sonnabend, J. A. and Friedman, R. M.**, Mechanisms of interferon action, in *Interferons and Interferon Inducers*, Finter, N. B., Ed., North-Holland, Amsterdam, 1973, 205.
108. **Marcus, P. I., Sekellick, M. J., Johnson, L. D., and Lazzarini, R. A.**, Cell killing by viruses. V. Transcribing defective interfering particles of vesicular stomatitis virus function as cell-killing particles, *Virology*, 82, 242, 1977.
109. **Hanson, R. P.**, The natural history of vesicular stomatitis, *Bacteriol. Rev.*, 16, 179, 1952.
110. **Baer, G. M., Shaddock, J. H., Moore, S. A., Yager, P. A., Baron, S. S., and Levy, H. B.**, Successful prophylaxis againts rabies in mice and Rhesus monkeys: the interferon system and vaccine, *J. Infect. Dis.*, 136, 286, 1977.
111. **Wiktor, T. J., Koprowski, H., Mitchell, J. R., and Merigan, T. C.**, Role of interferon in prophylaxis of rabies after exposure, in *Antivirals with Clinical Potential*, Merigan, T. C., Ed., University of Chicago Press, Chicago, 1976, 260.
112. **Flamand, A. and Bishop, D. H. L.**, Primary in vivo transcription of vesicular stomatitis virus and temperature-sensitive mutants of five vesicular stomatitis virus complementation groups, *J. Virol.*, 12, 1238, 1973.
113. **Youngner, J. S. and Preble, O. T.**, Evolution of virus populations in persistent infections of cell cultures by cytolytic RNA viruses, *J. Supramol. Struct.*, 2(Suppl.), 243, 1978.
114. **Stanners, C. P.**, personal communication.
115. **Lazzarini, R.**, personal communication.
116. **Marcus, P. I. and Sekellick, M. J.**, unpublished observations.
117. **Youngner, J. S., Preble, O. T., and Jones, E. V.**, Persistent infection of L cells with vesicular stomatitis virus: evolution of virus populations, *J. Virol.*, 28, 6, 1978.
118. **Sekellick, M. J. and Marcus, P. I.**, Persistent infection II. Interferon-inducing temperature-sensitive mutants as mediators of cell sparing:possible role in persistent infection by vesicular stomatitis virus, *Virology*, 95, 36, 1979.
119. **Bloom, B. R., Jimenez, L., and Marcus, P.I.**, A plaque assay for enumerating antigen-sensitive cells in delayed-type hypersensitivity, *J. Exp. Med.*, 132, 16, 1970.

Chapter 5

VIRUS VACCINES AND THERAPEUTIC APPROACHES

Tadeusz Wiktor

TABLE OF CONTENTS

I. INTRODUCTION

Rabies and vesicular stomatitis virus (VSV) are the only two members of the rhabdovirus group responsible for both human and animal disease. Since rabies causes such a dramatic symptomatology and a nearly 100% fatality rate, it attracted the curiosity of the first modern microbiologists, with the result that by the end of the 19th Century an effective vaccination procedure was developed.

Three important steps have been made since Pasteur's crucial advance in the conquest of rabies: the first is an increased understanding of the reservoir of virus in wild animals other than dogs; the second is the development of antirabies serum of human origin for passive/active immunization of humans; and the third is the development of a safe and potent antirabies vaccine from virus propagated in human diploid cell-tissue cultures, which has made possible a considerable reduction in the number of vaccine doses necessary for protection. Availability of this vaccine has also made possible investigation into the mechanism of postexposure rabies treatment.

II. HISTORICAL REVIEW

It has been known since ancient times that the saliva of rabid dogs is infectious. In 100 A.D. Celsus recommended the cauterization of animal bites with a hot iron, and this remained the treatment of choice for prevention of rabies until 1885, when Pasteur[1] introduced a rabies vaccine.

Experimental transmission of rabies by inoculation of saliva was first demonstrated in 1804 by Zinke.[2] In the last quarter of the century, Galtier's[3] introduction of rabbits as experimental animals for the study of rabies opened the way to a series of remarkable discoveries relating to the pathogenicity and prevention of rabies. In 1881 Pasteur and colleagues[4] concluded that the source of the virus is not solely in the saliva but that "the central nervous system material and especially the bulb which joins the spinal cord to the brain are particularly concerned and active in the development of the disease." Pasteur et al.[4] were able to transmit rabies in experimental animals by "submeningeally" inoculating them with saliva taken from naturally infected dogs or humans. The disease then could be transmitted serially in rabbits by intracerebral inoculation with a suspension of brain or bulb tissue taken from previously infected rabbits.

Two types of symptoms were observed by Pasteur et al.[5] in experimentally infected animals: paralytic or dumb rabies and furious rabies, a disease in which animals attacked everything. A small number of animals, however, were able to recover after showing initial signs of rabies and thus became resistant to further inoculations.

In 1884 Pasteur et al.[6] reported that the virulence of rabies virus could be modified by successive passage of virus in monkeys, and that such an attenuated virus could make dogs refractory to infection with fully virulent virus (street virus). Since rabies has a long incubation period in man, the possibility of using attenuated virus to immunize persons exposed to rabies was considered, but no steps were taken. It was not until October 26, 1885 that Pasteur[1] presented a report to the Academy of Science giving for the first time a detailed description of his procedure for prevention of rabies in humans. The following events led up to that report.

In 1882 Pasteur et al.[4] isolated a strain of virus from the brain of a cow and maintained it by intracerebral inoculation in rabbits for 90 consecutive passages. The original incubation period of 15 days became shorter with the number of passages and remained fixed at 7 days after about 50 passages (fixed virus). When spinal cords of rabbits showing signs of paralysis 7 days after infection were removed and suspended in dry air at room temperature, it was found that the virulence of such cords decreased rapidly, normally being completely lost after 15 days of dessication.

Pasteur immunized dogs by subcutaneously injecting suspensions of fragments of rabies virus-infected spinal cords, beginning with one dried long enough to be avirulent and using successively more virulent material until he finally reached a virulent cord. All of the more than 50 dogs submitted to this type of immunization resisted rabies infection when injected intracerebrally with virulent virus.

The first opportunity to apply this technique to the protection of man occurred in 1885. On July 6, a 9-year-old boy, Joseph Meister, who had been bitten 14 times by a rabid dog arrived in Paris from Alsace. The boy was examined by Doctors Vulpian and Grancher who thought the child had received a fatal inoculation of rabies virus. According to Pasteur, "the death of this boy seemed inevitable, and I decided, not without lively and cruel doubts, as one can believe, to try in Joseph Meister the method which had been successful in dogs." Apparently, Pasteur was not concerned that the 50 protected dogs had not been exposed to rabies before treatment because he had also been able to protect a number of animals treated after exposure. Consequently, 60 hr after being bitten, Joseph Meister received a subcutaneous inoculation consisting of a half syringe of the cord of a rabid rabbit preserved in a flask of dry air for 15 days. Twelve successive inoculations were made with cords of increasing virulence, for a total of 13 inoculations during a 10-day period. The boy not only did not develop rabies, but also escaped large quantities of highly virulent virus known to have been contained in the last five doses of vaccine. Meister remained in perfect health and grew up to become the janitor of the Pasteur Institute in Paris.

A second patient was treated on October 20, 1885, 6 days after having been bitten. No rabies developed. By February 25, 1886, Pasteur[7] reported that he had administered his 35th prophylactic treatment, and by April 12, 726 had been treated.[8] Of 688 cases treated after a dog bite, only one died, whereas three of 38 persons bitten by a wolf were not saved by the treatment. By October 31, 1886,[9] no less than 2490 persons had undergone a course of prophylaxis. Of this number, 1726 were treated after exposure in France or Algeria, and 10 were not protected. Treatment failures were observed mostly in children bitten on the face or head or receiving deep wounds of the extremities. Treatment for this group of patients was modified, being made more rapid and energetic. The typical regimen for intensive treatment is presented in Table 1.

The Pasteur method of treatment aroused very great interest in medical circles and, despite some antagonistic communications, was rapidly accepted. The Pasteur Institute of Paris was founded in 1888, and within a decade there were Pasteur Institutes throughout the world. Technical procedures were greatly simplified by Roux[10] who demonstrated that dessicated cords preserved in glycerol retained the appropriate degree of virulence for a prolonged period. Pasteur's original method has always been the method of choice in France and French colonies; in other parts of the world, however, numerous modifications have been introduced. The original Pasteur vaccination method has not been used in humans since 1953, when it was last employed at the Pasteur Institute in Paris.

III. NERVOUS TISSUE VACCINES

Of the several proposed modifications in vaccine preparation procedures we will mention only those that were widely used for human treatment.

A. Live Virus

1. Hoegyes' dilution method

In the procedure introduced by Hoegyes[11] in 1887 the virulence of fixed virus (spinal

TABLE 1

Schedule for Intensive Rabies Treatment (Pasteur[8])

Treatment day	Number of vaccine inocula-tions	Age of dessicated spinal cords (days)
1	3	12, 10, 8
2	3	6, 4, 2
3	1	1
4	3	8, 6, 4
5	2	3, 2
6	1	1
7	1	4
8	1	3
9	1	2
10	1	1

Note: In extreme cases of exposure a complete series of treatment could be given in a single day and repeated on following days. In contrast, the standard treatment at this time was inoculations over a 10-day period beginning with material dried for 14 days and concluding with the 5-day-old cord. Pasteur, L., *C. R. Acad. Sci. Paris,* 103, 777, 1886. With permission.

cords) was reduced by dilution rather than by dessication. Generally, treatment started with virus dilutions at 1/10,000 and finished with dilutions at 1/100. Treatment lasted 2 to 3 weeks. This method of treatment has been widely used in human practice (Bulgaria, Brazil, Istanbul, Madrid, and Riga) and has proved very efficacious.[12]

2. Fermi's phenol method

Inactivation of virus-infected brain tissue by phenol was proposed by Fermi.[13] Generally, 5% brain tissue suspensions were incubated in 1% phenol for 24 hr at room temperature (22 to 24°C). The resulting mixture was not completely inactivated, since a residual virulence was considered necessary for the vaccine's protective capacity. The Fermi type of rabies vaccine was widely used throughout the world and is the only live virus vaccine still in production, especially in Africa.

3. Etherized vaccines

Among several proposed procedures, Hempt's technique[14] is the only one still in use on a limited scale. Virus-infected brains are suspended in ether for a determined period of time, then emulsified in the presence of phenol. When phenol treatment is omitted the vaccine can contain varying amounts of live virus.

B. Inactivated Virus

Although usually harmless from the earliest days, antirabies treatment has occasionally been followed by severe and sometimes fatal accidents. It was found that the incidence of such reactions was considerably reduced by the use of phenolized vaccines completely free from live virus.

1. Semple's phenol method[15]

This vaccine is a slight modification of Fermi's method. Virus suspension phenolized at 1% is incubated at 37°C for 24 hr and kept at room temperature for at least 30 days before use. The resulting vaccine is completely free from infectious virus.

This vaccine has found by far the widest application, and even at the present time over 60% of all rabies vaccines used are of the Semple type.

2. Neonatal brain tissue vaccines

The myelin present in suspension of adult brain tissue is believed to be responsible for paralytic accidents connected with the use of rabies vaccines. Myelin-free vaccines prepared from neonatal mouse brains by Fuenzalida and Palacios[16] or from brains of immature rats by Svet-Moldavskaya and Svet-Moldavskij[17] or from suckling rabbit brains[18] have been described. Fuenzalida vaccine, which is presently used in all Latin American countries, was recently introduced in France.[19]

IV. NON-NERVOUS TISSUE VACCINE

A. Duck Embryo Vaccine

Duck embryo vaccine (DEV) was developed by Peck et al.[20] to avoid any allergic accidents caused by nervous tissue. Virus propagated in embryonated duck eggs is inactivated by beta-propiolactone and dessicated. Widely used in the U.S. for the last 25 years, DEV causes local reactions and can induce allergic reactions to avian antigen, but is neary completely devoid of encephalitogenic properties.

V. TISSUE CULTURE VACCINES

The solution to the problems of safety of rabies vaccine evidently lay in the development of vaccines prepared from rabies virus grown in tissue culture free of neural tissue. Ideally, to avoid foreign proteins, the tissue culture would be of human cells.

A. Primary Hamster Kidney Tissue Vaccine

Hamster kidney rabies vaccine has been thoroughly investigated since the early work done by Kissling.[21] The first vaccine prepared by Fenje[22] from Sad virus propagated in cells of hamster origin and inactivated by formaldehyde is still licensed in Canada for preexposure human immunization. A similar vaccine inactivated by phenol was developed in the U.S.S.R. by Selimov and Askenova[23] and extensively tested in pre- and postexposure immunization of man. The potency of this vaccine, however, is not higher than that of brain tissue vaccine. It is generally administered in 15 daily doses of 5 mℓ each.

B. Human Diploid Cell Tissue Vaccine

Production of human diploid cell vaccine (HDCV) was made possible by the development of the WI-38 normal human fibroblast cell line by Hayflick and Moorhead[24] and the adaptation of the Pitman-Moore strain of rabies virus to growth in WI-38 cells by Wiktor et al.[25] Virus grown in WI-38 cells is concentrated by ultrafiltration[26] or concentrated and purified by rate zonal centrifugation in a sucrose gradient[27] and inactivated by beta-propiolactone[26] or by tri(n-butyl)phosphate.[28]

HDCV vaccine containing concentrated and purified virus provided much better immunizing responses in rhesus monkeys than other vaccines such as DEV, suckling mouse brain, or brain tissue vaccine (modified Pasteur vaccine). Not only were antibody levels higher after immunization with the tissue culture vaccine than after inoculation with the other vaccines tested, but the antibody also appeared earlier.[29] The most spectacular result obtained in the course of these animal studies, however, was the demonstration,[29] for the first time, that a single injection of a vaccine given several hours after challenge with street virus could protect animals from rabies.

After extensive evaluation on laboratory animals, the vaccine was tested in several thousand human volunteers, first in the U.S.[30-34] then in Iran,[35] France,[36] Germany,[37-39] Great Britian,[40-41] India,[42] and Thailand.[43] The inoculations were well tolerated, and antibody testing indicated that three appropriately spaced doses of the new vaccine assured a rapid induction of virus-neutralizing antibody of much higher titer than that observed after administration of 14 to 21 daily doses of DEV or brain tissue vaccine.[33,42] From 1 to 6 years after primary vaccination, several hundred volunteers received a secondary vaccination.[33,37,38,41] All showed remarkable increases in antibody levels 1 or 2 weeks after booster doses, and none showed adverse reactions. Simultaneous inoculation of antirabies serum[35] or purified rabies gamma globulin of human origin[38,39,43] did not significantly interfere with active antibody production in human volunteers treated wih HDCV.

After 4 years of clinical studies in human volunteers, the new vaccine was used in humans exposed to rabies infection. Since the mortality rate of untreated people who suffer from severe wounds can be as high as 50%,[44] the postexposure trial conducted in Iran by Bahmanyar et al.[45] in humans severely bitten by rabid wolves and dogs is probably the most significant proof of efficiency of HDCV in treatment after exposure. None of the 45 severely bitten individuals died of rabies. They were treated with one dose of heterologous antirabies serum followed by four inoculations of HDCV given within 21 days and two additional booster inoculations administered 1 and 3 months later. All treated persons survived and are well more than 2 years after completion of treatment. All biting animals were proven to be rabid, and the severity of infection was such that 20 to 30% of the patients would have contracted rabies and died if they had not been treated. Parallelly, Kuwert et al.[39] reported the results of successful treatment of 108 people exposed to rabies infection; 44 of these individuals were bitten or severely hurt by wild animals or animals proven rabid. Eighteen were bitten by rabies-suspected animals in areas where rabies is endemic. These animals, however, escaped so that diagnosis could not be verified by laboratory examination. The remaining 46 patients came in close contact with the saliva of proven rabid animals of different species, but had only minor wounds at the site of the presumed virus contamination.

Data of postexposure treatment with the new vaccine in the U.S. have not yet been published. Available information indicates, however, that from 250 to 300 people have been treated with the HDCV. In about 10 to 15% of these cases, the biting animal was identified as being infected with rabies. The choice of patients to be treated with HDCV was dictated by known allergies to avian proteins present in DEV.

In conclusion, it is evident that an effective tool for protecting man against rabies has been developed. Not only can the number of inoculations used in postexposure treatment be decreased from 14 to 21 to 4 to 6, but in contrast to the vaccines used previously, the new HDCV is highly immunogenic and does not cause side effects. Extensive use of this vaccine may finally bring human rabies under control.

VI. EVALUATION OF ANTIRABIES TREATMENT IN HUMANS

Evaluation of antirabies treatment of man is complicated by several factors. The mortality from rabies in untreated persons bitten by rabid animals is, on the average, 15%; however, mortality figures vary directly with the amount of virus inoculated into the wound and the degree of trauma. Thus, in persons severely bitten by rabid wolves mortality can approach 50%.[44] Strains of street rabies virus probably vary in virulence and in antigenic structure. In addition, it is difficult to prevent development of rabies experimentally by injection of vaccine after infection.

Pasteur[8] recognized that his original method of vaccinating humans was founded solely on experiments with dogs vaccinated before challenge. Immediately after treating Meister, Pasteur initiated experiments with dogs infected prior to vaccination. Only partially successful results were obtained under the following conditions: "vaccination must be initiated in the first 24 hr after exposure, one must proceed rapidly and give a total series of vaccine inoculations in 24 hr. Treatment must be given every two hours and repeated once or twice."

In 1939 Webster[12] critically evaluated all publications related to postexposure protection of experimental animals. In only one instance from over 90 reports, was the vaccine shown to be protective, and in this case the total volume of vaccine injected was at least 50% of the body weight of the treated animals (rats or guinea pigs). It is difficult to conceive how this heroic treatment can be related to the vaccine regimen normally administered to humans.

A number of patients die of rabies despite the application of antirabies treatment. Approximately one third of such cases develop illness less than a month after the bite, one third between 1 and 2 months, and the remaining third after 2 months. It has been stated that antirabies treatment may actually bring about death from rabies after a short incubation period, the fixed virus inhibiting defense mechanisms and accelerating the replication of street rabies virus.

It has been suggested that rabies may develop in treated persons after an emotional shock, as if street virus were "latent" in the nervous system.

Information collected by the League of Nations from 1930 to 1946 was analyzed by McKendrick in 1940[46] and by Greenwood in 1945 to 1946.[47] Out of 1,670,848 cases of vaccine treatment, mortality from rabies was 0.15% in Europeans and 0.52% in non-Europeans. Even though untreated persons were not included in these statistics, results indicate that good protection was afforded by a vaccination procedure. No differences could be detected in the protective abilities of a variety of live and attenuated virus vaccines prepared by different laboratories. The lower incidence of fatal accidents after inactivated virus vaccines, rather than any superiority in their immunizing capacity, was a major factor in abandoning the use of living virus vaccines. McKendrick's observations led Webster[12] to conclude that "these vaccines are all equally effective or equally non-effective" — both disturbing conclusions.

Surprisingly few papers concerning the value of antirabies treatment in humans have included adequate observations of untreated control groups. The most convincing evidence supporting the efficacy of rabies vaccine is included in a series of reports of the Pasteur Institute of southern India in Coonoor.[48] In a group of 1221 persons bitten by dogs known to have caused at least one human rabies infection, the mortality rate of untreated persons was 56.6% as compared to only 7.0% in a similar group receiving complete treatment. In a group of 18,714 individuals bitten by dogs proven or presumed to be rabid, the mortality rate in nontreated was 16.4%, and in persons completely treated only 0.38%.

VII. REACTIONS AFTER HUMAN ANTIRABIES TREATMENT

A. General Systemic Reactions

The various minor disorders that may develop during or after a course of antirabies treatment include fever, headache, insomnia, palpitations, and diarrhea. Sensitization to proteins contained in vaccine can cause a sudden shock-like collapse usually towards the end of a course of treatment.

B. Local Reactions

Erythematous patches can develop about 1 week to 10 days after the beginning of antirabies treatment. Lesions appear a few hours after vaccine injection and fade in 6 to 8 hr, reappearing after the next vaccine inoculation.

C. Severe and Fatal Reactions

A patient may suffer from serious, often fatal, illness after antirabies treatment. Fortunately these accidents, which are of two types, have been extremely rare since the development of inactivated, nonnervous tissue vaccines.

Rage de Laboratoire—Rage de Laboratoire is a disease provoked by the living "fixed rabies virus" present in the vaccine.

Neuroparalytic Accidents—Neuroparalytic accidents present the greatest danger from rabies vaccination. All types of vaccine containing adult mammalian nervous tissue exhibit similar capacities for inducing neurologic reactions, regardless of whether they contain live virus or virus that has been inactivated by any of the various methods.

The accident usually develops between the 13th and 15th day of antirabies treatment and usually assumes one of three forms:

Landry type—The patient rapidly becomes pyrexial and suffers pain in the back. Flaccid paralysis of the legs sets in, and within the next day the arms become paralyzed. Later the infection spreads to the face, tongue, and other muscles supplied by the bulb. The fatality rate is about 30%; in the remaining 70% recovery usually occurs rapidly.

Dorsolumbar type—Less severe than the Landry type, this is the commonest form of neuroparalytic accident. Clinical features are explicable by the presence of dorsolumbar myelitis. The patient may be febrile and feel weak and the lower limbs become paralyzed, usually with diminished sensation and sphincter disturbances. The fatality rate is not over 5%.

Neuritic Type—The patient may be pyrexial and usually shows a temporary paralysis of the facial, oculomotor, glossopharyngeal, or vagus nerves.

Neuroparalytic accidents are caused by "allergic encephalomyelitis" attributable specifically to sensitivity to adult nervous tissue antigen (myelin). The incidence of these reactions varies widely from 0.017%[45] to 0.44%.[49] The incidence of neurologic reactions is definitely lower in persons receiving DEV and is reduced when properly manufactured vaccine of newborn rodent brain is administered.

VIII. SERUM AND VACCINE TREATMENT

The ability of treatment with brain tissue vaccine alone to protect persons exposed to wolf bites was analyzed by Baltazard and Ghodssi at the Pasteur Institute of Teheran in 1953.[50] The findings were considered alarming. In patients bitten by confirmed rabid wolves and treated with vaccine, the overall mortality was 25%, regardless of the severity and place of the wounds; among individuals wounded in the head and face mortality was 42%. Comparing these figures with statistically comparable cases exposed to rabid wolves but untreated for various reasons, the authors came to the conclusion that postexposure treatment of severely exposed individuals with vaccine alone offers insignificant protection.

The use of antirabies serum in postexposure prophylaxis goes back many years; in 1891 Babes and Cerchez[51] treated 12 persons severely bitten by a rabid wolf with whole blood from vaccinated humans or dogs. Each person received four to six injections of immune blood in addition to the Pasteur treatment, and all survived. In addition, another person bitten by the same wolf, who did not present himself for treatment, died of rabies. Between this date and the 1940s, a number of publications appeared on the use of rabies immune serum or serum concentrates in the prevention of experimental rabies in animals or in the postexposure treatment of man. Results ranged from complete protection to no protection at all.

The unquestioned efficacy of antirabies serum in postexposure prophylaxis was finally established through the studies of the World Health Organization Committee on Rabies. The superior results obtained experimentally by Habel[52] and by Koprowski et al.[53] were confirmed in a field trial in Iran in 1954.[54] Twenty-nine people bitten by the same rabid, infectious wolf, 18 of whom had very severe wounds of the head and neck, arrived in Teheran less than 36 hr after having been bitten. There they were divided into two groups: some received treatment with phenolized vaccine alone; others received vaccine therapy associated with one or several injections of antirabies serum. Three of five patients treated with vaccine alone contracted rabies and died, whereas only one of 13 subjects treated with vaccine and antirabies serum did so. Thus, the addition of serum therapy to vaccine seems to improve the results of treatment to such an extent that there can be little doubt of the crucial role played by passive immunization in the prevention of rabies.

Many experimental studies in humans have established that the combined use of a single dose of antirabies serum given at the start of treatment and a course of at least 14 daily doses of vaccine is the best specific prophylaxis available today. However, it has been found that the antibody tends to inhibit the antigenic effect of the vaccine. To overcome this inhibition, it is now recommended that two booster doses of vaccine be given 10 and 20 days after the regular vaccine schedule.

A. Human Rabies Immune Globulin

To eliminate the serum sickness reactions that follow inoculation with animal rabies serum, volunteers were hyperimmunized with rabies vaccine and bled, and gamma globulin was prepared from their plasma. Human rabies immune globulin (HRIG) is now commercially available and, although expensive, will probably replace animal rabies serum for use in postexposure treatment. As an homologous antiserum, HRIG persists longer in the circulation of the host, but may have an even greater dampening effect on active immunization. Thus Hattwick et al.[55] found that 23 doses of DEV were needed to overcome the suppressive effect of 15 to 40 units of HRIG per kilogram.

IX. RABIES VACCINATION OF ANIMALS

From the beginning of his involvement with rabies, Pasteur recognized that protection of man could be effectively achieved through the vaccination of dogs. Even though dogs were used for most of the experimental data on protection obtained from 1884 to 1885, it was not until the early 1920s that a practical vaccine for dogs was developed and used successfully in practice.

The first vaccine used for mass vaccination of dogs was a modified Semple type prepared by Umeno and Doi in Tokyo.[56] It proved effective in controlling rabies in dogs in Japan and in other countries that began to produce and use this type of vaccine. The quality of vaccines improved greatly with Habel's introduction[57] of a standard mouse potency test for Semple-type vaccine, which assured the use of potent vaccines in mass vaccination programs.

Johnson[58] demonstrated that a single dose of a potent, phenol-inactivated vaccine can protect dogs against a challenge with street rabies virus for a period of over 1 year. From then on, this was virtually the only type of vaccine used for control of rabies in dogs, cats, and other domestic animals.

The modified live-virus rabies vaccine was introduced by Koprowski and Cox.[59] Successive passages of a strain of virus of human origin, first in 1-day-old chicks then in embryonated hen's eggs, resulted in the loss of pathogenicity for dogs (Flury low egg passage; LEP)[60] providing an attenuated strain safe for use in dogs. Further passages of Flury strain virus in embryonated hen's eggs resulted in a vaccine no longer infective for adult laboratory animals, yet lethal for newborn mice (Flury high egg passage; HEP).[61] Both Flury LEP and Flury HEP are still widely used in all types of domestic animals.

More recently, another attenuated strain of rabies virus, ERA, was introduced by Canadian workers.[62] The ERA vaccine was shown to provide excellent immunity lasting for at least 3 years. Several new rabies vaccines of tissue culture origin containing live or inactivated virus have been introduced during the last 10 years and are in use in several European and American countries.

Immunization of dogs against rabies is still the most important measure in a local rabies control program. The breaking of the chain of transmission in canine rabies can be considered the greatest success story in the battle against zoonotic disease.

X. POSSIBLE MECHANISMS OF POSTEXPOSURE PROTECTION FROM RABIES

The role of the immune response in rabies virus infection and prophylaxis remains obscure. Virus-neutralizing antibodies (VNA) can be detected in animals that are inoculated with virulent or attenuated virus a few days after infection; the presence and titers of these antibodies, however, are unrelated both to the time required for symptoms to appear and to the outcome of the disease. In humans infected with rabies, VNA usually appear a few days after the onset of symptoms,[63] and their titers increase over the course of the disease. High levels of antibody have been reported in cases of protracted disease,[61] and in two reported instances of recovery from rabies,[64,65] the levels of VNA oberved were exceptionally high. When VNA are induced by vaccination or acquired through the administration of an antirabies serum before exposure to rabies virus, protection against subsequent challenge is usually assured.[29] However, only vaccine has a protective effect when administered postexposure; antirabies serum alone given after challenge only prolongs the incubation time and fails to protect from death.[29] The manner in which antibodies protect when vaccine is administered after exposure is poorly understood.

The study of the mechanism involved in protection from rabies infection by means of vaccination after exposure became possible a few years ago when it was finally demonstrated[29] that a single dose of a potent, experimentally produced, inactivated rabies vaccine of tissue culture origin could protect monkeys and other animals treated several hours after experimental exposure to street rabies virus. The simultaneous administration of antiserum against rabies virus with vaccine in this experimental situation did not improve the results obtained with vaccine alone; treatment with the antiserum alone was less effective than treatment with vaccine alone.[29] No correlation could be established between the level of VNA in individual animals and protection against death, nor could the time at which antibody appeared in the animal's serum be correlated with protection.

Since concentrated rabies vaccine of tissue culture origin was shown to induce high levels of circulating interferon in hamsters and to protect them from rabies infection if given shortly before or after challenge with rabies virus,[66] the hamster model was used to determine whether similar activity could be obtained with other inactivated viral vaccines. Treatment with inactivated vaccines of influenza B virus and Kern Canyon virus (another rhabdovirus) rapidly induced interferon in hamsters and was as effective as the potent interferon-inducing rabies vaccine in protecting hamsters treated 24 hr before or after infection with rabies virus. However, when vaccination preceded challenge by 5 days or more, only rabies vaccine (and not heterologous vaccine) rendered the animals immune and resistant to infection. All rabies vaccine preparations that induced interferon-protected hamsters treated after exposure. Moreover, vaccine preparations that did not induce detectable levels of interferon in serum did not protect animals treated after exposure, although they were able to protect hamsters when administered several days before challenge.

In an experiment with monkeys,[67] the simultaneous inoculation of vaccine and interferon inducers (polyribocytidylic acid homopolymer pair or Newcastle disease virus) did not improve the results obtained with vaccine alone, which indicates that factors other than interferon and antibody may play an important role in the treatment of rabies after exposure.

Clinical observations suggest that cell-mediated immunity (CMI) plays a role in promoting recovery from infections with herpesviruses, poxviruses, and some paramyxoviruses and arboviruses, as well as in resistance to virus-induced tumors (reviewed by Woodruff and Woodruff[68]). It was demonstrated previously[69] that clinically inapparent disease caused by attenuated strains of rabies virus can be converted into lethal infection by immunosuppression of mice or by the use of thymus-deprived nude mice as experimental animals.

Wiktor et al.[70] recently reported that a strong, specific cell-mediated cytotoxic (CMC) response is generated in mice exposed to an attenuated rabies virus or injected with an inactivated rabies virus vaccine. This reaction is a function of thymus-derived immune lymphocytes and is severely depressed by the prior inoculation of mice with antirabies antibody. It was further reported[71] that only very low levels of CMC were detected in spleen cells of animals infected with virulent strains of virus and that such animals were not only deficient in generating CMC against rabies virus-infected target cells, but were also unable to develop CMC against an unrelated influenza virus. No inhibition of CMC against influenza virus could be observed in mice that had been infected with an attenuated strain of rabies virus.

It has recently been reported by Wiktor[72] that in mice resisting infection after intracerebral inoculation with attenuated strains of rabies virus, infectious virus and fluorescent viral antigens were detected in the brains for only short periods (postinfection days 3 to 9).[68] In mice infected with virulent strains of rabies, infectious virus and viral antigens were detectable throughout the course of the disease, despite the presence of circulating and brain-associated VNA. Infectious virus and viral antigens disappeared from the brain tissue of mice infected with attenuated strains of virus at the same time that "killer" cells appeared. These cells may have been specifically programmed to destroy infected cells in the brain. The released virus would then have been inactivated by neutralizing antibody. The clearing from brain tissue of virus and viral antigen appears to be absolute, as the virus cannot be reisolated either by classical methods or by culturing of brain tissue, cocultivation, or fusion with indicator cells.[74]

The generation of CMI after infection with attenuated strains of virus was directly correlated with protection from rabies death. Similarly, the generation of CMC in vaccine-treated animals correlated with protection from street rabies virus infection. The timing of vaccine administration was very important in the experimental model studied. Vaccine was efficient when given before or at the time of challenge; however, when treatment was delayed until 24 hr after infection with street virus, the immunosuppressing activity of the virus could not be overcome, the CMC reaction was absent, and vaccine failed to protect.

A high level of CMC could be maintained by repeated inoculations of vaccine, and the presence of VNA did not interfere with the secondary stimulation of sensitized lymphocytes. This observation provided experimental justification for the long established practice of repeated vaccine inoculations in postexposure treatment.

Finally, it was observed that challenge inoculations with street rabies virus can induce a secondary CMC response in vaccinated animals and thus can contribute to the mechanism of protection. The same dose of street virus given to nonvaccinated animals was lethal and suppressed CMC induction. These experiments clearly demonstrate that induction of CMC must play an important role in the mechanism of postexposure protection from rabies.

The methodology now available for studying cellular level reactions in mice have provided new insights into the possible mechanism of postexposure protection from rabies infection. More work is needed to answer the many questions that still remain. For instance, the role of interferon induction by rabies vaccine during the early stages of the postexposure treatment needs to be clarified. Evidence has been presented that interferon may act as a positive regulator of fundamental cellular processes, including enhancement of specific CMC responses.[73]

REFERENCES

1. **Pasteur, L.,** Methode pour prevenir la rage après morsure, *C. R. Acad. Sci.,* 101, 765, 1885.
2. **Zinke, M.,** *Neue Ansichten der Hundswut,* C. E. Gabler, 1804, 180; cited by **van Rooyen, C. E. and Rhodes, A. J.,** Virus Diseases of Man, Oxford University Press, London, 1948.
3. **Galtier, M.,** Deuxieme note sur la rage, *C. R. Acad. Sci.,* 89, 444, 1879.
4. **Pasteur, L., Roux, E., Chamberland, C., and Thuillier, L.,** Sur la rage, *C. R. Acad. Sci.,* 92, 1259, 1882.
5. **Pasteur, L., Roux, E., Chamberland, C., and Thuillier, L.,** Nouveaux faits pour servir a la connaissance de la rage, *C. R. Acad. Sci.,* 95, 1187, 1882.
6. **Pasteur, L., Chamberland, C., and Roux, E.,** Sur la rage, *C. R. Acad. Sci.,* 98, 1229, 1884.
7. **Pasteur, L.,** Resultats de l'application de la methode pour prevenir la rage après morsure, *C. R. Acad. Sci.,* 102, 459, 1886.
8. **Pasteur, L.,** Nouvelle communication sur la rage, *C. R. Acad. Sci.,* 103, 777, 1886.
9. **Pasteur, L.,** Note complementaire sur les results de l'application de la methods de prophylaxie de la rage après morsure, *C. R. Acad. Sci.,* 102, 835, 1886.
10. **Roux, E.,** Note sur un moyen de conserver les moelles rabiques avec leur virulence, *Ann. Inst. Pasteur Paris,* 1, 87, 1887.
11. **Hoegyes, A.,** *Orv. Hetil.,* 31, 121, 1887; cited by **van Rooyen, C. E. and Rhodes, A. J.,** *Virus Diseases in Man,* Oxford University Press, London, 1948, 861.
12. **Webster, L. T.,** The immunizing potency of antirabies vaccines. A critical review, *Am. J. Hyg.,* 30, 113, 1939.
13. **Fermi, C.,** Über die Immunisierung gegen Wutkrankheit, *Z. Hyg. Infektionskr.,* 58, 233, 1908.
14. **Hempt, A.,** Sur la methode rapide de traitement antirabique, *Ann. Inst. Pasteur Paris,* 39,632, 1925.
15. **Semple, D.,** The preparation of a safe and efficient antirabic vaccine, *Scientific Memoirs of Medical and Sanitary Department, India,* No. 44, 1911.

16. **Fuenzalida, E. and Palacios, R.**, Rabies vaccine prepared from brains of infected suckling mice, *Biol. Inst. Bacteriol.*, 8, 3, 1955.
17. **Svet-Moldavskaya, I. A. and Svet-Moldavskij, G. Y.**, Antirabies vaccine not producing postvaccinal neuroparalytic complications, *Vopr. Virusol.*, 7, 68, 1962.
18. **Gispen, R. Schmittmann, G. J. P., and Saathof, B.**, Rabies vaccine derived from suckling rabbit brain, *Arch. Gesamte Virusforsch.*, 15, 366, 1965.
19. **Gamet, A., Atanasiu, P., Dodin, A., Tsiang, H., and Vialat, C.**, Les vaccinations antirabiques en France en 1972, *Ann. Microbiol.*, 124 B, 411, 1973.
20. **Peck, F. B., Powell, H. M., and Culbertson C. G.**, Duck-embryo rabies vaccine...study of fixed virus vaccine grown in embryonated duck eggs and killed with beta-propiolactone, *JAMA*, 162, 1373, 1956.
21. **Kissling, R E.**, Growth of rabies virus in non-nervous tissue culture, *Proc. Soc. Exp. Biol. Med.*, 98, 223, 1958.
22. **Fenje, P.**, A rabies vaccine from hamster kidney tissue cultures; preparation and evaluation in animals, *Can. J. Microbiol.*, 6, 605, 1960.
23. **Selimov, M. A. and Askenova, T. A.**, Tissue culture antirabic vaccine for human use, *Symp. Ser. Immunobiol. Stand.*, 1, 377, 1966.
24. **Hayflick, L. and Moorhead, P. S.**, The serial cultivation of human diploid cell strains, *Exp. Cell Res.*, 25, 585, 1961.
25. **Wiktor, T. J., Fernandes, M. V., and Koprowski, H.**, Cultivation of rabies virus in human diploid cell strain WI-38, *J. Immunol.*, 93, 353, 1964.
26. **Wiktor, T. J., Sokol, F., Kuwert, E., and Koprowski, H.**, Immunogenicity of concentrated and purified rabies vaccine of tissue culture origin, *Proc. Soc. Exp. Biol. Med.*, 131, 799, 1969.
27. **Majer, M., Hilfenhaus, J., Mauler, R., and Hennessen, W.**, Zonal-centrifuged human diploid cell rabies vaccine, *Dev. Biol. Stand.*, 37, 267, 1977.
28. **Tint, H., Dobkin, M. B., and Rubin, B. A.**, A new tissue culture (WI-38) rabies vaccine, inactivated and disaggregated with Tri-(n)-butyl phosphate, *Symp. Immunobiol. Stand.*, 21, 132, 1974.
29. **Sikes, R. K., Cleary, W. F., Koprowski, H., Wiktor, T. J., and Kaplan, M. M.**, Effective protection of monkeys against death by street virus by post-exposure administration of tissue culture rabies vaccine, *Bull. W.H.O.*, 45, 1, 1971.
30. **Wiktor, T. J., Plotkin, S. A., and Grella, D. W.**, Human cell culture rabies vaccine, *JAMA*, 224, 1170, 1973.
31. **Kaplan, M. M.**, Preliminary results on antibody profiles of humans receiving concentrated inactivated vaccine prepared in human diploid cells, *Symp. Ser. Immunobiol. Stand.*, 21, 226, 1974.
32. **Cabasso, V., Dobkin, M. B., Roby, R. E., and Hammar, A. H.**, Antibody response to a human diploid cell rabies vaccine, *Appl. Microbiol.*, 27, 553, 1974.
33. **Plotkin, S. A., Wiktor, T. J., Koprowski, H., Rosanoff, E. I., and Tint, H.**, Immunization schedules for the new human diploid-cell vaccine against rabies, *Am. J. Epidemiol.*, 103, 75, 1976.
34. **Garner, W. R., Jones, D. O., and Pratt, E.**, Problems associated with rabies preexposure prophylaxis, *JAMA*, 235, 1131, 1976.
35. **Bahmanyer, M.**, Results of antibody profiles in man vaccinated with the HDCS vaccine with various schedules, *Symp. Ser. Immunobiol. Stand.*, 21, 231, 1974.
36. **Soulebot, J. P.**, *Resultats Serologiques de l'Immunisation et de Hyperimmunisation de l'Homme avec un Nouveau Vaccin Antirabique Obtenu sur Cultures de Cellules Diploides Humaines WI-38*, La Rage, Colloque, Paris, 173, Expansion Scientifique Francaise, Publie sous la Direction de Pr. J. B. Dureaux, 1974, 149.
37. **Cox, J. H. and Schneider, L. G.**, Prophylactic immunization of humans against rabies by intradermal inoculation of human diploid cell culture vaccine, *J. Clin. Microbiol.*, 3, 96, 1976.
38. **Kuwert, E., Marcus, I., and Hoher, P. G.**, Neutralizing and complement-fixing antibody responses in pre- and postexposure vaccines to a rabies vaccine produced in human diploid cells, *J. Biol. Stand.*, 4, 249, 1976.
39. **Kuwert, E. K., Marcus, I., Werner, J., Scheiermann, N., Hohner, P. G., Thraenhart, O., Hierholzer, E., Wiktor, T. J., and Koprowski, H.**, Postexposure use of human diploid cell culture rabies vaccine, *J. Biol. Stand.*, 37, 274, 1977.
40. **Aoki, F. Y., Tyrrell, D. A., and Hill, L. E.**, Immunogenicity and acceptability of a human diploid-cell culture rabies vaccine in volunteers, *Lancet*, 1, 660, 1975.
41. **Turner, G. S., Aoki, F. Y., Nicholson, K. G., Tyrrell, D. A., and Hill, L. E.**, Human diploid cell strain rabies vaccine. Rapid prophylactic immunization of volunteers with small doses, *Lancet*, 1, 1379, 1976.
42. **Shah, U., Jaswal, G. S., Mansharamani, H. J., Plotkin, S. A., and Wiktor, T. J.**, Trial of human diploid-cell rabies vaccine in human volunteers, *Br. Med. J.*, 1, 997, 1976.

43. **Nelson, K. E., Vithayasai, V., Makornkawkayoon, S., Rosanoff, E., Tint, H., Plotkin, S. A., and Wiktor, T. J.**, Studies of a human diploid cell strain (HDCS) rabies vaccine used with human rabies immune globulin (RIG), *Conf. of Antimicrobial Agents and Chemotherapy, Am. Soc. Microbiol.*, Chicago, No. 97(abstr.), 1976.
44. **Gremliza, L.**, Kasuistik zum Lyssa-Problem, *Z. Tropenmed. Parasitol.*, 4, 382, 1953.
45. **Bahmanyar, M., Fayaz, A., Nur-Salehi, S., Mahamdei, M., and Koprowski, H.**, Successful protection of humans exposed to rabies infection postexposure treatment with the new human diploid cell rabies vaccine and antirabies serum, *JAMA*, 236, 2751, 1976.
46. **McKendrick, A. G.**, A ninth analytical review of reports from Pasteur Institutes on the results on anti-rabies treatment, *Bull. W.H.O.*, 9, 31, 1940.
47. **Greenwood, M.**, Tenth report on data of anti-rabies treatment supplied by Pasteur Institutes, *Bull. W.H.O.*, 12, 301, 1945.
48. **Veeraraghavan, N.**, "Annual Report", Director Pasteur Institute of Southern India, Coonoor, Diocesan Press, Madras, India, 1970.
49. **Shiraki, H. and Othani, S.**, *Allergic Encephalomyelitis*, Charles C Thomas, Springfield, Ill., 58, 1959.
50. **Baltazard, M. and Ghodssi, M.**, Prevention de la rage humaine, *Rev. Immunol.*, 17, 366, 1953.
51. **Babes, V. and Cerchez, T.**, *Ann. Inst. Pasteur Paris*, 10, 625, 1891.
52. **Habel, K.**, Seroprophylaxis in experimental rabies, *Public Health Rep.*, 60, 545, 1945.
53. **Koprowski, H., van der Scheer, J., and Black, J.**, The use of hyperimmune antirabies serum concentrates in experimental rabies, *Am. J. Med.*, 8, 412, 1950.
54. **Baltazard, M. and Bahmanyar, M.**, Essai pratique du serum antirabique chez les mordus par loups enrages, *Bull. W.H.O.*, 13, 747, 1955.
55. **Hattwick, M. A. W., Rubin, R. H., Music, S., Sikes, R. K., Smith, J. S., and Gregg, M. B.**, Postexposure rabies prophylaxis, with human rabies immune globulin, *JAMA*, 227, 407, 1974.
56. **Umeno, S. and Doi, Y.**, The study on the anti-rabic inoculation of dogs and the results of its practical application, *Kitasato Arch. Exp. Med.*, 4, 89, 1921.
57. **Habel, K.**, Evaluation of a mouse test for the standardization of the immunizing power of antirabies vaccines, *Public Health Rep.*, 55, 1473, 1940.
58. **Johnson, H. N.**, Proc. 49th Annu. Meet., USA Livestock Sanitary Assoc., 1945, 99.
59. **Koprowski, H. and Cox, H. R.**, Studies on chick embryo adapted rabies virus, *J. Immunol.*, 60, 533, 1948.
60. **Koprowski, H. and Black, J.**, Studies on chick embryo adapted rabies virus. II. Pathogenicity for dogs and use of egg-adapted strains for vaccination purposes, *J. Immunol.*, 64, 185, 1950.
61. **Koprowski, H.**, Biological modification of rabies virus as a result of its adaptation to chicks and developing chick embryos, *Bull. W.H.O.*, 10, 709, 1954.
62. **Abelseth, M. K.**, An attenuated rabies vaccine for domestic animals produced in tissue cultures, *Can. Vet. J.*, 5, 279, 1964.
63. **Hattwick, M. A. W., and Gregg, M. B.**, The disease in man, in *The Natural History of Rabies*, Vol. 2, Academic Press, New York, 1975, 281.
64. **Hattwick, M. A. W., Weis, T. T., Stechschulte, J., Baer, G. M., and Gregg, M. B.**, Recovery from rabies, *Ann. Intern. Med.*, 76, 931, 1972.
65. **Porras, C., Barboza, J. J., Fuenzalida, E., Lope Adaros, H., Oviedo de Diaz, A. M., and Furst, J.**, Recovery from human rabies, *Ann. Intern. Med.*, 85, 44, 1976.
66. **Wiktor, T. J., Postic, B., Ho, M., and Koprowski, H.**, Role of interferon in protective activity of rabies vaccines, *J. Infect. Dis.*, 126, 408, 1972.
67. **Wiktor, T. J., Koprowski, H., Mitchell, J. R., and Merigan, T.**, Role of interferon on prophylaxis of rabies after exposure, *J. Infect. Dis.*, 113, 260, 1976.
68. **Woodruff, J. F. and Woodruff, J. J.**, T lymphocytes interaction with viruses and virus-infected tissues, *Prog. Med. Virol.*, 19, 120, 1975.
69. **Kaplan, M. M., Wiktor, T. J., and Koprowski, H.**, Pathogenesis in immunodeficient mice, *J. Immunol.*, 114, 1761, 1975.
70. **Wiktor, T. J., Doherty, P. C., and Koprowski, H.**, *In vitro* evidence of cell-mediated immunity after exposure of mice to both live and rabies virus, *Proc. Natl. Acad. Sci. U.S.A.*, 74, 334, 1977.
71. **Wiktor, T. J., Doherty, P. C., and Koprowski, H.**, Suppression of cell-mediated immunity by street rabies virus, *J. Exp. Med.*, 145, 1617, 1977.
72. **Wiktor, T. J.**, Cell-mediated immunity and postexposure protection from rabies by inactivated vaccines of tissue culture origin, *Dev. Biol Stand.*, 1978, 40, 171, 1978.
73. **Heron, I., Berg, K., and Cantel, K.**, Regulatory effect of interferon on T cells in vitro, *J. Immunol.*, 117, 1370, 1976.
74. **Wiktor, T.**, unpublished observations.

Chapter 6

DROSOPHILA SIGMA VIRUS

Danielle Teninges, Didier Contamine, and Gilbert Brun

TABLE OF CONTENTS

I. INTRODUCTION

Sigma virus is a rhabdovirus (Figure 1) which infects the fruitfly *Drosophila*. Exposure of infected flies to carbon dioxide gas induces paralysis, while a similar treatment of uninfected flies anesthetizes them. This CO_2 sensitivity was discovered in 1937 by L'Héritier and Teissier[1] as a unique physiological feature of a *Drosophila* strain. Following this observation, the CO_2 sensitivity was studied as a genetic character[2,3] and regarded as an example of nonchromosomal heredity, since the CO_2 sensitivity was not transmitted by mere contact between flies. The infectious nature of the agent inducing CO_2 sensitivity was demonstrated later by experiments involving organ transplantation and injection of acellular fly extracts into larvae and imagoes. The agent was then named sigma virus.[4,5] It was only in 1965 that sigma virus was identified as a member of the Rhabdoviridae family from electron microscopy in the laboratories of Berkaloff and associates.[6]

Since 1937, a large amount of information has been obtained concerning the relationship that exists between sigma virus and its host organism. Different aspects of this relationship have been discussed in previous reviews.[7-11] It was discovered in early studies that there was a difference in the hereditary transmission of sigma virus by male and female flies. Thus female *Drosophila* were recognized to transmit the viral genetic information to all their descendants with few exceptions. These descendants were found to exhibit the same transmission pattern as the parental strain no matter which male was used in the cross (except for males having certain nonpermissive alleles of *Drosophila ref* genes). By contrast, male flies only transmit the CO_2 sensitivity to a fraction of their progeny. The infected descendants from such males exhibit quite different transmission patterns when compared to either male or female parents of the original strain. Thus the progeny males do not transmit the virus, while progeny females transmit it to only a fraction of their progeny.

Following the discovery that inoculation of *Drosophila* by acellular extracts from infected flies could induce the CO_2 sensitivity after a suitable period of incubation, the offspring of such inoculated flies have also been studied. It was found that the transmission capabilities of inoculated flies were analogous to those of flies infected by the paternal gamete: i.e., progeny males did not transmit the agent while females transmitted it to only a fraction of their progeny. Such transmission occurs in a time-dependent fashion,[12,13] as shown in Figure 2, in which the kinetics of accumulation of infectious material in the inoculated flies is also plotted.

An important step in the understanding of sigma virus transmission was the discovery of a sigma virus mutant (the mutation g^-) which does not permit the transmission of the agent to progeny of inoculated females, or females infected by the paternal gamete. Wild-type sigma virus is g^+.[14] The g^- virus strains can be perpetuated through any number of generations (as can the g^+ strains) by females from "sigma virus stabilized maternal lines" — a term which has replaced the expression "pure CO_2-sensitive strains" which is found in the early literature.

It is believed that in an adult female *Drosophila*, the germ cells are protected from viral infection throughout oogenesis by a barrier which arises from cell differentation and which prevents the propagation of g^- virus strains, but permits that of g^+ strains. The concept and nature of this barrier will be discussed further, but it allows us to understand how the transmission of sigma virus in stabilized maternal lines differs from that obtained from inoculated females both quantitatively and qualitatively. This barrier does not need to be overcome for females from a stabilized maternal line whose germ cells were infected from their origins. The male germ line cells appear to be totally protected against *de novo* infection. We have no data allowing analysis of the nature of this latter protection, but it is likely to also arise from cell differentiation.

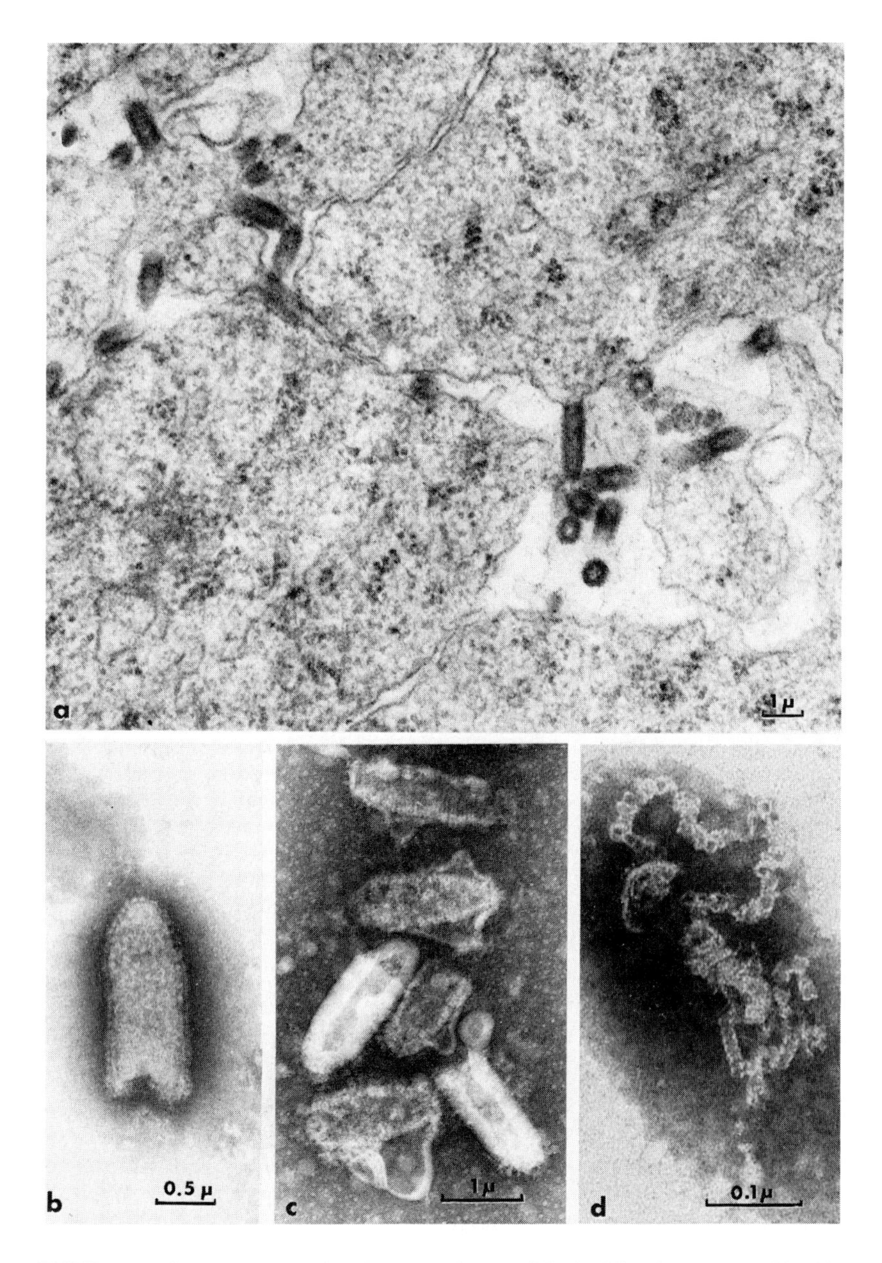

FIGURE 1. Sigma virus: a rhabdovirus. (a) Virus particles budding from spermatids (thin section of testis from males transmit sigma virus to their progeny); (b) a sigma virion from infected cell culture supernatant fluids seen in negative contrast. The particle's diameter, including the surface spikes is estimated to the 75 nm, while its total length is estimated to be 200 nm; (c) thermolabile 9 mutants; virions are seen losing their envelope and revealing a coiled nucleocapsid which has 32 turns; (d) an unwound viral nucleocapsid.

The infected eggs of females inoculated with g^+ virus give birth to flies of several types, and in varying proportions, including some females which establish new stabilized maternal lines. The type of progeny obtained from females inoculated with g^+ virus appears to depend on the number of viral genomes present in the embryo at the beginning of cell differentiation, and therefore depends on both the time of oocyte infection[15,16] and on the rate of virus multiplication. If, for example, an egg becomes

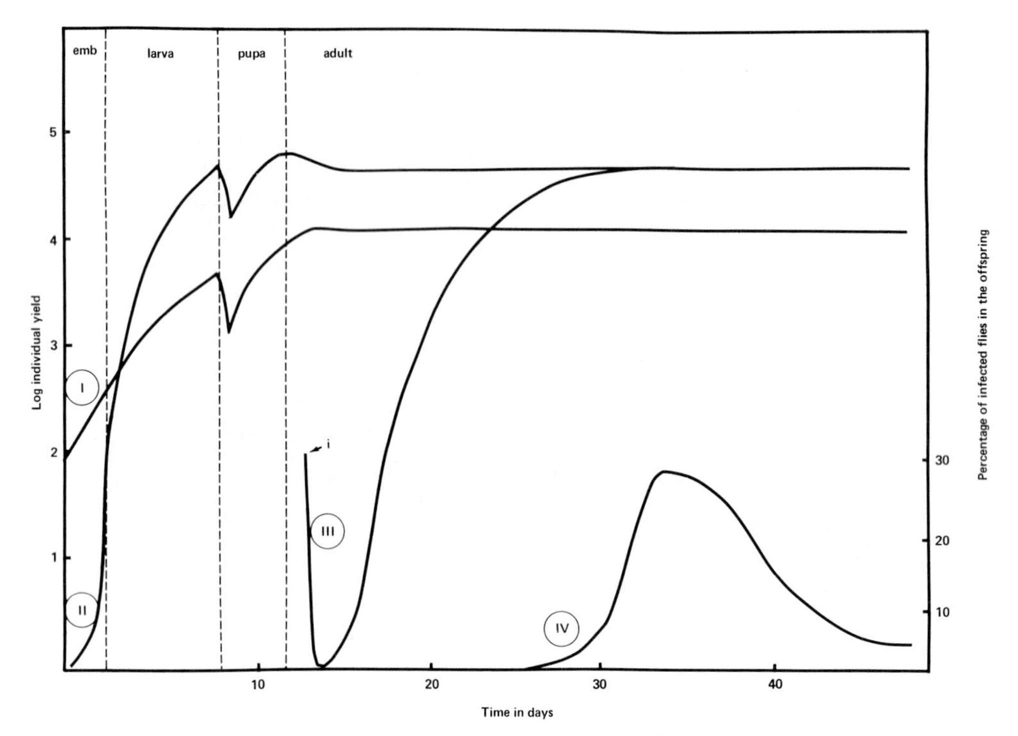

FIGURE 2. Kinetics of accumulation of infectious virus in *Drosophila*. (Line I) *Drosophila* from a stabilized maternal line; (line II) *Drosophila* infected by the paternal gamete; (line III) inoculated adult *Drosophila;* (line IV) time-dependent variation of the percentage of infected individuals among the offspring of inoculated flies.[12,79]

infected upon entry of the spermatozoon material, then competition between invasion of the embryo by the viral information and the production of cell differentiation barriers could result in heterogeneity among the progeny. Such heterogeneity could vary both according to the virus and the host genotypes.[17,18] This explanation might also reconcile the data of Williamson,[19] who observed that male *Drosophila* species of another genotype which received viral information from the paternal spermatozoon, were in fact capable of transmitting sigma virus, in contrast to the results described above.

Another important step to our present understanding of the relationship of sigma virus with *Drosophila*, was the discovery of maternal lines stabilized for defective viruses.[20,21] Defective viruses, which are unable to direct the synthesis of infectious material, can be perpetuated in a stabilized maternal line in a manner which parallels that of infectious viruses. These observations indicate that persistent viral derived infections do not require agents that are infectious per se. Also they indicate that the association of viral genetic information with cells can be maintained through cell divisions. The carrier state nature of the sigma virus-*Drosophila* cells association has been amply demonstrated as discussed in Section III.

Although comparisons have been made between the sigma virus-*Drosophila* association to that of lysogeny,[8,21] the total absence of linkage of the viral genetic information with the chromosomes of *Drosophila* precludes the use of this term.

Sigma virus is widespread in the natural fruitfly populations of several continents: Western Europe, Middle East, South America,[22] North America,[19,23] and Central America.[24] A survey by Fleuriet[25] of natural fruitfly populations in France showed that on average 10 to 20% of the flies are infected. In North America, species other than *D. Melanogaster,* for example *D. affinis* and *D. athabasca,* carry viruses which exhibit similar characteristics (symptoms, hereditary transmissions, etc.[19]) to those reported for Sigma virus of *D. melanogaster*. Still, these viruses may be distinct. Indeed, hereditary transmission as the only propagation process of a virus implies that in different hosts the viruses will experience genetic drift.

The paralysis of sigma virus infected flies is specifically induced by CO_2 and no other gas. L'Héritier and Teissier[1,7] described a relationship between the temperature of exposure and the CO_2 concentration necessary for inducing an irreversible paralysis which killed the flies. It is probable that this relationship depends on the strain of CO_2-sensitive flies which are used. The CO_2-induced paralysis occurs when the fly's thoracic nervous center is infected,[26] although infection of the fly's cephalic nervous center, while it does not cause a visible paralysis, also can cause death of the insect a few days after exposure — probably by paralyzing the mouth and/or sense organs.[27]

Incomplete paralysis occurs when the CO_2 concentration is lower than the lethal threshold, or when the symptom is not completely expressed in the inoculated fly. Partially paralyzed flies (which walk sideways and/or have uncoordinated movements) may recover. It may be that elimination of virus material is involved in the repair of nervous system lesions. This is suggested by the protective effect of a prior exposure to sublethal CO_2 concentrations upon subsequent CO_2 exposure,[7] and by the more recent observations that when virus synthesis is reduced under restrictive conditions (temperature or host genotype), daily repeated exposures to CO_2 totally inhibit expression of CO_2 sensitivity.[28,29]

Exposure of flies to CO_2 does not affect virus infectivity since it does not modify the yield of virus obtained from infected flies. From studies involving certain defective virus strains, it is possible that CO_2 sensitivity is related to the insertion into nerve cell membranes of those viral proteins that contain neutralizing determinants.[30]

Paralysis (immediate sensitivity), or death after a few days (delayed sensitivity), is also observed in *Drosophila* following inoculation of several rhabdoviruses belonging to the vesiculovirus group, including two strains of VSV Indiana, one strain of VSV New Jersey, VSV Argentina, VSV Brazil, Piry, and Chandipura viruses,[31-33] as well as by two fish rhabdoviruses (Spring viremia of carp virus, SVCV, and Pike fry rhabdovirus, PFRV).[28,34]

The CO_2 sensitivity symptom appears to be specific to *Drosophila* infection by rhabdoviruses. Togaviruses, such as Sindbis or Semliki forest virus, which are also capable of multiplying in the fruitfly, do not induce a CO_2 sensitivity.[35,36] The sensitivity appears to be unique to rhabdoviruses, and it is not the same as the premortal sensitivity to anoxia induced by the *Drosophila* X virus.[37]

The time between inoculation and the first detection of infectious progeny virus in sigma inoculated flies is approximately 24 hr. The invasion of a fly by sigma virus at any temperature is much slower than that observed with VSV Indiana, New Jersey, or Piry virus.

In natural conditions, sigma virus is not known to cause any deleterious effect on the insect, although for some laboratory strains of *Drosophila* a deleterious effect can be observed, resulting in a slow oogenesis and even death of the embryos.[38,39] Innocuous variants of sigma virus have probably been selected in nature due to the disadvantage of having pathogenic strains involved in the hereditary transmission cycle.

II. TECHNIQUES FOR THE STUDY OF SIGMA VIRUS

A. Test for the CO_2-Sensitivity Symptom

Experimental conditions which are used to demonstrate CO_2 sensitivity induced by sigma virus involve exposing the flies for 15 min to a 100% CO_2 atmosphere at 12°C and determining whether the anesthetized *Drosophila* awake within 15 min at 25°C after the CO_2 treatment. Those flies which fail to awake are considered CO_2 sensitive.

B. Assay for Sigma Virus

1. Infectious Units

Infectious sigma virus is assayed by the end point dilution method using the induction of CO_2 sensitivity in susceptible flies as the test for the presence of virus. Infected fly extracts are assayed by first homogenizing flies in a saline solution, which is afterwards clarified by low speed centrifugation for 15 min at 1000 **g**. Serial dilutions of the infectious material (that is present in the supernatant fluids[12]), are injected intra-abdominally into batches of flies. Devices for the successful injection of the insects have been described by L'Héritier.[40] At the dilution end point of infectivity, only a fraction of the flies in a batch acquires sensitivity to CO_2. Extracts of the insensitive flies in that batch have been shown to lack infectious material.

If the infectious virus particles are independent in their probability to infect a *Drosophila* fly, and the flies are homogeneous with respect to their ability to be infected, then the average number of infectious units per volume injected (n) is related to the fraction (p) of uninfected flies by the formula:

$$p = e^{-n} \qquad (1)$$

This is the zero term of the Poisson distribution. When the values y = log(−Lp) are plotted against the logarithm of the inoculum concentration (x = log c), a straight line with a slope of 1 is obtained. The statistical analyses of experimental data indicate that the experimentally obtained regression lines do not differ significantly from a linear function having a slope of 1.[21] Several factors can contribute to experimental variation in the results one obtains: for example, fluctuations and inaccuracies in the volumes injected, and the physiological conditions of the flies (such as their age and nutrition). Attention to the control of these factors increases the precision of the assay.

While the estimate of the number of infectious units in a virus suspension is proportional to the estimates obtained of the number of physical virus particles, the ratio can vary depending on the virus strains, genotype of the flies, temperature of incubation, etc.

2. Infectious Centers

An infectious center is a virus-cell complex arising from the infection of one cell of a fly by one infectious unit. They are assayed by their ability to induce the symptom of CO_2 sensitivity within the same time period as that obtained by one infectious virus particle. The mean number of infectious centers per fly is determined in exactly the same way as the number of virus infectious units, per volume injected.

The maximum time allowed for a single infectious center to induce the CO_2 sensitivity syndrome is 30 days at 20°C, after the end of heat treatment.[41]

3. Clones

Cloning of sigma virus can be achieved by injecting flies at the end point dilution of a virus suspension such that 10% or less of the flies in a batch become CO_2 sensitive. Under these conditions, the average number of infectious units injected in each fly is 0.1, and the probability that each sensitive fly has been infected by a single infectious unit is 95%. Recovery of virus from the few paralyzed flies allows clones of virus to be obtained.

4. Mean Incubation Time

One parameter in the study of sigma virus which can be exploited experimentally is the incubation time that precedes the appearance of CO_2 sensitivity in inoculated flies. This mean incubation time is related to several factors: the size of the inoculum, the incubation temperature, the host and virus genotypes.

The incubation time is inversely proportional to the logarithm of the number of infectious units injected. A standard curve, relating the mean time of incubation to the number of infectious units determined by the end point dilution method, can be obtained and used for an approximate assay of the number of infectious units in a virus suspension.[19,21] The mean incubation time is used to detect, locate, and characterize host genes that partially or totally restrict virus expression,[42] to characterize differences among virus clones at a fixed temperature,[21,43] and to detect thermosensitive mutants.[41]

III. THE VIRAL CARRIER STATE

A. A Noncytocidal Persistent Infection

The fact that persistent infection of flies is produced by sigma virus is shown by the fact that the amount of virus recovered from infected insects is constant throughout the lifetime of the fly (Figure 2). The plateau that is obtained results from an equilibrium between continuous production and inactivation of infectious virus particles within the fly. This has been demonstrated by taking infected flies in which the maximum amount of sigma virus has been produced, and subjecting them to either heat treatment at 30°C or protein starvation.[44,45] Both treatments induce a significant drop in the infectious titer of sigma virus. The virus titers, however, return to their previous levels after return of the insect to normal conditions.

There is no evidence which indicates that infected flies have shorter life spans than uninfected flies. The uninterrupted hereditary transmission of sigma virus without loss of fecundity implies that the persistent infection of the germ line cells of stabilized females does not affect their physiology.

Persistent infections can be observed at the level of dividing cell clusters by transplanting pieces of infected larval imaginal discs into adult flies (technique of Hadorn[46]). Such blastemas grow in the adult host, can be recovered, dissociated, and retransplanted into secondary hosts. For blastemas originating from infected larvae, or those which acquire an infection following transplantation into an infected host, it was shown that they can remain infected throughout successive transplantations, even when the secondary receiver flies are nonpermissive for the sigma virus strain in the implant.[47,48]

Drosophila cells cultured in vitro also maintain a sigma virus infection throughout successive cell transfers (Figure 3), and without any evidence of cytopathic effect.[49,50]

B. Infection Maintained in the Absence of Infectious Virus Production

Stabilized *Drosophila* lines perpetuating a sigma virus strain defective for the production of infectious virus can be selected from normal stabilized *Drosophila* strains. These strains are called ultra-rho. Flies from these carrier strains are not CO_2 sensitive, and their extracts are devoid of any infectious material that induces CO_2 sensitivity. The only expression for the presence of a virus is an immunity to superinfection by other sigma virus strains. This immunity has exactly the same inheritance patterns as a virus which induces CO_2 sensitivity. Its hereditary transmission can be interrupted by heat treatment and by restrictive alleles of certain refractory genes introduced into the descendants' genotypes, etc. in the same manner that these agents can interrupt wild-type sigma virus transmission. Such observations indicate that not only is the immunity viral in origin, but also that virus genetic information can be perpetuated throughout the divisions of the germ line cells without a complete virus cycle producing infectious agents.[21]

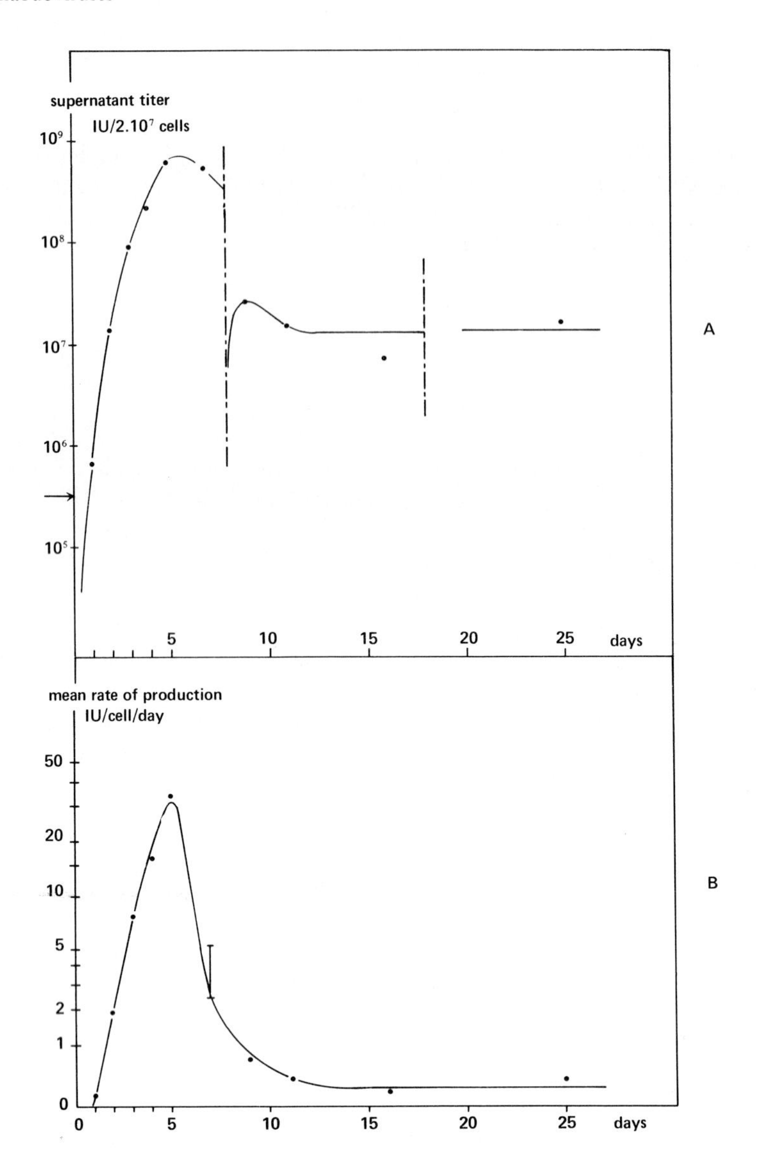

FIGURE 3. Sigma virus production in *Drosophila* cell cultures. (A) Kinetics of virus obtained from 2 × 10^7 cells. Each piont represents the titration of virus in the supernatant fluids. By cell counting, the mean number of infectious virus producer per cell per day was estimated. (B) Vertical bars correspond to cell transfers, the arrow to the inoculum titer.[80]

Similar observations have been made with temperature-sensitive mutants defective in maturation functions. Thus flies stabilized with sigma *ts* 9 do not show a CO_2 sensitivity when they are incubated at a nonpermissive temperature. Also, they do not yield infectious particles although the hereditary transmission of the virus proceeds in a manner similar to that observed with wild-type virus.[41]

The perpetuation of sigma ultra-rho virus has also been observed during successive transfers of imaginal wing discs,[51] whereby the presence of ultra-rho viral genomes can be revealed by a total resistance of the blastema to superinfection by wild-type sigma viruses.

These observations should not be taken to imply that successive reinfections do not play a role in the persistent infection, but that the process of virus production and reinfection of cells is not obligatory for the perpetuation of virus information.

C. Cytoplasmic Localization of the Viral Genetic Information

As discussed previously, the hereditary transmission of sigma virus is totally independent of the *Drosophila* chromosomes.[2,3] This led to the concept that the CO_2 sensitivity was a model of cytoplasmic heredity.[7] This was especially obvious for a virus strain *(v−)* which cannot be transmitted by male flies, but can be transmitted by females to their entire progeny.

It has been shown that the female germ line cells can be cured by heat or ethylmethane sulfonate (EMS) treatment. Using different times of heat treatment, or different doses of EMS, Brun and Diatta[52] showed that the number of viral genomes in the oogonia of flies varied from 10 to 40 according to the virus strain and the breeding temperature. From the EMS treatment experiences, the number of viral genomes per cell at different stages of oogenesis seemed to increase in the same proportion as the cytoplasmic volume. From the distribution of the frequency of spontaneously cured germ cells, it was deduced that the viral genomes replicate as independent units, and are randomly distributed in daughter cells at each cell division. Other examples of spontaneous cures of virus infection have also been reported.[51,53,54] Their probability depends on both the virus strain and the host cells. Cured cells are permissive to reinfection by the sigma virus strain from which they have been freed.

Evidence obtained from studies of sigma *ts* mutants (such as *ts* 4 whose hereditary transmission can be interrupted by incubation at a nonpermissive temperature) indicates that the expression of viral functions are necessary for the maintenance of viral genomes in germ line cells.[41] Other studies with sigma mutant *ts* 4 suggest that the genetic defect of this mutant concerns a protein involved in the virus replication processes (see below).

D. Cell-Specific Immunity to Superinfection by Homologous Virus

As indicated above, insect cells carrying sigma virus resist superinfection by the homologous strain. This has been shown to be true for adult somatic cells, infected imaginal wing discs, infected tissue culture cells, and infected female germ line cells. To prove the latter, superinfection attempts were made by mating females stabilized for defective virus strains with males transmitting a normal homologous, infectious virus.

The immunity to superinfection appears to be a direct consequence of the presence of virus genomes, or virus-related information, since spontaneous or induced loss of the virus is accompanied by simultaneous loss of immunity. Immunity is strictly virus specific, since cells infected by sigma virus can be superinfected by heterologous viruses such as VSV or Sindbis.[31,55]

E. The Regulation Process

The fact that small numbers of active genomes have been detected in infected insect cells (for example, in oogonia[52]), and the lack of cytopathic effects in carrier state cells, is taken to imply that virus replication is regulated. The factors which are involved in such regulation are not known. However, carrier states have been observed with other RNA viruses which also induce a specific resistance to superinfection (for example, Sindbis virus[35,56] and rhabdoviruses[31]), suggesting that this may be a generalized property of infection of a dipteran cell by an arbovirus.[57-61]

Whether the regulation of viral replication in insects parallels that obtained in persistently infected tissue culture cells is not known. The latter, as discussed in Chapter 3 of this volume, probably represents either perpetuation of viral genomes through cell divisions without the necessary involvement of infectious virus particles and/or reinfection of cells freed of virus information, by superinfecting extracellular virus. These two factors may be in perpetual balance, varying in their relative importance according to factors such as the relative speed of viral and cellular multiplication, etc. To what extent a cell regulates the viral replication process, particularly in different differentiated cells within an insect, is not known. Likewise, the effect on the virus of maintaining restricted replicaton is also unknown. The possibility as discussed elsewhere in this book, that it results in the selection of temperature-sensitive mutants and/or defective viruses has been suggested.

Infection by defective viruses or temperature-sensitive mutants in restrictive conditions seem necessary to establish a carrier state in several vertebrate cells.[62,63] It is possible that such variants act in the maintenance of carrier states in dipteran cells, but they are not required for its establishment.

It is important to know when regulation of viral replication occurs and whether it is an induced or a constitutive function of the cells. From tissue culture experiments, it has been suggested that regulation occurs a few hours after the initiation of an infection. For several arboviruses, the published kinetic data of virus production in insect cell cultures (see Figure 3) may not show evidence of such an early regulation due to the fact that the input multiplicities of infection are usually less than one. However, when the infection of cells is performed in a one-step growth situation, as described by Davey and Dalgarno[58] for Semliki forest virus, it can be shown that virus production increases for 12 to 15 hr postinfection, and then decreases to very low levels.

Similar results have been obtained for VSV Indiana in *Drosophila* cells, as shown in Figure 4A, where it can also be seen that the number of infected cells, compared to the number found in mock-infected cultures, shows only a slight decrease between 11 and 30 hr postinfection (Figure 4B). In these VSV-infected *Drosophila* cultures, 95% of the cells were immunofluorescence positive by 11 hr postinfection; however, the percentage dropped to 10% by 30 hr postinfection and remained at that level for the next 40 hr.[64] Which step of the virus cycle is controlled and what host cell factors are involved in such regulation remains to be determined.

IV. SIGMA VIRUS GENETICS

A. Temperature-Resistant Variants

Most laboratory strains of sigma virus do not multiply in *Drosophila* when incubated at 30°C. However, in an early study several temperature-resistant strains were obtained (named Tr) which had acquired the capability of growing at 30°C in *Drosophila*.[8,65] In view of the various selection conditions that were used to obtain such temperature-resistant variants, it is possible that a variety of mutational changes might induce such variants. Of the Tr variants that were selected by serial inoculation into flies, most have lost one central property of sigma virus: they fail to infect female germ line cells. Even those which succeed are unable to maintain a continuous association with these cells.[66]

B. Temperature-Sensitive Mutants

Several temperature-sensitive mutants have been isolated from a clone of wild-type sigma virus (clone 23 of the strain M_2). This sigma virus clone was chosen in view of its ability to be perpetuated in female germ line cells, as well as its ability to infect them upon inoculation. Four *ts* mutants have been isolated, three (*ts* 4, *ts* 6, and *ts* 9) after mutagenesis in vivo with 5-fluorouracil (5-FU), and one as a spontaneously occurring temperature-sensitive mutant in the wild-type sigma virus clone 23.[41] In addition to these mutants, which were selected directly for their temperature-sensitive phenotype, some host range mutants *(hap)* have been obtained which are also thermosensitive.

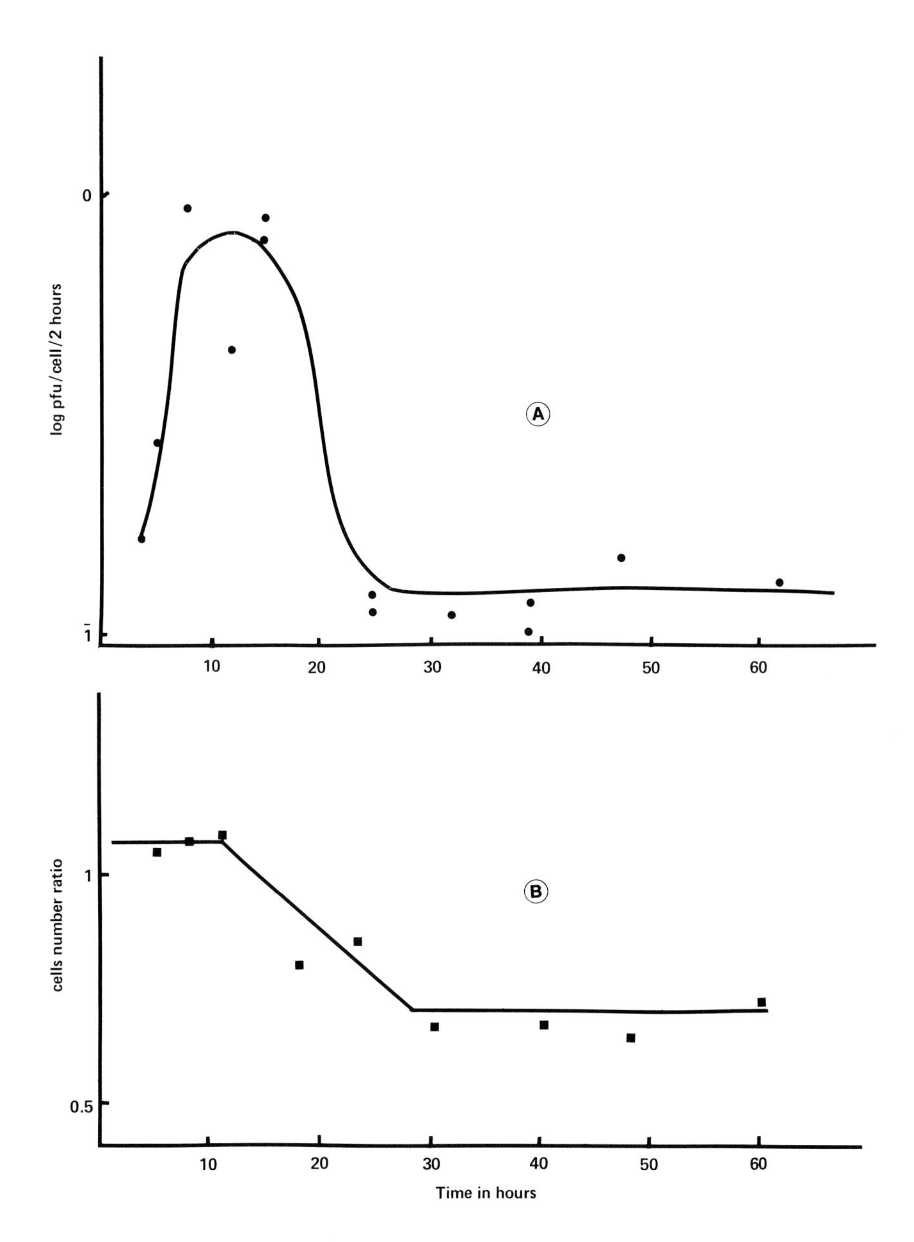

FIGURE 4. Kinetics of VSV Indiana production in *Drosophila* cells at 30°C. (A) Some 4×10^6 *Drosophila* cells in 3 cm dishes were inoculated with 2.6×10^8 PFU of VSV Indiana (equivalent to 3 fluorescence-forming units (FFU) per cell). After 1 hr adsorption, the inoculum was removed. The medium was changed at various intervals and aliquots taken 2 hr later for PFU assay in chick cells. The cells in the dish were counted at each time point. Each point represents the average number of PFU produced per cell, per 2 hr. (B) At various intervals the cells in infected and in mock-infected cultures were counted and the ratio determined.[64]

The different temperature-sensitive mutants have been characterized by the variations of the mean incubation time as a function of the incubation temperature (Figure 5).

As shown in Table 1, certain physiological characteristics of the temperature-sensitive mutants have been analyzed, and criteria established for the classification of these mutants as described below.

TABLE 1

ts Mutants Classification Based on Physiological Characteristics

Viral clone	23 *ts*⁺	*ts*4	*hap*7	*hap*27	*ts*9
Hereditary transmission at 30°C	+	−	+	+	+
Expression of the CO_2 sensitivity symptom at 31°C[a]	+	+	+	−	−
Thermostability of the virions[b]	+	+	+	+	−
Early temperature-sensitive period[c]	No	Yes	Yes	No	No
Late temperature-sensitive period[c]	No	Yes	No	Yes	Yes

[a] When the flies' thoracic ganglia are infected.
[b] The reference for thermostability is that of 23 ts^+ at 31°C.
[c] Non permissive temperature = 28°C.

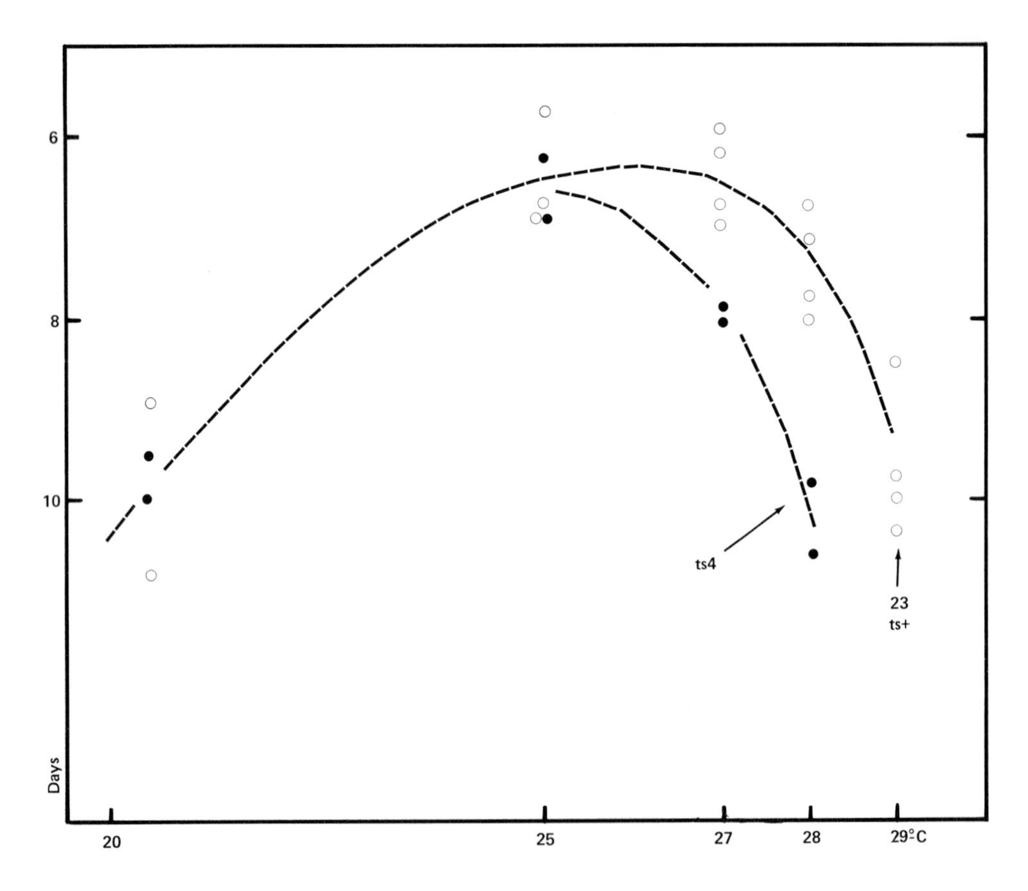

FIGURE 5. Growth of temperature-sensitive sigma virus mutants. The mean incubation time preceding the appearance of CO_2 sensitivity in inoculated flies is plotted as a function of incubation temperature for sigma wild-type 23 (ts^+) and *ts*4.[41]

1. The hereditary Transmission of ts Mutants at 30°C

Two categories of temperature-sensitive mutants have been defined according to their hereditary transmission abilities at 30°C. Some mutants (e.g., *ts* 4) have impaired hereditary transmission capabilities indicating that a function necessary for virus genome replication in germ cells is defective. Other mutants (e.g., *ts* 9), maintain the hereditary transmission capability of wild-type virus at 30°C.

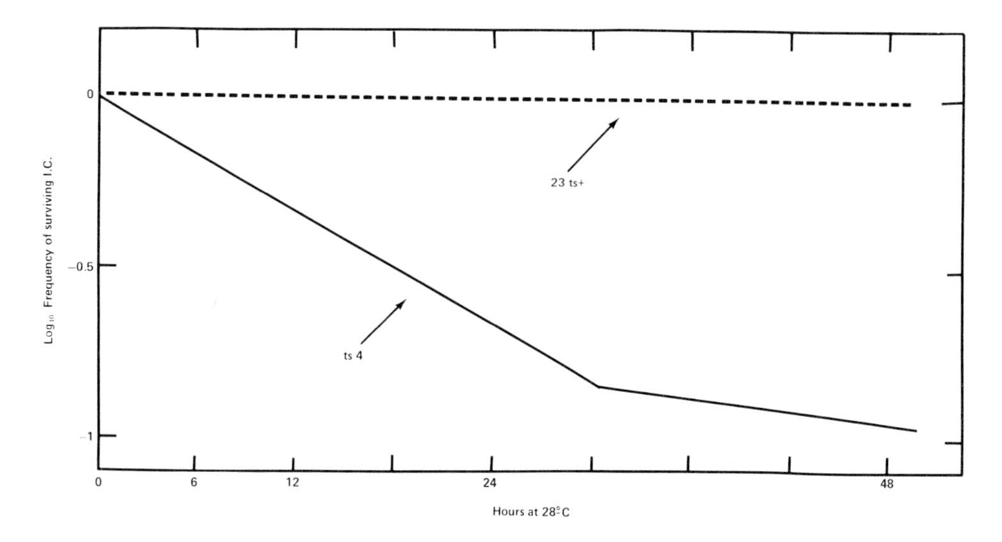

FIGURE 6. Kinetics of infectious centers decay at 28°C. The decay of infectious centers of sigma wild-type 23 (*ts*+) and *ts*4 is plotted as a function of time of exposure at 28°C prior to inoculation.

2. The Decay of Infectious Centers at Nonpermissive Temperatures

The survival of infectious centers is another characteristic which has been used to categorize temperature-sensitive mutants. The wild-type infectious centers resist inactivation when incubated at 28°C (Figure 6), while infectious centers infected with certain mutants are heat sensitive. An infectious center is primarily an infected cell from which infectious material emanates and causes the infection of the whole fly. Essentially, therefore, the infectious center assay allows one to analyze the productive infection cycle of a single cell. The kinetics of the decay of infectivity of infectious centers on exposure to heat have been shown to be exponential and biphasic, with two apparent slopes when plotted on a semilogarithm graph. Since one of these slopes dissects the origin, it suggests that the target of that reaction is a single entity. The fact that *ts* 4 virus particles are as heat stable as the wild-type virus suggests that the heat-sensitive structure cannot be the virus particle itself but a structure arising after the initial adsorption and penetration process. It may be postulated that if replication of viral genomes is permitted (e.g., for a preincubation at 20°C), infectious centers will be more heat-resistant in shift-up experiments. Such a result is obtained with *ts* 4 mutant after an initial incubation for 15 hr at 20°C.

The second kinetic decay of the infectious centers (Figure 6) may correspond to viral genomes that have already performed replications competently in the infected cells before their inactivation. That this is the probable explanation is substantiated by the fact that the virus obtained after shift down to 20°C is temperature-sensitive in phenotype, and not revertant or pseudorevertant.

Two periods of thermosensitivity have been defined. Early-function thermosensitivity, occurring during the first 24 hr of the virus cycle at 20°C (*ts* 3, *ts* 4, *ts* 6, *hap* 7), and late-function thermosensitivity (*ts* 9, *hap* 27), for which the proportions of surviving infectious centers are the same at the nonpermissive temperature (28°C) either with or without pretreatment for 24 hr at 20°C. For sigma mutant *ts* 9 and other mutants affected in late functions, the second part of the infectious center kinetic decay has a negative slope. This can be seen more clearly at 29.5°C as shown in Figure 7. At 29.5°C, not only is *ts* 4 temperature sensitive in its early function, but the wild-type virus also appears to be somewhat thermosensitive early in its infection. For mutant *ts* 4, the second part of the infectious centers decay kinetics also has a negative slope, which may indicate that the temperature-sensitive function of *ts* 4 is also expressed late in the infection cycle.

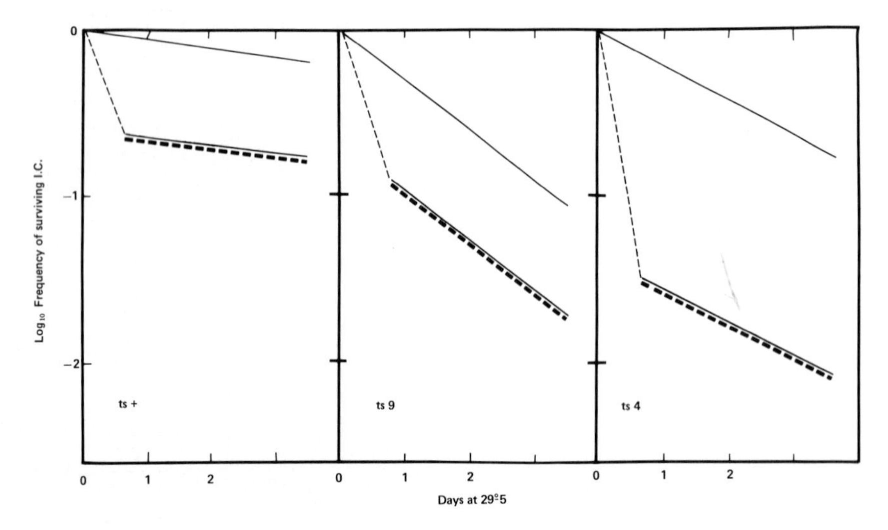

FIGURE 7. Kinetics of infectious centers decay at 29.5°C. The decay of infectious centers of sigma wild-type virus (*ts*⁺), sigma mutants *ts* 9, and sigma mutant *ts* 4 is plotted. The first part of the inactivation kinetics is not known in these experiments (broken line), but is estimated from the distance between the two parallel straight lines. The solid line is the decay following 24-hr pretreatment at 20°C, while the broken/solid line is without pretreatment at 20°C. The differences in the distance between the parallel lines for *ts*⁺ and *ts* 9 is not significant. The most temperature-stable infectious centers are much more labile for *ts* 4 and *ts* 9 than for *ts*⁺(in the case of *ts*⁺, the slope of the parallel lines can be regarded as zero).[41]

Although sigma mutants *ts* 3, *ts* 4, and *ts* 6 are affected in an early function, and are not transmitted hereditarily at nonpermissive temperatures,[41] the mutant *hap* 7, which is also affected in an early function at 28°C, is transmitted hereditarily even at 30°C (Table 1).[67] This situation is also true for the sigma wild-type 23 strain when it is used to infect flies incubated at 29.5°C or above. Detailed studies with *hap* 7 indicate that the time during which the temperature-sensitive defect is expressed, is within the first 9 hr of an infection cycle at 20°C — which contrasts the 15 hr observed with sigma *ts* 4. Therefore, it appears that sigma *hap* 7 may be affected in an earlier function than that of *ts* 4. One interpretation is that the temperature-sensitive defect of *hap* 7 occurs prior to genome replication, and that when several virus genomes are present in a cell (i.e., after viral genome replication or in the female germ cells of stabilized maternal lines), the deficiency of the function affected by *hap* 7 does not appear.

From analyses of the fate of mutants in infectious centers temperature shift-up and shift-down experiments, it can be concluded that mutants which have defective viral genome replication processes are the mutants which are not transmitted hereditarily.

C. Host-Range Mutants

Certain genes of *Drosophila melanogaster* were detected by their specific action on sigma virus: some of their alleles are restrictive for some strains of sigma virus. These genes are called refractory genes (*ref*). At the present time five such genes are known: *ref* (1) H, *ref* (2) M, *ref* (2) P, *ref* (3) O, and *ref* (3) D. The number in parenthesis refers to the *Drosophila* chromosome which carries the gene, while the capital letter is the initial letter of the name of the *Drosophila* strain in which the restrictive allele of this gene was first observed. The different alleles are noted in lower case letters exponents (e.g., *ref* (2) P^p, is the allele p of the *ref* (2) P gene). These five genes operate during the general process of virus invasion of the fly following inoculation.

There are some genes which are specifically expressed in certain differentiated tissues as, for example, *ref* (3) V. For this gene the different alleles are characterized by their effect on the transmission of virus infection by spermatozoa. The different allelic forms of the *ref* genes have been obtained from natural *Drosophila* populations or from laboratory strains of the fly, and do not appear to affect the physiology of the insect.[42]

The Paris strain of *Drosophila* was the strain for which the first restrictive allele was identified. It carries the allele *ref* (2) P^p.[68] The wild-type sigma virus strains which are restricted in the Paris strain, are termed P^-. Sigma variants that are not restricted in the Paris strain have been obtained and are termed P^+.[69,70] These early results proved that the sensitivity of a sigma virus to a restrictive allele of a host gene could be modified by mutations.

Depending on whether the initial virus strain is restricted (or not) by a particular host gene allele, and according to the direction in which the mutation modifies the virus, both host-adapted *(ha)* and host-restricted *(hr)* mutants can be distinguished. For practical reasons in order to analyze the virus functions which are affected by the products of the *ref* genes, the mutants which become both host-adapted and temperature-sensitive by a single mutational event are of particular interest.

Starting with the sigma virus clone 23, which is restricted by the P^p allele of *ref* (2) P, mutants have been derived by mutagenesis in vivo with 5-FU, and selected in nonpermissive hosts.[67] Twenty-four mutants were obtained (*hap* mutants). Of these mutants, 6 out of 24 were also found to be thermosensitive while among the unselected clones only 4 out of 122 clones were thermosensitive. Of the six *hap* temperature-sensitive mutants, the temperature sensitivity of five was found to be directly dependent on the particular *ref* allele (*ref* (2) P) that was present in the host. The sixth mutant was a different type (Table 1). These observations suggest that viral proteins interact directly in some way with the *ref* (2) P gene product.

Information on the role of the host gene products in the virus infection cycle can be derived by comparative physiological studies of these mutants at permissive and restrictive temperatures in permissive and restrictive hosts. For the sigma mutant *hap* 7, the kinetics of infectious centers decay at 28°C have been studied in both permissive and restrictive hosts. The following observations have been made:

1. The kinetics of adsorption/penetration are the same in both hosts.
2. The infectious centers are more thermolabile in restrictive hosts than in permissive hosts.
3. The proportion of temperature-resistant infectious centers is smaller in nonpermissive hosts (Figure 8).
4. In the permissive hosts the period in which the temperature-sensitive defect is effective is in the first 9 hr, while in the restrictive host at 20°C, it is in the first 32 hr (Figure 9).
5. The viruses produced in both hosts maintain their genotypes.

It has been suggested from these observations that the viral genome-protein complexes which are liberated from the parental virion after penetration interact at the beginning of the infectious cycle with the *ref* (2) P gene product. This interaction is expressed as a modification in the expression of the viral genome at 28 and 20°C, according to the host genotype.

These experiences may be extended to other host genotypes whereby the stability of the infectious centers can be used as a screen for other host genes whose products act on the viral cycle at its initiation.

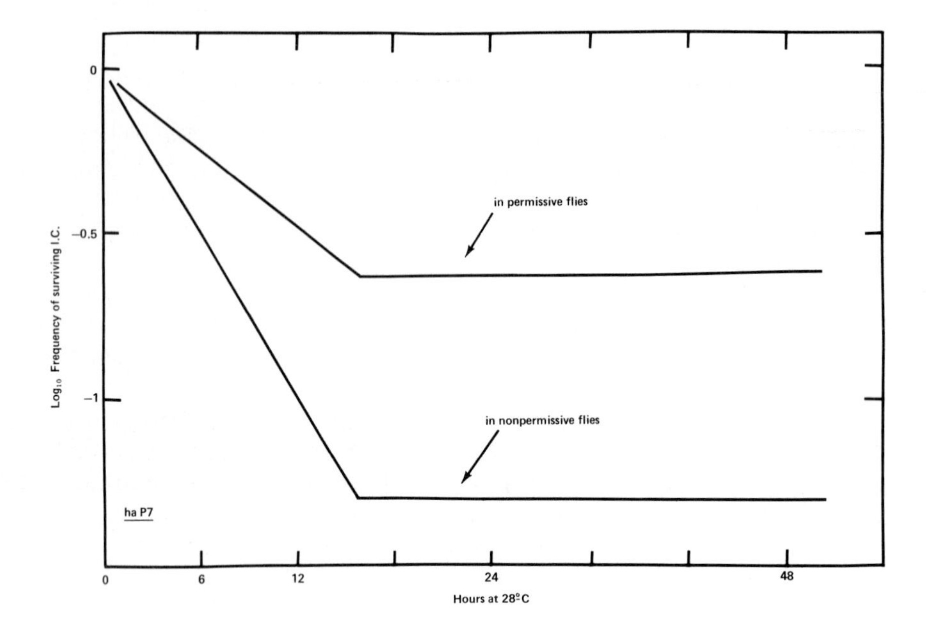

FIGURE 8. Kinetics of infectious centers decay for *hap*7 in permissive and in nonpermissive hosts.

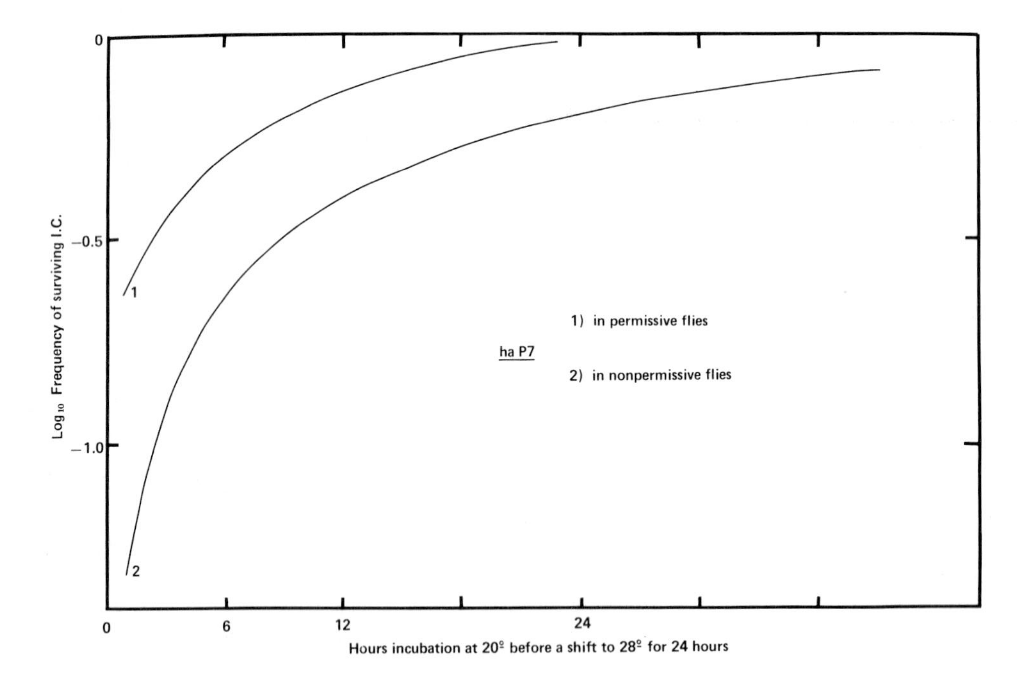

FIGURE 9. Determination of the temperature-sensitive period for *hap* 7 in permissive and in nonpermissive hosts. In these experiments, the frequency of temperature-stable infectious centers is estimated for each pretreatment at 20°C. The curves presented are adjusted to experimental data, assuming that the decay of temperature-labile infectious centers and the transformation of temperature-labile into temperature-stable infectious centers occurs at a constant rate. In these experiments, as in those presented in Figure 8, four infectious units defined in the permissive host are equivalent to one infectious unit in the nonpermissive host.

Until recently, no viral mutations producing a host restriction had been observed. The reasons for this in part reflects the inappropriateness of the screening procedure used for the selection of *ha* mutants, to the selection of *hr* mutants (i.e., selection of survivors in a nonpermissive host). In addition, the difficulties in obtaining and analyzing each clone of sigma virus are so laborious that they do not allow a random screening of a large number of clones.

However, an indirect selection of *hr* mutants is possible. If we assume that several host-gene products interact with the same viral protein, then a mutation affecting the interaction of one of these products may possibly affect the interaction with others. On this hypothesis one might expect that the frequency of modification of the sensitivity toward a *ref* gene allele other than the *ref* (2) P^p allele, would be high among *hap* mutants. This has been found to be true for the gene *ref* (1) H. The different known alleles of that gene are permissive for the wild-type clone 23. Eighteen *hap* mutants have been studied and two of them found to be restricted in flies homozygous for the H^h allele of *ref* (1) H. In addition, the temperature-sensitivity of one of these mutants, *hap* 23, is *ref* (1) H allele dependent. We cannot exclude the possibility that these results are coincidental; however, they do support the premise that the viral proteins which interact with the *ref* (2) P gene product also interact with the *ref* (1) H gene product. By pursuing this line of investigation for genes other than *ref* (1) H, and *ref* (2) P, it will be of interest to determine whether the products of other *ref* genes interact with the same viral proteins.

D. Mutants for Adaptation to Differentiated Cells

The first mutants of sigma virus which were detected differed from the original virus type by their particular expression in particular tissues of the fly. These mutants were detected because of their altered hereditary transmission characteristics. It was found that certain viral strains (g^-) derived from the wild-type virus (g^+) by a single mutational event are never transmitted to the progeny of inoculated females, while they are perpetuated in the normal manner in stabilized maternal lines. The g^+ character is a specific adaptation of the virus for overcoming a barrier raised by cell differentiation at the oocyte level. Heat-exposure experiments on egg-laying flies indicate that the oocytes receive the infection from a particular cell type. The infection from these cells is exhausted as layings proceed.[21] Bregliano[71] analyzed the differences between females inoculated with a g^- virus strain and those inoculated with a g^+ revertant derived from the former strain. He observed that the sheath cells of ovarioles are the cells which produce the virus which infects the oocyte's cysts, and that the g^- virus is not absorbed by the follicle cells.

Following inoculation of a mixture of g^+ and g^- virus mutants into female flies, Duhamel and Plus[72] observed a transmission of infection to the progeny whereby the descendants either carried the g^- or g^+ virus type. Since the g^- progeny virus were also all temperature resistant (Tr) as were the g^- parents, it is unlikely that the g^+ progeny came from g^- mutations to g^+ phenotypes since neither the g^+ progeny or parents carried the Tr marker. These results suggest that mutant complementation must have occurred, and to date it is the only example of such complementation in flies.

A similar situation exists for the sigma variants (v^+, v^-) which can be differentiated by their ability to be hereditarily transmitted by males obtained from a stabilized maternal line.[73] For v^+ virus, a large percentage of the progeny from these males are infected (30 to 90%), while infected progeny are almost never obtained from males carrying v^- mutants derived from v^+ virus. In *Drosophila melanogaster*, the only time that sigma virus infects the male germ line cells is when the carrier state is already established at the egg stage, and in germ cell primordia. This carrier state has to be maintained throughout the fast cell divisions of spermatogenesis and in spite of the profuse elimination of cytoplasm occuring during spermiogenesis. A comparative electron microscopy study of males originating from v^+- and v^--stabilized maternal lines have shown a large number of virus particles maturing exclusively in spermatids of v^+-infected males (Figure 1), but not at all in the germ cells of v^--infected males. Therefore, in v^--stabilized males the viral genetic information is probably lost before spermiogenesis, during the fast cell divisions of spermatogenesis.[74,75] It is probable that in the case of the v^- character, viral replication is the step which is inhibited or reduced to the extent that the replication of genetic information does not keep pace with cell division.

It has been observed that the adaptation to the Paris *Drosophila* strain of a virus which was originally unable to multiply in it (P^-), and yet which could be transmitted by males (v^+), was systematically accompanied by a transformation of the v^+ character to v^-.[70] This observation suggests a mechanism for tissue specificity, at least in the case of the v^+, v^- character, similar to that proposed for host specificity. For example, if in certain differentiated tissues, a particular host gene product assumes the function which in other cells is undertaken by the *ref* (2) P or *ref* (3) D gene products, then in those differentiated tissues for an adapted variant P^+ the interaction is changed and the tropism is transformed (becoming v^-). While other explanations may be conceived, the tissue tropism relationship of a virus is one which is of particular interest and bears further attention.

E. Defective Mutants

Drosophila flies with stabilized maternal lines carrying defective viruses do not yield infectious virus particles. Most of them do not exhibit a CO_2 sensitivity, but they are immune to superinfection by normal sigma virus, and this immunity is transmitted exactly according to the same inheritance patterns as that found for lines carrying virus which induce CO_2 sensitivity. The defective viruses cannot be propagated unless they appear in dividing carrier cells, like, for instance, the germ cells of a stabilized female. The defective viruses may be defective for any viral function which is not necessary for the replication of the viral genome.

Two types of *Drosophila* strains carrying spontaneous defective virus have been described by Brun.[21] These are the rho-types, in which reverse mutations can be obtained (frequencies around 10^{-4}), and the ultra-rho-types for which the total absence of reverse mutations suggests that the defect may involve a major genetic deficiency (such as a genome deletion or multiple mutations).

Other types of defective virus have been isolated: (1) IC_4 (which was obtained following in vivo mutagenesis with EMS of the viral genomes present in the germ cells of stabilized females — reverse mutations for this type occur at a lower frequency than that of the rho type, around 10^{-7}); (2) 62S, a spontaneous mutant which still retains an ability to induce the sensitivity syndrome.

Physiologically, the thermosensitive sigma mutant *ts* 9 (which is affected in a late function), behaves at 30°C like a defective mutant of the IC_4 type in that its hereditary transmission is normal and the flies are not CO_2 sensitive.

F. Dimensional-Type Mutants

When a batch of flies is inoculated with an extract of a fly previously inoculated with one infectious unit of virus, and homogenized in 0.1 mℓ after a given time of incubation, the mean time for the development of a CO_2 sensitivity characterizes the "dimensional type" of the virus inoculum. This character has also been termed the "plaque size" by analogy to the plaque size characteristic of cytopathic viruses whereby a short incubation time may produce large plaque size viruses, while a long incubation time may produce small plaques. The "dimensional type" character probably reflects the overall growth rate of the virus and can be influenced by several viral functions. While some virus clones are quite stable in the expression of this character, others appear to be unstable.[21,43] The reason for this instability which has also been observed for the rho character, is not known.

V. CONCLUSIONS

Although Sigma virus has long been regarded as a very particular virus, evidence obtained recently with other arboviruses concerning their transovarial transmission capabilities in insects, suggest that the particular relationship of sigma to *Drosophila* may have parallels in other systems. Whether this is the case will have to await further investigation.

The induction of CO_2 sensitivity is a particular property that sigma shares with other rhabdoviruses, and it has allowed significant exploration of the virus-insect relationship as described above. In fact, without this particular characteristic, it is doubtful if sigma virus would have been discovered. This leads one to reflect on how many other arboviruses exist which are totally subliminal to our present methods of identification.

The phylogenetic relationship of sigma virus with other rhabdoviruses is very likely in view of the electron microscopy evidence, and the identity of the symptoms induced in *Drosophila* by the various rhabdoviruses that have been studied. Biochemical characterization and comparison of sigma virus to other rhabdoviruses is in progress, and this should substantiate the phylogenetic relationship.

The biology of sigma virus in *Drosophila* is a particularly interesting model for understanding the relationship of those rhabdoviruses which are also arboviruses. Transovarial transmission of VSV Indiana in sandflies has been reported.[76] Although a detailed study of the hereditary transmission of VSV in sandflies similar to that realized with sigma virus in *Drosophila* is, at the present time, not possible, the question of whether the mechanisms are analogous is raised. The hereditary transmission of sigma virus in *Drosophila* is a complex process with at least two overall mechanisms in operation. One involves infection of germ cells by the virus, while the other involves carrier state with or without reinfection. To what extent these processes are involved in the transmission of other rhabdoviruses which infect insects, or for other arboviruses families, is not known. However, the perpetuation of arboviruses during epizootic intervals probably involves several generations in an invertebrate host in particular ecological niches. Such perpetuation may be subject to selection pressures which are quite different to those in mammalian hosts and contribute to the diversity that exists among the different arboviruses — even among different isolates of the same strain of the same serotype virus such as VSV Indiana.[77]

It is of interest to record that no transmission of VSV Indiana has been observed to the progeny of inoculated *Drosophila*. It is possible that the different strains tested behave like the g^- sigma virus mutants. However, the efficiency of replication in *Drosophila* of some viruses such as VSV New Jersey and Piry, as well as results of the VSV Indiana inoculations in *Drosophila* eggs,[78] suggests an alternate reason. If infection of oocytes causes the death of the descendants, then transmission will not be possible. Certain sigma virus strains have been described[39] that do induce a lethal effect on *Drosophila* embryos, and it is noteworthy that such effects are often associated with sigma virus strains which have been selected for their multiplication efficiency in *Drosophila*.[66]

Carrier states observed for VSV Indiana and Sindbis virus both in *Drosophila* and in mosquitoes cells may resemble that produced by sigma virus in *Drosophila* cells. The few cases in which carrier states of VSV and Sindbis have been produced in vertebrate cells do not seem to parallel the sigma-*Drosophila* situation. For example, in the vertebrate cell carrier state not only is there massive initial cytolytic destruction of the cells, but also after the establishment of the carrier condition the infected cells undergo periods of crisis and reestablishment. This does not occur in sigma-infected *Drosophila* cells, nor does it occur in arbovirus-infected dipteran cells.

In conclusion, it is worth emphasizing the value of sigma virus and its host *Drosophila* as a system for the genetic analysis of the role of host gene products in the biology of a virus. In view of the parasitic nature of viruses, there is little reason to doubt that host gene products play significant roles in the overall replication achieved by a virus. Thus, the sigma-*Drosophila*- system may well prove extremely useful as a model for understanding virus-host protein interactions.

REFERENCES

1. **L'Héritier, Ph. and Teissier, G.,** Une anomalie physiologique héréditaire chez la *Drosophile, C. R. Acad. Si.,* 205, 1099, 1937.
2. **L'Héritier, Ph. and Teissier, G.,** Un mécanisme héréditaire aberrant chez la *Drosophile, C. R. Acad. Sci.,* 206, 1193, 1938.
3. **L'Héritier, Ph. and Teissier, G.,** Transmission héréditaire de la sensibilité au gaz carbonique chez la *Drosophile, C. R. Acad. Sci.,* 206, 1683, 1938.
4. **L'Héritier, Ph. and Hugon de Scoeux, F.,** Transmission of the carbon dioxyde sensitivity of *Drosophila* by grafting, *Nature (London),* 157, 729, 1946.
5. **L'Héritier, Ph. and Hugon de Scoeux, F.,** Transmission par greffe et injection de la sensibilité héréditaire au gaz carbonique chez la *Drosophile, Bull. Biol. Fr. Belg.,* 81, 70, 1947.
6. **Berkaloff, A., Bregliano, J. C., and Ohanessian, A.,** Mise en évidence de virions dans des *Drosophiles* infectées par le virus hereditaire sigma, *C. R. Acad. Sci.,* 260, 5956, 1965.
7. **L'Héritier, Ph.,** Sensitivity to CO_2 in *Drosophila.* A review, *Heredity,* 2, 325, 1948.
8. **L'Héritier, Ph.,** The hereditary virus of *Drosophila, Adv. Virus Res.,* 5, 195, 1958.
9. **Seecof, R.,** The sigma virus infection of *Drosophila melanogaster, Curr. Top. Microbiol. Immunol.,* 42, 59, 1968.
10. **L'Héritier, Ph.,** *Drosophila* viruses and their role as evolutionary factors, in *Evolutionary Biology,* Vol. 4, Dobzhansky, Th., Hecht, M. K., and Steere, W. C., Eds., Meredith Corp., New York, 1970, chap. 7.
11. **Brun, G.,** Sigma virus and general aspects of the multiplication of certain vertebrate rhabdoviruses and togaviruses in *Drosophila,* in *The Genetics and Biology of Drosophila,* Vol. 2, Ashburner M. and Novitski, E., Eds., Academic Press, New York, in preparation.
12. **Plus, N.,** Etude de la multiplication du virus de la sensibilité au gaz carbonique chez la *Drosophile, Bull. Biol. Fr. Belg.,* 88, 248, 1954.
13. **Plus, N.,** Etude de la sensibilité héréditaire au CO_2 chez la *Drosophile.* I. Multiplication du virus σ et passage à la descendance après inoculation ou transmission héréditaire, *Ann. Inst. Pasteur Paris,* 88, 347, 1955.
14. **Duhamel, C.,** Etude de la sensibilité héréditaire à l'anhydride carbonique. Description de quelques variants du virus, *C. R. Acad. Sci.,* 239, 1157, 1954.
15. **Brun, G. and Sigot, A.,** Etude de la sensibilité héréditaire au gaz carbonique chez la *Drosophile.* II. Installation du virus σ dans la lignée germinale à la suite d'une inoculation, *Ann. Inst. Pasteur Paris,* 88, 488, 1955.
16. **Bregliano, J. C.,** Etude de l'infection de la lignée germinale chez des *Drosophiles* femelles injectées avec le virus sigma. I. Evolution au cours du temps des diverses catégories de descendants et de cystes infectés, *Ann. Inst. Pasteur Paris,* 118, 741, 1970.
17. **Bregliano, J. C. and Fleuriet, A.,** Contribution à l'étude de l'infection de la lignée germinale femelle par le virus sigma chez *Drosophila melanogaster.* II. Mise en évidence d'un passage au germen tardif chez les femelles qui n'ont reçu le virus que par le gamète paternel, *Ann. Microbiol. (Inst. Pasteur),* 126B, 491, 1975.
18. **Brun, G.,** A propos de l'infection de la lignée germinale femelle de la *Drosophile* par le virus sigma, *Ann. Microbiol. (Inst. Pasteur),* 128A, 119, 1977.
19. **Williamson, D.,** Carbon dioxide sensitivity in *Drosophila affinis* and *Drosophila athabasca, Genetics,* 46, 1053, 1961.

20. **L'Héritier, Ph.**, The CO_2 sensitivity problem in *Drosophila, Cold Spring Harbor Symp. Quant. Biol.*, 16, 99, 1951.
21. **Brun, G.**, *Etude d'une association du virus σ et de son hôte la Drosophile:* l'état stabilisé, These, Paris-Orsay, 1963, 254.
22. **Kalmus, H., Kerridge, J., and Tattesfield, F.**, Occurrence of susceptibility to carbon dioxide in *Drosophila melanogaster* from different countries, *Nature (London)*, 173, 1101, 1954.
23. **Williamson, D.**, Incidence of CO_2 sensitivity in several *Drosophila* species, *Drosophila Information Service*, 31, 169, 1957.
24. **Felix, R., Guzman, J., and de Garay-Arellano.**, Distribution of CO_2 sensitivity in an urban population of *D. melanogaster* from Mexico City, *Drosophila Information Service*, 47, 105, 1971.
25. **Fleuriet, A.**, Répartition et fréquence du virus σ dans des populations naturelles et expérimentales de *D. melanogaster, C. R. Soc. Biol.*, 166, 598, 1972.
26. **Bussereau, F.**, Etude du symptôme de la sensibilité au CO_2 produit par le virus sigma chez la *Drosophile.* I. Influence du lieu d'inoculation sur le delai d'apparition du symptôme, *Ann. Inst. Pasteur Paris*, 118, 367, 1970.
27. **Bussereau, F.**, Etude du symptôme de la sensibilité au CO_2 produit par le virus sigma chez la *Drosophile.* II. Evolution comparée du rendement des centres nerveux et des divers organes après inoculation dans l'abdomen et dans le thorax, *Ann. Inst. Pasteur Paris*, 118, 626, 1970.
28. **Bussereau, F.**, The CO_2 sensitivity induced by two rhabdoviruses, Piry and Chandipura in *Drosophila melanogaster, Ann. Microbiol. (Inst. Pasteur)*, 126B, 389, 1975.
29. **Deneubourg, A. M.**, personal communication, 1975.
30. **Diatta, F.**, personal communication.
31. **Bussereau, F.**, Etude du symptôme de la sensibilité au CO_2 produit par le virus de la stomatite vésiculaire chez *Drosophila melanogaster.* I. VSV de sérotype New Jersey et le virus Cocal, *Ann. Inst. Pasteur Paris*, 121, 223, 1971.
32. **Bussereau, F.**, Etude du symptôme de la sensibilité au CO_2 produit par le virus de la stomatite vésiculaire chez *Drosophila melanogaster.* II. VSV de sérotype Indiana, *Ann. Inst. Pasteur Paris*, 122, 1029, 1972.
33. **Bussereau, F.**,Etude du symptôme de la sensibilité au CO_2 produit par le virus de la stomatite vésiculaire chez *Drosophila melanogaster.* III. Souches de différents serotypes, *Ann. Microbiol. (Inst. Pasteur)*, 124A, 535, 1973.
34. **Bussereau, F., de Kinkelin, P., and Le Berre, M.**, Infectivity of fish rhabdoviruses for *Drosophila melanogaster, Ann Microbiol. (Inst. Pasteur)*, 126A, 389, 1975.
35. **Herreng, F.**, Etude de la multiplication de l'arbovirus Sindbis chez la *Drosophile, C. R. Acad. Sci.*, 264, 2854, 1967.
36. **Herreng, F.**, Personal communication.
37. **Teninges, D., Ohanessian, A., Richard-Molard, Ch., and Contamine, D.**, Isolation and biological properties of the *Drosophila* X virus, *J. Gen. Virol.*, 62, 241, 1979.
38. **Jupin, N., Plus, N., and Fleuriet, A.**, Action d'une souche de virus sigma sur la fertilité des *Drosophiles* femelles, *Ann. Inst. Pasteur Paris*, 114, 577, 1968.
39. **Seecof, R.**, Deleterious effects on *Drosophila* development associated with the sigma virus infection, *Virology*, 22, 142, 1964.
40. **L'Héritier, Ph.**, A convenient device for injecting large number of flies, *Drosophila Information Service*, 26, 131, 1952.
41. **Contamine, D.**, Etude de mutants thermosensibles du virus sigma, *Mol. Gen. Genet.*, 124, 233, 1973.
42. **Gay, P.**, Les gènes de la *Drosophile* qui interviennent dans la multiplication du virus sigma, *Mol. Gen. Genet.*, 159, 269, 1978.
43. **Vigier, Ph.**, Contribution à l'étude de l'instabilité génétique du virus σ de la *Drosophile, Ann. Genet.*, 9, 63, 1965.
44. **De Lestrange, M. T.**, Contribution à l'étude du virus héréditaire de la *Drosophile.* Action de la température sur le contenu en virus des tissus somatiques de l'hôte, *Ann. Genet.*, 6, 39, 1963.
45. **Printz, P.**, Influence de facteurs inhibant les synthèses sur le rendement final en virions de *Drosophiles* propageant le virus σ à l'état stabilisé, *C. R. Acad. Sci.*, 258, 378, 1964.
46. **Hadorn, E.**, Differenzierungsleitungen Wiederholt Fragmentierter Teilstücke Männlicher Genitalscheiben von *Drosophila melanogaster* nach Kultur in vivo, *Dev. Biol.*, 7, 617, 1963.
47. **Bernard, J.**, Etude de la propagation du virus héréditaire, σ, de la *Drosophile*, au niveau de disques imaginaux transplantés, *C. R. Acad. Sci.*, 262, 2102, 1966.
48. **Bernard, J.**, Contribution à l'étude du virus héréditaire de la *Drosophile*, sigma, sa propagation au niveau de disques imaginaux transplantés dans des adulte, *Exp. Cell Res.*, 50, 117, 1968.
49. **Ohanessian, A. and Echalier, G.**, Multiplication of *Drosophila* hereditary virus (σ virus) in *Drosophila* embryonic cells cultivated in vitro, *Nature (London)*, 212, 1049, 1967.
50. **Ohanessian, A.**, Sigma virus multiplication in *Drosophila* cell lines of different genotypes, *Curr. Top. Microbiol. Immunol.*, 55, 230, 1971.

51. **Bernard, J.**, Propagation d'un virus sigma défectif au niveau de disques imaginaux d'aile implantés dans des *Drosophiles* adultes, *J. Gen. Virol.*, 8, 209, 1970.
52. **Brun, G. and Diatta, F.**, Guérison germinale de femelles de *Drosophile* stabilisées pour le virus sigma, à la suite d'un traitement par la température ou par l'éthylméthanesulfonate, *Ann. Microbiol. (Inst. Pasteur)*, 124A, 421, 1973.
53. **Iconomidis, J. and L'Héritier, Ph.**, Les relations du virus σ avec son hôte *Drosophile*. Etude de la perte spontanée de l'état stabilisé, *Ann. Genet.*, 2, 53, 1961.
54. **Richard-Molard, Ch.**, Etude de la multiplication du virus sigma dans plusieurs cultures primaires et dans une lignée continue de cellules de *Drosophiles* issues d'embryons perpétuant un virus sigma défectif, *C. R. Acad. Sci. Ser. D.*, 277, 212, 1973.
55. **Bras-Herreng, F. and Ohanessian, A.**, personal communication, 1970.
56. **Bras-Herreng, F.**, Multiplication du virus Sindbis dans des cellules de *Drosophile* cultivées *in vitro*, *Arch. Ges. Virusforsch.*, 48, 121, 1975.
57. **Peleg, J.**, Studies on the behaviour of arboviruses in an *Aedes aegypti* mosquito cell line, *Arch. Gesamte Virusforsch.*, 37, 54, 1972.
58. **Davey, M. W. and Dalgarno, L.**, Semliki forest virus in cultured *Aedes albopictus* cells: studies on the establishment of persistence, *J. Gen. Virol.*, 24, 453, 1974.
59. **Artsob, H. and Spence, L.**, Persistent infection of mosquito cell lines with vesicular stomatitis virus, *Acta Virol.*, 18, 331, 1974.
60. **Stollar, V., Igarashi, A., and Koo, R.**, Properties of *Aedes albopictus* cell cultures persistently infected with Sindbis virus, in *Microbiology*, Schlessinger, D., Ed., American Society for Microbiology, Washington, D.C., 1977, 456.
61. **Riedel, B. and Brown, D. T.**, Role of extracellular virus in the maintenance of the persistent infection induced in *Aedes albopictus* (Mosquito) cells by Sindbis virus, *J. Virol*, 23, 554, 1977.
62. **Holland, J. J. and Villareal, L. P.**, Persistent non-cytocidal vesicular stomatitis virus infections mediated by defective T particles that suppress virion transcriptase, *Proc. Natl. Acad. Sci., U.S.A.*, 71, 2956, 1974.
63. **Youngner, J. S., Dubovi, E. J., Quagliana, M. O., Kelly, M., and Preble, O. T.**, Role of *ts* mutants in persistent infections initiated with VSV, *J. Virol.*, 19, 90, 1976.
64. **Richard-Molard, Ch. and Teninges, D.**, unpublished data, 1972.
65. **Ohanessian-Guillemain, A.**, Etude de facteurs génétiques contrôlant les relations du virus σ et de la *Drosophile* son hôte, *Ann. Genet.*, 5, 1, 1963.
66. **Courtin, M. F.**, Etude d'un Variant Génétique du Virus σ Modifiant ses Relations Avec la *Drosophile*, son Hôte, *Diplome d'Etudes Superieures*, Paris-Orsay, 1965.
67. **Contamine, D.**, manuscript in preparation, 1978.
68. **Guillemain, A.**, Découverte et localisation d'un gène empêchant la multiplication du virus de la sensibilité héréditaire au CO_2, *C. R. Acad. Sci.*, 236, 1085, 1953.
69. **Ohanessian-Guillemain, A.**, Sensibilité héréditaire à l'anhydride carbonique chez la *Drosophile*. Etude d'une mutation du virus σ modifiant son spectre d'activité, *C. R. Acad. Sci.*, 243, 1922, 1956.
70. **Gay, P.**, Adaptation d'une population virale à se multiplier chez un hôte réfractaire, *Ann. Genet.*, 11, 98, 1968.
71. **Bregliano, J. C.**, Contribution à l'étude des relations entre le virus σ et son hôte. Etude du caractère g^-, *Ann. Inst. Pasteur Paris*, 117, 325, 1969.
72. **Duhamel, C. and Plus, N.**, Phénomène d'interférence entre deux variants du virus de la *Drosophile*, *C. R. Acad. Sci.*, 242, 1540, 1956.
73. **Goldstein, L.**, Contribution à l'étude de la sensibilité au gaz carbonique chez la *Drosophile*. Mise en évidence d'une forme nouvelle du génoide, *Bull. Biol. Fr. Belg.*, 83, 177, 1949.
74. **Teninges, D.**, Etude de la localisation du virus sigma au cours de la différenciation des cellules germinales de *Drosophiles* mâles stabilisées, *Ann. Inst. Pasteur Paris*, 122, 541, 1972.
75. **Teninges, D.**, Le virus sigma et la cellule germinale mâle des *Drosophiles*, *Ann. Inst. Pasteur Paris*, 122, 1183, 1972.
76. **Tesh, R. B.**, Chianolis, B. N., and Johnson, K. M., Vesicular stomatitis virus (Indiana serotype) transovarial transmission by Phlebotomine sandflies, *Science*, 175, 1477, 1972.
77. **Clewley, J. P., Bishop, D. H. L., Kang, C. Y., Coffin, J., Schnitzlein, W. M., Reichmann, M. E., and Shope, R.**, Oligonucleotide fingerprints of RNA species obtained from rhabdoviruses belonging to the VSV subgroup, *J. Virol.*, 23, 152, 1977.
78. **Ohanessian, A. and Zalokar, M.**, personal communication, 1977.
79. **Bregliano, J. C.**, Contribution à l'étude de l'infection de la lignée germinale femelle par le virus sigma chez *Drosophila melanogaster*. I. Action du traitement thermique sur la transmission héréditaire du virus par les femelles qui ont reçu le virus sigma uniquement par le gamète paternel, *Ann. Microbiol. (Inst. Pasteur)*, 124A, 393, 1973.
80. **Ohanessian, A.**, personal communication.

Chapter 7

RHABDOVIRUSES INFECTING PLANTS

R. I. B. Francki and J. W. Randles

TABLE OF CONTENTS

I. INTRODUCTION

The detection of large bacilliform particles in cells of plants infected with maize mosaic virus (MMV)[1] was followed by the discovery of numerous other plant viruses with similar morphology. Working with lettuce necrotic yellows virus (LNYV), Harrison and Crowley[2] were the first to observe that plant viruses with large bacilliform particles showed remarkable structural similarities to viruses of vertebrates such as vesicular stomatitis virus (VSV). All these viruses are now included in the family Rhabdoviridae.[3]

As can be seen from the various chapters in the three volumes, our knowledge of rhabdoviruses infecting vertebrates, and particularly VSV, is very extensive; unfortunately, the same cannot be said of rhabdoviruses infecting plants. Although knowledge of the structure of plant rhabdoviruses and their relationship with host cells, as studied in the electron microscope, is as advanced as that of the vertebrate members of the family, biochemical studies of the infection process and virus multiplication are only in their infancy. This reflects, in some measure at least, the difficulties associated with biochemical studies on viruses in plant tissues as compared to those of animals; the plant virologist has nothing approaching the advantages of his colleagues who can use vertebrate cell cultures. Even the painstaking development of insect cell lines in which some plant viruses have been grown are, to date, rather disappointing in that they have not contributed much to our knowledge of rhabdovirus replication.

Research concerning plant rhabdoviruses has been reviewed in some detail by Francki,[4] and several other articles contain much relevant material regarding these viruses.[5-12] In this chapter we will briefly summarize the present knowledge of plant rhabdoviruses, discussing in more detail information published after the appearance of the review by Francki.[4] Where appropriate, emphasis will be placed on the comparison between the properties of the plant rhabdoviruses and those infecting vertebrates.

II. MEMBERS OF THE FAMILY RHABDOVIRIDAE INFECTING PLANTS

Comprehensive lists of plant rhabdoviruses have been published in reviews by Francki,[4] Knudson,[7] Russo and Martelli,[8] and Martelli and Russo[12] who enumerated 16, 25, 35, and 41 viruses, respectively. More recently Jackson et al.[11] have listed 11 viruses infecting hosts within the family Gramineae alone. By updating these lists, a new one with about 45 entries could be compiled. However, many entries on such a list would be based on reports in which only rhabdovirus-like particles had been observed by electron microscopy in thin sections of plant cells or leaf-dip preparations without any virus transmission or other data. Because of the scanty data available, there would be no assurance that the same virus had not been entered under different names just because it has been observed in cells of a different host plant. It is true that in many descriptions of rhabdoviruses, differences in reported particle dimensions are given as evidence that two viruses are distinct. However, it is our opinion that measurements of rhabdovirus particles can be misleading. It is well documented that in leaf-dip preparations preparative artifacts occur, and this topic has been discussed at some length elsewhere.[4,13]

We consider that the morphology of two rhabdoviruses in negatively stained preparations can be taken as distinct only when significant differences are evident in a mixture of the viruses. Figure 1 illustrates such a difference where the morphological types characteristic of *Digitaria* striate virus (DSV)[14] and cereal chlorotic mottle virus (CCMV)[15] are still observed when the two viruses are mixed before staining.

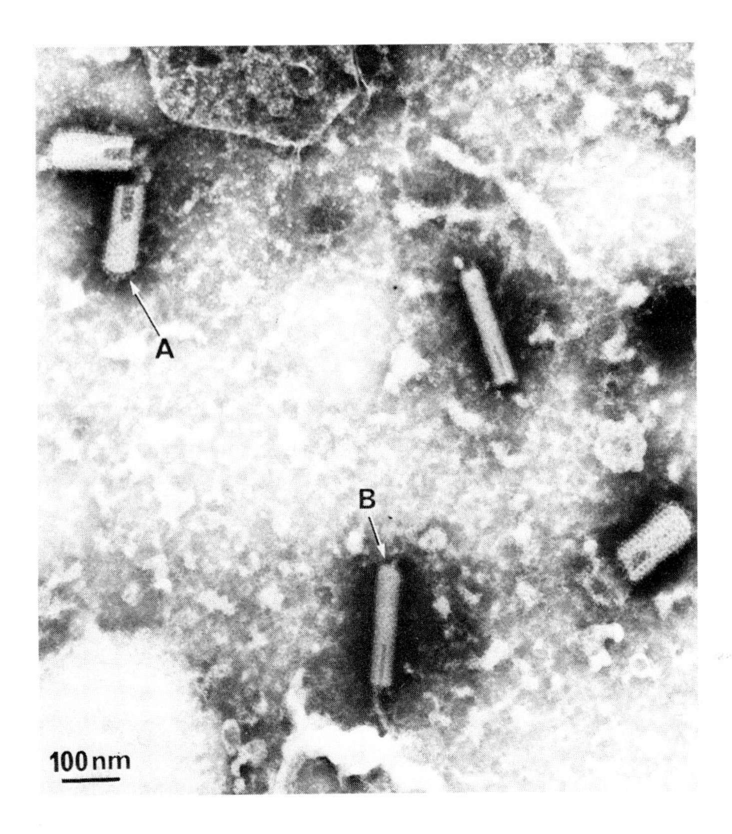

FIGURE 1. Morphological differences between cereal chlorotic mottle virus (A) and *Digitaria* striate virus particles (B) in a mixture of leaf extracts from virus-infected plants negatively stained with phosphotungstic acid. Micrograph supplied by R. S. Greber.

Although the properties of rhabdovirus particles in thin sections of infected cells are less prone to artifacts, there are still problems which require clarification. For example, the dimensions of maize mosaic virus (MMV) have been reported to be 242 × 48 nm and 300 × 75 nm in different laboratories.[16,17] Under no circumstances should measurements from thin sections and negatively stained preparations be compared as the two are seldom similar. For example, the same author reports the dimensions of MMV to be 242 × 48 nm in sections and 225 × 90 nm in leaf-dip preparations.[16] Virus taxonomy would be better served if much more care were taken in describing new rhabdoviruses and far more caution exercised in introducing new virus names.

Thus, in presenting a list of rhabdoviruses infecting plants (Table 1), we have limited ourselves to including those for which transmission data are available, whether by vector or sap inoculation. Even among these 23 viruses, there are many about which knowledge is scanty and it may turn out that some entries are synonymous or relatively closely related strains of the same virus.

In addition to viruses with typical rhabdovirus morphology, several smaller bacilliform particles have been observed in diseased plants, and it has been suggested that they are related to true rhabdoviruses. Of these, orchid fleck virus (OFV) has been transmitted by sap inoculation[38] and *Rubus* yellow net virus (RYNV) by an aphid vector.[39] It is interesting that OFV can infect both orchids (monocotyledons) and *Nicotiana* spp. (dicotyledons).[38] To our knowledge, no true rhabdovirus is known to have hosts among both dicotyledons and monocotyledons. Similar particles to those of OFV have been observed in plants showing symptoms of coffee ringspot[40] and citrus le-

TABLE 1

Some Rhabdoviruses Infecting Plants

Virus	Vectors	Sap transmission	Distribution
Aphid-Transmitted Viruses			
1. Lettuce necrotic yellows virus (LNYV)[18]	*Hyperomyzus lactucae* (L.), *Hyperomyzus carduellinus* (Theob),	+	Australia, New Zealand
2. Broccoli necrotic yellows virus (BNYV)[19]	*Brevicoryne brassicae* L.	+	Great Britian, Australia
3. *Sonchus* yellow net virus (SYNV)[20]	*Aphis coreopsidis* (Thomas)	+	North America
4. Sowthistle yellow vein virus (SYVV)[21]	*Hyperomyzus lactucae* (L.), *Macrosiphum euphorbiae* (Thomas)	—	North America, Europe
5. Strawberry crinkle virus (SCV)[22]	*Chaetosiphon fragaefolii* (Cock), *Chaetosiphon jacobi* (H.R.L.)	—	Americas, Europe, South Africa, Australia, New Zealand
6. Raspberry vein chlorosis virus (RVCV)[23]	*Aphis idaei* v.d.G.	—	Europe, North America, New Zealand
7. Lucerne enation virus (LEV)[24]	*Aphis craccivora* Koch	—	France
Leafhopper Transmitted Viruses			
8. Potato yellow dwarf virus (PYDV)[25]	*Aceratagallia sanguinolenta* (Provancher), *Aceratagallia lyrata* (Baker), *Aceratagallia obscura* Oman, *Aceratagallia curvata* Oman, *Agallia constricta* van Dusee, *Agallia quadripunctata* (Provancher), *Agalliopsis novella* (Say)	+	North America
9. Maize mosaic virus (MMV)[16]	*Peregrinus maidis* (Ashm)	—	Americas, Caribbean, Hawaii, India, Mauritius, Africa
10. (American) wheat striate mosaic virus (WSMV)[26]	*Endria inimica* (Say)	—	North America
11. Rice transitory yellowing virus (RTYV)[27]	*Nephotettix apicalis* (Motsch), *Nephotettix cincticeps* (Uhler), *Nephotettix impicticeps* Ish.	—	Taiwan
12. Northern cereal mosaic virus (NROV)[28]	*Laodelphax striatellus* Fallen, *Ribantodelphax albifascia* (Mats.), *Unkanodes sapporanus* (Mats.), *Muellerianella fairmairei* Perris	—	Japan

TABLE 1 (continued)

Some Rhabdoviruses Infecting Plants

Virus	Vectors	Sap transmission	Distribution
13. Barley yellow striate mosaic virus (BYSMV)[29]	*Laodelphax striatellus* Fallen	—	Europe
14. (Russian) winter wheat mosaic virus (WWMV)[30]	*Psamotettix striatus* (L.) *Psamotettix alienus* (Dhlb.)	—	Europe
15. Digitaria striate virus (DSV)[14]	*Sogatella kalophon* (Kirkaldy)	—	Australia
16. Oat striate mosaic virus (OSMV)[31]	*Graminella nigrifrons* (Forbes)	—	North America
17. Cereal chlorotic mottle virus (CCMV)[15]	*Nesoclutha pallida* (Evans)	—	Australia
18. Wheat chlorotic streak virus (WCSV)[32]	*Laodelphax striatellus* Fallen	—	France
19. Bobone disease virus (BDV)[33]	*Torophagus proserpina* (Kirk)	—	Pacific Islands
Viruses With No Known Vector			
20. Eggplant mottled dwarf virus (EMDV)[34]	?	+	Italy
21. *Gomphrena* virus (GV)[35]	?	+	South America
22. *Sonchus* virus (SV)[36]	?	+	South America
23. *Cynara* virus[37]	?	+	Spain

prosis.[41] Particles of these viruses look similar to rhabdovirus nucleocapsids (Figure 2). However, any suggestions that these viruses are envelope-defective rhabdoviruses will have to be substantiated by more data.

Of the seven rhabdoviruses which are transmitted by aphids (Table 1), LNYV is the most extensively studied and will be referred to frequently in this article. Both broccoli necrotic yellows virus (BNYV) and *Sonchus* yellow net virus (SYNV) are also sap-transmissible but have not been studied as extensively as LNYV, although some excellent work has recently been done on SYNV.[20,42] The other four viruses vectored by aphids are not sap-transmissible and hence are more difficult to work with. Of these, SYVV has been studied fairly extensively, both with respect to its structure and vector relationship which will be discussed later. However, relatively little information is available on strawberry crinke virus (SCV), raspberry vein chlorosis virus (RVCV), and lucerne enation virus (LEV).

It is interesting to note that there has been some confusion on the identity of RVCV. Rhabdovirus-like particles were first observed by Putz and Meignoz[43] in tissues from raspberries showing symptoms of raspberry mosaic virus (RMV). However, Stace-Smith and Lo[44] were unable to detect such particles in plants infected with RMV but did find rhabdovirus-like particles in plants infected with RVCV. It would thus appear that Putz and Meignoz[43] were working with plants infected with both RMV and RVCV. This raises an important problem regarding the determination of disease etiology.

Detection of virus-like particles in cells of diseased plants and the transmission of the disease should never be considered adequate evidence that the observed particles

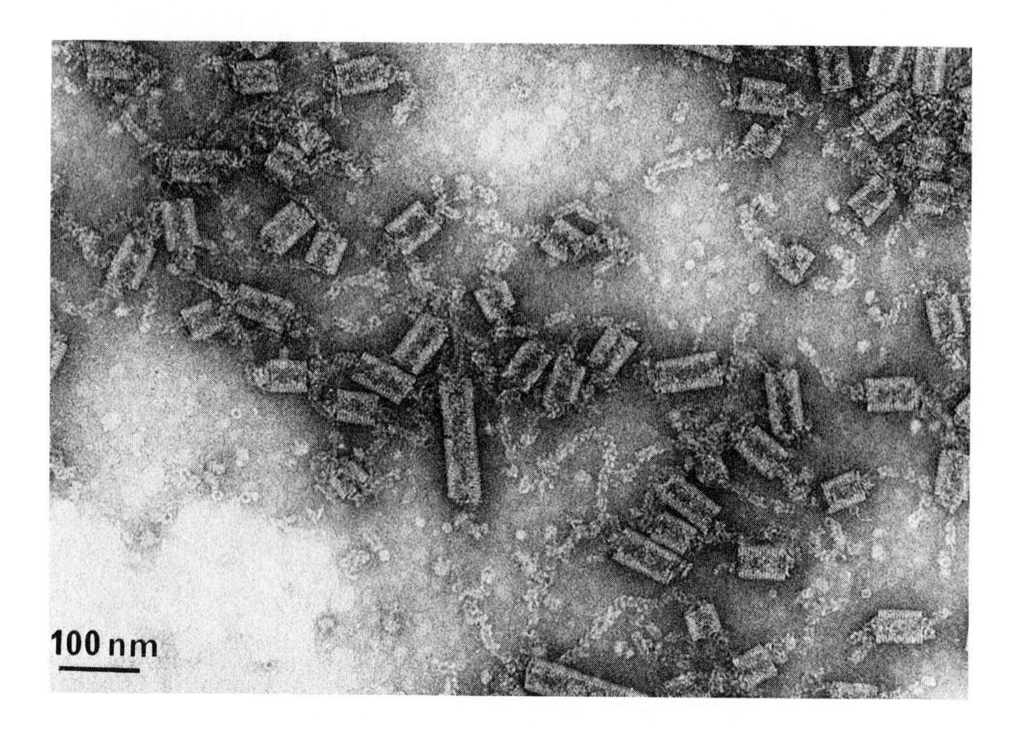

FIGURE 2. Purified preparation of orchid fleck virus negatively stained with ammonium molybdate. Micrograph supplied by D. Lesemann.

are the causal agents. It is always possible that the disease may be caused by another agent which defied microscopic detection. Infection by rhabdoviruses may readily be involved in confusions of this sort since they are easy to detect with the electron microscope under conditions where many smaller viruses may have gone unobserved. There are a number of examples where rhabdoviruses, together with other pathogens, have been shown to infect plants. For example, Lesemann[45] detected a rhabdovirus together with *Atropa* mild mosaic and belladonna mottle viruses in plants of *Atropa;* Maramorosch et al.[46] examined diseased pigeon pea plants and found them to be infected with both a rhabdovirus and a mycoplasma-like agent without being able to conclude whether either or both agents were pathogenic. Such situations are not confined to plants as we have recently observed leafhoppers with small polyhedral virus-like, rhabdovirus-like and reovirus-like particles (Figure 3); the insects were from colonies which induced wallaby ear disease when transferred to maize plants. The experimental evidence for a number of diseases with rhabdovirus etiology described in the literature, including some of those listed in Table 1, are not convincing and require confirmation by more thorough investigations.

Of the twelve viruses transmitted by leafhoppers listed in Table 1, only potato yellow dwarf virus (PYDV) is sap-transmissible; it is also the only one which infects dicotyledons. It has been investigated thoroughly over many years, and our knowledge about the molecular biology and biology of the virus can only be rivaled by LNYV. The information about the other 11 leafhopper-borne viruses varies greatly but is far less extensive. With further studies, it is conceivable that some of them may be shown to be synonymous or relatively closely related. Jackson et al.[11] have mentioned the results of unpublished data indicating serological relationships between oat striate mosaic virus (OSMV) and (American) wheat striate mosaic virus (WSMV), and between MMV and barley yellow striate mosaic virus (BYSMV). Relationships could also be expected between northern cereal mosaic virus (NCMV), BYSMV, and wheat chlorotic streak

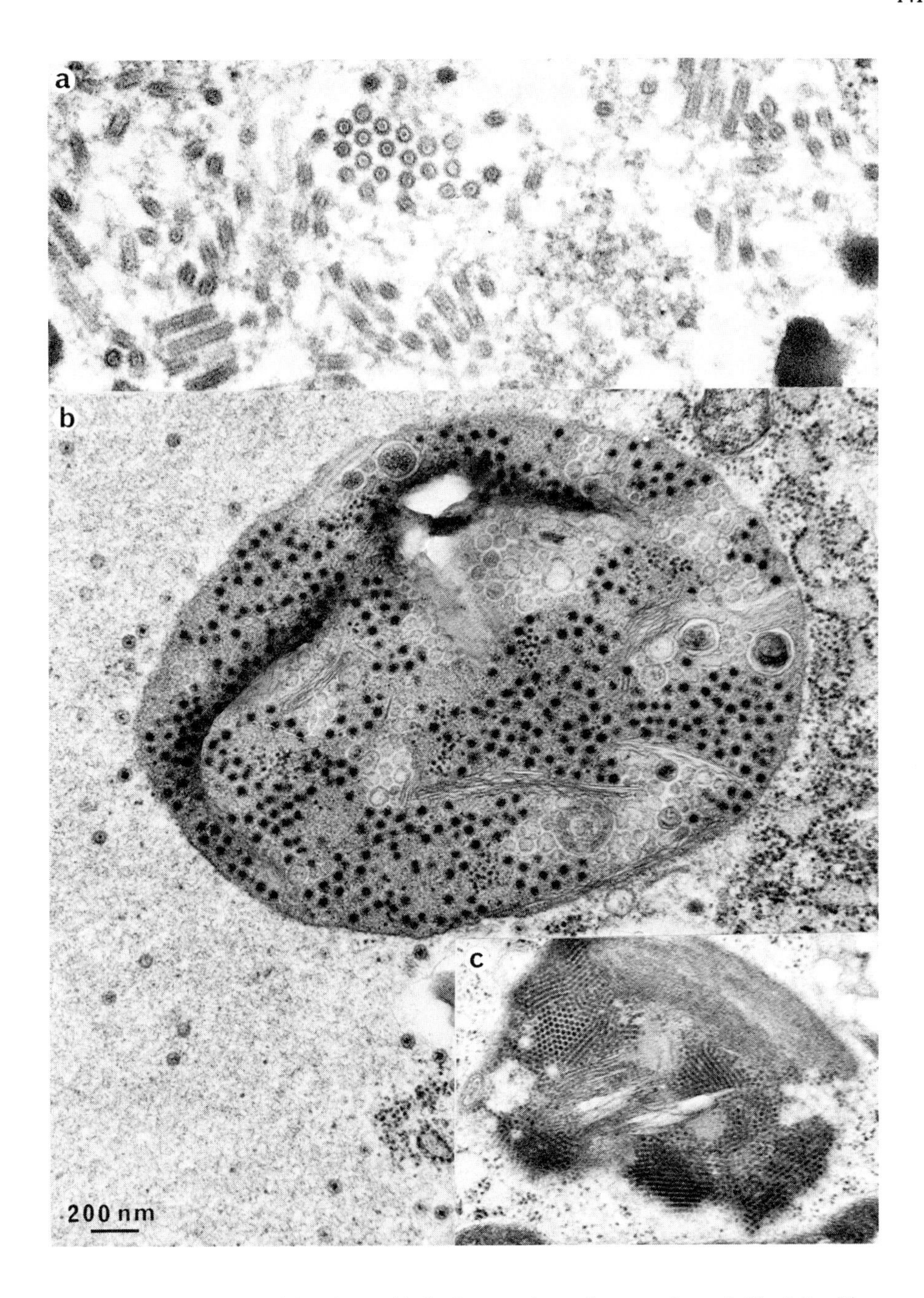

FIGURE 3. Various particles observed in leafhopper tissues from a colony of *Cicadulina bimaculata* (Evans) capable of transmitting maize wallaby ear disease. Typical rhabdovirus-like particles are seen in (a), reoviruslike particles in (b), and crystalline arrays of small polyhedral viruslike particles in (c). Micrographs supplied by T. Hatta.

virus (WCSV) which are vectored by the same species of leafhopper. Furthermore, the cytopathology of plants infected by NCMV[28] and BYSMV[29] show striking similarities.

Of the four viruses which have been transmitted by sap inoculation but which have no known vector (Table 1), only eggplant mottled dwarf virus (EMDV) has been investigated to any extent. More work is required on the remaining three to fully justify their description as distinct viruses. Vega et al.[36] have actually pointed out that *Sonchus*

TABLE 2

Plants Used for Propagation and Assay of Sap-Transmissible Rhabdoviruses

Virus	Propagation plant	Local lesion assay plant
Lettuce necrotic yellows virus (LNYV)[18]	*Nicotiana glutinosa* L.	*Nicotiana glutinosa* L.
Potato yellow dwarf virus (PYDV)[25]	*Nicotiana rustica* L.	*Nicotiana rustica* L.
Sonchus yellow net virus (SYNV)[20]	*Nicotiana clevelandii* Gray, x *Nicotiana glutinosa* L. Hybrid	*Chenopodium quinoa* Willd
Eggplant mottled dwarf virus (EMDV)[34]	*Solanum melongena* L., *Nicotiana glutinosa* L., *Nicotiana tabacum* L.	*Nicotiana tabacum* L.
Broccoli necrotic yellows virus (BNYV)[19]	*Datura stramonium* L.	

virus (SV) shares many properties with LNYV, and hence comparative studies between the two viruses should be very interesting.

From what has been said above, it becomes amply clear that the existing descriptions of rhabdoviruses are far from satisfactory. Any list of members which are assigned as belonging to the Rhabdoviridae and attempts at classification into subgroups must be tentative and will have to remain so until much more data become available.

III. GROWTH AND ASSAY OF THE VIRUSES

Of the 23 viruses listed in Table 1, only nine can be transmitted by sap inoculation. Sap transmission of the viruses offers an advantage to the researcher in that the growing and assay of virus is reasonably easy. However, only five of the sap-transmissible rhabdoviruses have been investigated in any detail; they are listed in Table 2 together with their propagation and assay hosts. With the exception of BNYV, these viruses induce local lesions on assay plants providing another advantage in that relative infectivity determinations can be done with reasonable precision.[47,48] However, it must be remembered that virus strains vary considerably[49] and not all strains will necessarily multiply satisfactorily in any particular host species or may fail to induce suitable local lesions on others. For example, most strains of LNYV will infect *N. glutinosa* and replicate satisfactorily but only one strain, the S.E.3 isolated by Stubbs and Grogan,[50] will induce on this plant lesions which are suitable for assay work; it is with this strain that most of the work on LNYV has been done to date. Symptoms produced by SYNV on leaves of its propagation and assay plants are illustrated in Figure 4A and B, respectively; and those of LNYV (Strain S.E.3), which can be propagated and assayed on the same species, in Figure 4C.

Virus growth curves have been determined in plants inoculated with LNYV,[51] PYDV,[52] and SYNV.[20] Although the exact kinetics are dependent on environmental conditions, the concentration of each virus increases rapidly after an initial latent pe-

FIGURE 4. Plants infected with two rhabdoviruses. The leaf shown in (a) is from a *N. glutinosa* × *N. clevelandii* hybrid plant systemically infected with SYNV which can be used as a source of virus for purification. The leaf of *C. quinoa* in (b) was inoculated with SYNV and the induced local lesions make the plant suitable for virus assays. In (c), the *N. glutinosa* plant on the left is healthy, and the one on the right infected with LNYV shows local lesions on inoculated leaves (I) and systemic symptoms on the younger leaves; (s) indicates the first leaf to develop systemic symptoms. Photographs in (a) and (b) supplied by A. O. Jackson.

riod of 4 to 6 days, reaching a maximum about 10 to 12 days after inoculation, and declining thereafter. These observations are based on the estimation of virus by infectivity assays, and hence the rapid decline in the amount of virus could be due to an artifact such as a buildup of virus inhibitors in the plant at late stages of infection, breakdown of virus *in situ*, or its inactivation.

Although experimentation with viruses that can only be transmitted by their insect vectors is much more involved, several have been studied to some degree; they include SYVV,[53] WSMV,[54] NCMV,[28] and MMV.[1] Although no growth curves of these viruses in their plant hosts have been determined, electron micrographs of infected tissues show that virus particles accumulate in large numbers.

The assay of viruses which can only be transmitted by vectors is both laborious and inaccurate. Virus preparations are either injected into virus-free insects or the vectors are fed on the virus through a membrane[55,56] and then tested by caging on healthy plants to see if they transmit the virus. By using virus at increasing dilutions for the injections or membrane feeding, the dilution end point can be determined, which is a function of virus concentration in the inoculum. However, the errors involved in such assays are many and require large numbers of insect and plant replicates.[57]

During the past decade there have been striking advances in the development of insect-culture techniques, and cultures suitable for infection by some viruses are available.[58] However, such methods have been attempted with only two rhabdoviruses, PYDV and SYVV in leafhopper and aphid cells, respectively.[52,59] In the case of aphid cells, infection of only primary cultures has been achieved and considerable improvements in culturing techniques are needed before they can be used for virus work with any degree of success.[60] Results obtained with leafhopper cells are much more promising, and growth curves of PYDV in vector cell monolayers have been reported. They demonstrate that after a latent period of about 10 hr there is exponential virus growth for about 20 hr, after which a plateau is reached.[61] There was no suggestion of any drop in virus concentration after the maximum was reached, as apparently occurs with the same virus in cells of intact plants.[52] PYDV does not appear to induce any clear cytopathic effects in leafhopper cell monolayers, but infection can readily be detected and assayed by immunofluorescent techniques.[58] It has been demonstrated that PYDV can be assayed with greater sensitivity and precision on insect cell monolayers than on plants by the local lesion method.[62]

IV. VIRUS PURIFICATION

Purification of viruses from plant tissues requires efficient methods of clarifying crude extracts before the virus is concentrated. The most effective methods make use of organic solvents,[63] and more recently, treatment with nonionic detergents has been used.[64] Since rhabdoviruses contain lipoprotein envelopes, both the use of organic solvents and detergents is precluded, and hence other, less effective methods of clarification must be used. The best methods involve filtration through pads of celite (diatomaceous earth) under pressure with or without prior addition of adsorbants such as cacium phosphate gel, charcoal, diethylaminoethyl (DEAE) cellulose, or bentonite.[4,63] Efficient selective adsorption of host plant materials is dependent on the pH and ionic composition of the extract.[63]

Concentrating rhabdoviruses is relatively simple since they sediment readily at low centrifugal forces. However, unless the extracts are clarified efficiently before sedimentation, the pelleted virus can be difficult to resuspend. In the case of some rhabdoviruses, polyethylene glycol precipitation has proved very satisfactory for their concentration; however, the optimum salt and polymer concentrations for precipitation can vary from virus to virus and even between strains of the same virus.[65]

Some rhabdoviruses have been obtained in reasonably pure preparations by subjecting concentrated virus to rate-zonal and quasi-equilibrium centrifugation.[20] In other instances, sucrose density-gradient electrophoresis and column chromatography on calcium phosphate gel have been used; the details of these techniques are summarized by Francki.[4] The electrophoretic method has the great disadvantage for use with labile viruses, which includes most rhabdoviruses, in that it requires periods of at least 16 hr for efficient separation of virus and contaminating materials. The chromatographic method is very quick but has the disadvantage that it is difficult to prepare or obtain commercially calcium phosphate gels with similar absorptive properties. It has become

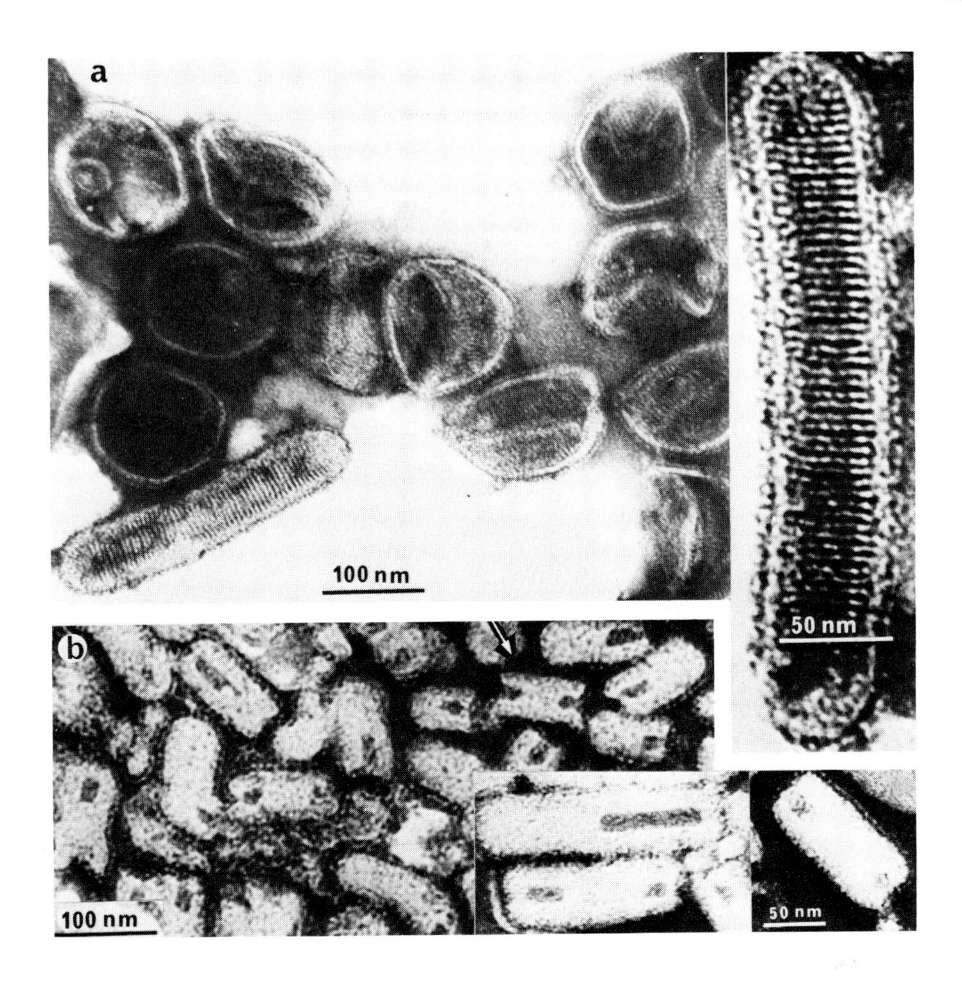

FIGURE 5. Purified preparation of LNYV negatively stained with uranyl acetate (a) and phosphotungstic acid (b). In (a) most of the particles are pleomorphic in shape but some are bacilliform. In (b) most particles are bullet-shaped but a few are "blunt" at both ends (arrowed and bottom right corner). Micrographs supplied by T. C. Chambers and B. S. Wolanski.

quite clear that no one method is suitable for purifying all rhabdoviruses from plants. For example, whereas BNYV was purified satisfactorily[66] by a method developed for LNYV,[67] the method was quite unsuitable for the preparation of SYNV.[20]

V. VIRUS STRUCTURE AND COMPOSITION

Bullet-shaped and bacilliform, as well as pleomorphic particles have been observed in negatively stained preparations of rhabdoviruses from plants examined by electron microscopy (Figure 5). The bullet-shaped particles are similar to those seen in preparations of most vertebrate rhabdoviruses. It is generally accepted that the bullet-shaped and pleomorphic particles seen in micrographs of plant rhabdoviruses result from preparative artifacts and that the unaltered particle is bacilliform, like that seen in thin sections of infected cells (Figure 6). Conversely, it is also accepted that particles of the vertebrate viruses are bullet shaped.[9] However, recent work by Orenstein et al.[68] suggests that although the VSV nucleocapsid is bullet-shaped, the virion is bacilliform, although it would appear that preservation of this structure during extraction and preparation for electron microscopy requires careful manipulation involving chemical fixation.

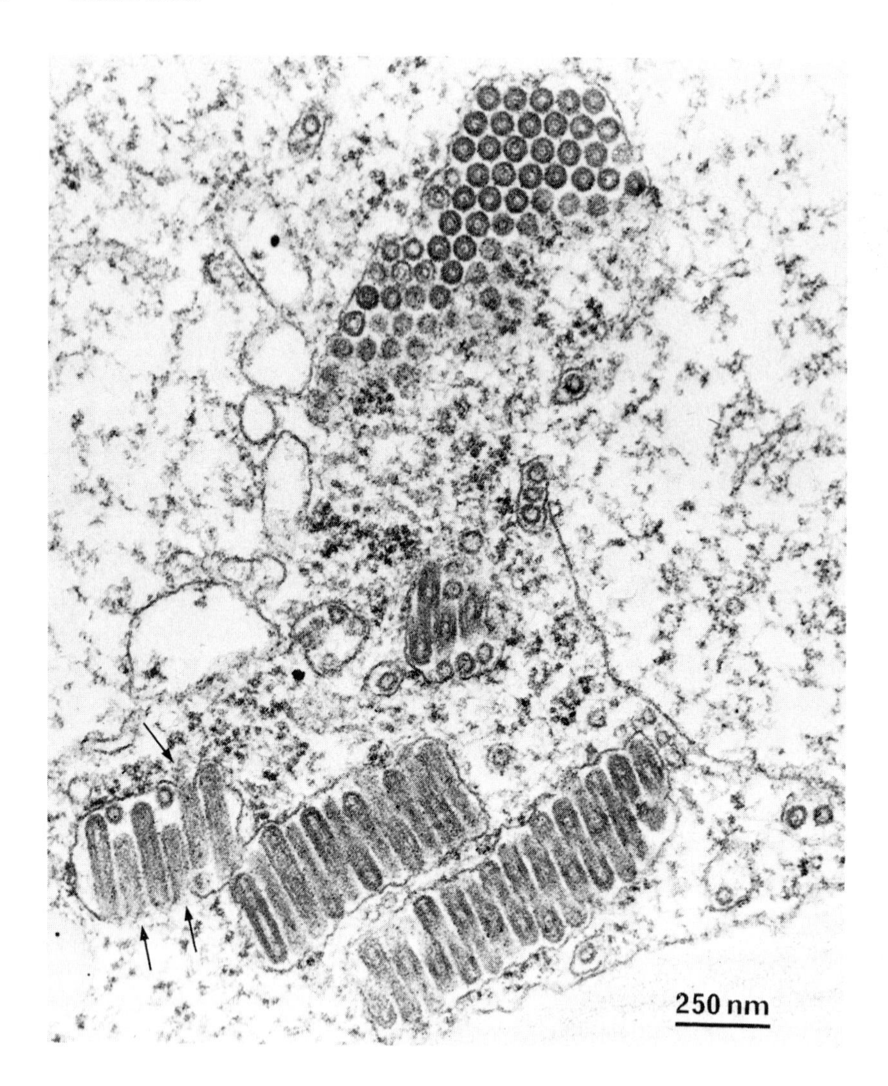

FIGURE 6. Thin section of plant leaf cell infected with BNYV showing aggregates of virus particles seen in transverse (above) and longitudinal section (below) associated with the endoplasmic reticulum. Some particles (arrowed) appear to be budding from the membranes of the endoplasmic reticulum. Micrograph supplied by R. G. Garrett.

The effects of various conditions of handling plant rhabdoviruses during purification and preparation for electron microscopy on their morphology have been studied with several viruses, and the results have been discussed in some detail elsewhere.[4,13,63] Preparative artifacts almost certainly occur with all rhabdoviruses but it would appear that the original structure of some, such as PYDV,[69] are more difficult to retain than that of others such as LNYV.[70,71] It has already been stressed that extreme caution is required in interpreting electron microscopic data regarding the morphology of rhabdoviruses. Nevertheless, our present concept of rhabdovirus structure is that of a bacilliform particle consisting of a viral envelope enclosing a long strand of nucleoprotein wound into a helix of low pitch (Figure 7).

Although the physical and chemical properties of only a few plant rhabdoviruses have been investigated, they all appear to be similar to those of rhabdoviruses from vertebrates. The most significant differences are in their size and sedimentation properties (Table 3). The protein compositions of rhabdoviruses from plants and vertebrates are similar (Table 4), for which a common nomenclature has been proposed.[72]

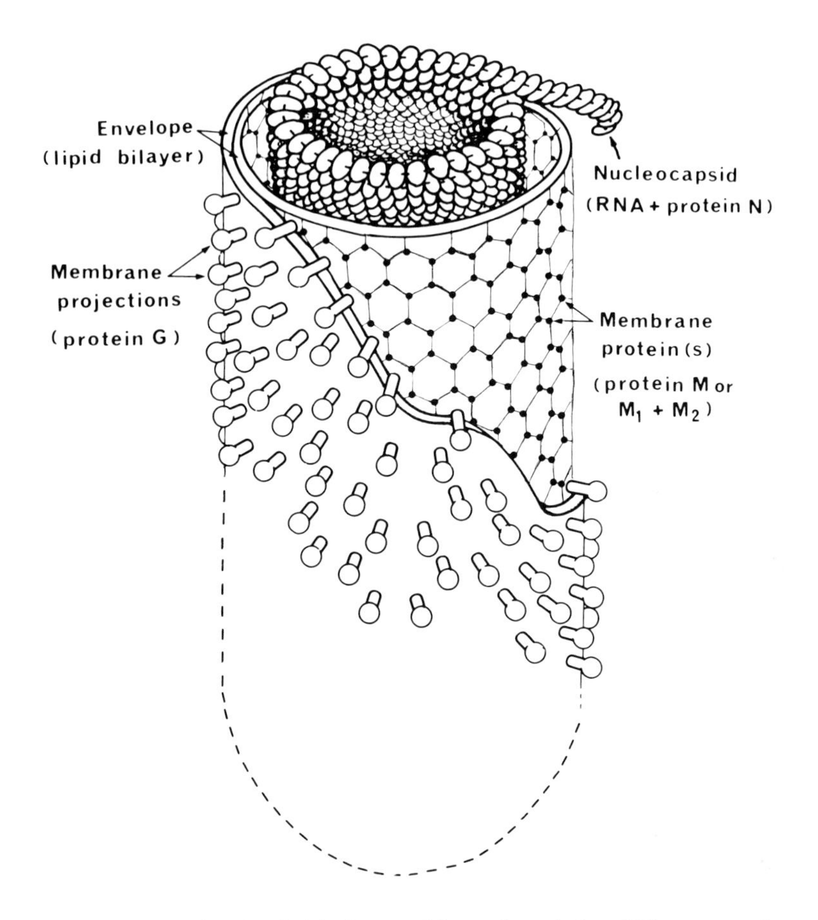

FIGURE 7. Model of a rhabdovirus particle cut through the middle to expose its internal structure (see text for details).

The envelopes of all the rhabdoviruses tested can be removed with nonionic detergent to leave a nucleocapsid which, in most cases, retains some of its infectivity. Two or three of the major virion proteins are released with the envelope, the glycosylated G protein constituting the surface projections and the M or M_1 + M_2 proteins associated with the envelope matrix. Some rhabdoviruses, like LNYV and VSV, have only one envelope matrix protein (M), whereas others like SYNV and Rabies virus contain two M_1 + M_2 (Table 4). The nucleocapsid consists of the viral RNA and protein which can be dissociated with sodium dodecyl sulfate (SDS) to yield single-stranded RNA of about 40S, one major protein (N), and up to two minor proteins (NS and L). In the case of VSV, it has been demonstrated that nucleocapsids capable of transcribing the viral RNA in vitro (see also Section VI) have the N, NS, and L proteins. Although the major proteins G, N, and M (or M_1 + M_2) have been identified in all the plant rhabdoviruses which have been analyzed, minor proteins have been detected only in some (Table 4).

The location of proteins in the virions of only two rhabdoviruses from plants have been investigated in detail, those of SYVV[81] and SYNV,[42] both of which contain M_1 and M_2 proteins. The G protein of both viruses was the one most readily labeled with ^{125}I in vitro indicating its exposed location, thus confirming that it is the polypeptide

TABLE 3

Physical Properties of Plant and Vertebrate Rhabdoviruses

Property	LNYV[18,73-75]	PYDV[25,76]	SYNV[20]	WSMV[26,77,78]	BNYV[19]	VSV[4,9]	Rabies[4,9]
Virions							
Dimensions (nm)							
In sections	360 × 52	380 × 75	—	250 × 75	297 × 64	—	—
In negative stain	227 × 66	—[a]	250 × 94	245 × 75	275 × 75	175 × 68	180 × 75
Sedimentation coefficient(s)	945	880	1045	875	874	625	600
Buoyant density (gm/cm^3)	1.20	1.17	1.18	1.22	1.19	1.18	1.20
Nucleocapsids							
Sedimentation coefficient(s)	260	250	250	—	—	140	200
Infectivity	Yes	Yes	Yes	—	—	Yes	No
Ribonucleic acid							
Sedimentation coefficient(s)	43	45	44	32	—	36—45	45
Strandedness	Single	Single	Single	Single	—	Single	Single
Infectivity	No	No	No	No	—	No	No
Molecular weight (× 10^6 daltons)	4.2	4.6	4.4	2.2	—	3.6—4.5	4.6

[a] No data available.

TABLE 4

Structural Proteins of Plant and Vertebrate Rhabdoviruses

Protein[72]	PYDV[79]	LNYV[80]	SYNV[20,42]	SYVV[81]	WSMV[82]	VSV[9]	Rabies[9]
L	+[a]	171	+	150	145	161	–
G	78	71	77	83	92	64	69 / 60
N	56	56	64	60	59	52	60
NS	–[b]	38	–	–	–	42	45
M_1	33		45	44			29
M_1 / M_2		19			25	24	
M_2	22		39	36			24

[a] Detected but molecular weight not estimated accurately.
[b] Not detected.

associated with the envelope spikes. Ziemiecki and Peters[81] demonstrated that the M_1 and M_2 polypeptides of SYVV were also labeled strongly, and it was concluded that they are constituents of the envelope. Protein N was the most difficult to iodinate which is consistent with the notion that it is located internally to the envelope and hence is a constituent of the nucleocapsid. Jackson[42] came to similar conclusions regarding the G and N proteins of SYNV. However, he showed that M_1 protein appears to be more tightly complexed to the virion than M_2 under low salt and low pH conditions, and his data from lactoperoxidase iodination of intact virions indicate that the tyrosine residues of M_1 are deeper into the envelope membrane than those of the M_2 protein. It would appear that M_2 has all the characteristics of a membrane matrix protein whereas those of M_1 are not altogether clear. Unlike the L protein of VSV, high molecular weight protein of SYNV appears to be located near the virion surface.[42] However, this protein has only been detected in very small amounts, and the possibility has not been eliminated that it may be either host plant protein contaminating the virus preparations or aggregates of some of the major viral proteins.

The lipids of three plant rhabdoviruses, PYDV,[83] NCMV,[84] and WSMV,[78] have been investigated. From the data on PYDV and WSMV, it would appear that the lipid accounts for about one quarter of the virion weight, and both the viruses have been shown to contain sterols and free fatty acids. It was demonstrated that the proportion of the various fatty acids in PYDV was not the same as that in extracts of healthy plants.[83] Toriyama[84] analyzed the sterol composition of NCMV and demonstrated that the proportions of the four sterols differed from those in extracts from plants. These studies indicate that lipid is an integral part of the virion, but more detailed studies are needed to determine how it becomes associated with viral-specified proteins to produce the viral envelope.

Three of the four rhabdoviruses from plants whose RNAs have been examined in any detail, contain ss-RNA of molecular weight between 4.2 to 4.6×10^6 daltons (Table 3). However, it has been reported that the ss-RNA of WSMV has a molecular weight of only 2.2×10^6 daltons.[77] We feel that these data need confirmation before they are accepted since it can be calculated that in order to code for the four WSMV proteins,[82] an RNA with molecular weight of at least 2.8×10^6 daltons would be required.

All published data support the conclusion that particles of rhabdoviruses from plants are longer than those of vertebrates. Although we are aware of the problems in drawing conclusions from particle measurements,[4,13,63] the sedimentation data (Table 3) support the view that there are real differences in size of the plant and vertebrate viruses. However, the reported sizes of the plant and vertebrate viral RNAs are similar, ranging from 4.0 to 4.6×10^6 daltons, as are the molecular weights of the structural

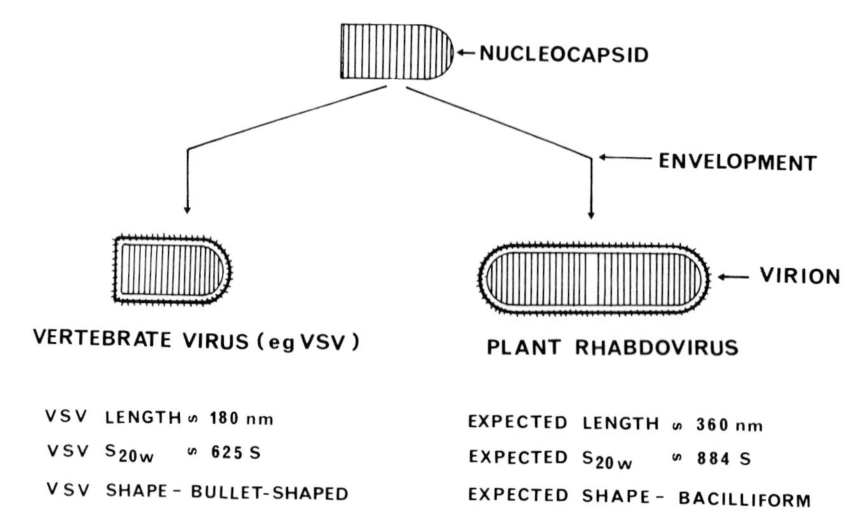

FIGURE 8. Diagrammatic representation of possible structural differences in the assembly of rhabdoviruses infecting vertebrates and plants.

proteins (Table 4). Thus, it is difficult to account for the striking differences in the sedimentation coefficients of the viruses. A suggestion has been put forward by Francki,[4] and the view was supported by Wagner,[9] which could account for this apparent impasse. Whereas a particle of a vertebrate virus such as VSV consists of bullet-shaped nucleocapsid surrounded by an envelope, one of a plant virus such as LNYV may consist of two similar nucleocapsids positioned end-to-end and enveloped together (Figure 8). If we assume that the sedimentation coefficient of VSV is 625S as quoted by Wagner,[9] a particle containing two nucleocapsids would be expected to have a sedimentation coefficient of approximately 884S which is not widely different from that reported for some of the plant rhabdoviruses (Table 3). The acceptance or rejection of this hypothesis will have to await experimental evidence.

VI. VIRUS-ASSOCIATED TRANSCRIPTASE

Rhabdovirus RNAs lack demonstrable infectivity (Table 3). This, together with the observation that m-RNAs capable of synthesizing VSV protein have nucleotide sequences complementary to the viral RNA,[85] prompted Baltimore et al.[86] to test for the presence of an RNA-dependent RNA polymerase associated with the virions. This search was successful and an enzyme (transcriptase) capable of transcribing the viral RNA in vitro has been characterized and shown to be an integral part of the VSV virion.[9,87] Subsequently, similar transcriptases were detected in preparations of several other rhabdoviruses from vertebrates, but not in others such as rabies virus.[87]

Although attempts have been made to detect transcriptases in preparations of several rhabdoviruses from plants, enzymatic activity has only been demonstrated in preparations of LNYV.[74,75,88] However, negative results have been obtained from work with PYDV, SYVV, and SYNV.[89] Both PYDV and SYNV have many characteristics in common with LNYV (Table 29.3). Thus, it would appear that either these viruses contain transcriptases which have defied detection, or they must depend for their replication on enzymes from their host plants. Future research must take into account both these possibilities. We favor the former since the viruses are very similar to LNYV, but the latter possibility cannot be dismissed as there is evidence of RNA-dependent RNA polymerases in virus-free plant tissues.[90] It is conceivable that these host-plant enzymes could be utilized by the viruses for their replication.

Since preparations of rhabdoviruses from plants are difficult to obtain in a highly purified state (see Section IV), difficulties in detection of virus-associated transcriptases can be anticipated. In order to detect enzyme activity, it is necessary to have virus preparations in which both the RNA template and the enzyme are undamaged and which are free of other enzymes capable of degrading either template and product (RNases) or the transcriptase itself (proteases). In the case of SYVV, it has been demonstrated that virus preparations contained proteolytic activity in spite of an elaborate purification procedure.[91] Fragments of cellular membranes are probably the most difficult contaminants to remove from rhabdovirus preparations, and it has been demonstrated that such material interferes with the assay of the LNYV-associated transcriptase and that this interference is at least in part due to RNase activity associated with the membrane fragments.[92] If cellular membranes contaminating virus preparations interfere with the assay of the virus-associated transcriptases, then it may be more fruitful to isolate purified viral nucleocapsids to test for the presence of the enzyme.

Studies on the transcription of LNYV in vitro, although less complete than those on VSV, indicate that the two systems are very similar.[13] The LNYV transcriptase was active when assayed under the conditions used for the assay of the VSV enzyme, except that the incubation temperature was 25°C. The optimal Mg^{2+} concentration was similar for both enzymes, and both systems were sensitive to RNase but were unaffected by DNase or actinomycin D. Analysis of the LNYV transcriptase product by sucrose density-gradient centrifugation and polyacrylamide-gel electrophoresis indicate that the viral RNA is transcribed into at least three molecules.[75] However, critical comparisons to the transcripts of VSV have not been made.

VII. SEROLOGY

Relatively little attention has been paid to serological studies with rhabdoviruses from plants, probably because of the unavailability of reasonable quantities of adequately purified viruses for use as immunogens. However, antisera to several of the viruses have been prepared, and it would appear from the work already done (Table 5) that serological studies hold considerable potential for the investigation of a number of problems; perhaps the most pressing is that of serotyping the viruses within the group. From the limited studies done so far, there appears to be no serological relationship between LNYV and SYVV,[93] and no positive reactions were detected between preparations of SYNV and antisera to LNYV, BNYV, SYVV, or PYDV.[20] In most of these tests relatively low-titered antisera were used, and hence distant relationships could have escaped detection.

Immunodiffusion techniques are widely used by plant virologists in spite of the disadvantage of being insensitive compared to a number of other serological methods. Their ability to resolve antigenic differences and simultaneously test for host plant antigen-antibody reactions probably account for their popularity. A disadvantage of these tests when dealing with rhabdoviruses is that intact virions are too large to diffuse into agar gels and hence cannot form precipitin lines with homologous antiserum.[94] However, this problem can be circumvented by using chemically or enzmatically disrupted virus preparations which react satisfactorily and have the advantage that both internal and external antigens can be examined. Perhaps the most satisfactory agents for disrupting rhabdoviruses prior to immunodiffusion tests are either nonionic or ionic detergents; their action on rhabdoviruses has aleady been mentioned (see Section V). However, caution is required when using ionic detergents such as SDS which under some conditions can produce nonspecific precipitin lines.

The rhabdoviruses appear to be antigenically complex and two or more antigens have been identified in LNYV,[97] PYDV,[79] BNYV,[66] and WSMV.[82,99] The work of

TABLE 5

Antisera Prepared Against Rhabdoviruses From Plants

Antiserum to:	Antiserum used for:	Ref.
Potato yellow dwarf virus (PYDV)	Virus strain differentiation[a]	94,95
	Immunofluorescent detection of virus in vector cells[a]	96
	Basic immunochemical studies[a]	79
Lettuce necrotic yellows virus (LNYV)	Detection of virus[b]	2
	Basic immunochemical studies[c]	97
Wheat striate mosaic virus (WSMV)	Detection of virus in plants and vectors[b]	98
	Basic immunochemical studies[c]	99
	Analysis of viral antigens in vector[a]	100
	Basic immunochemical studies[a]	82
Sowthistle yellow vein virus (SYVV)	Testing possible relationship to a vertebrate virus[a]	101
	Immunofluorescent detection of virus in vector cells[a]	59
Broccoli necrotic yellows virus (BNYV)	Basic immunochemical studies[c]	66
Northern cereal mosaic virus (NCMV)	Virus identification[c]	102
Maize mosaic virus (MMV)	Virus identification[a]	103
Sonchus yellow net virus (SYNV)	Testing relationships among plant rhabdoviruses[b]	20

[a] No tests reported of tests for freedom of host plant-specific antibodies in antisera.
[b] Host plant-specific antibodies were detected in antisera.
[c] Tests failed to detect host plant-specific antibodies in antisera.

Knudson and MacLeod[79] is particularly interesting in that it demonstrates that when rabbits were immunized with two strains of PYDV, distinct antibodies specific to the proteins G, N, M_1, and M_2 were elicited. Furthermore, it appears that whereas proteins N and M_2 are serotype specific, G and M_1 were shown to be antigenically similar from the two virus strains studied.

Antisera prepared against some of the rhabdoviruses from plants have been shown to react not only with preparations of homologous antigens, but also with soluble antigens specific to infected plants.[79,97,99] The work of Knudson and MacLeod[79] indicates that soluble antigen in PYDV-infected plants has the specificity of N protein, the major protein constituent of nucleocapsids. Since this protein has been shown to be serotype specific,[79] it may be very useful in rhabdovirus serotyping. Because of the problems associated with purifying intact rhabdovirus particles, it may be easier to prepare antisera to N protein only for which preparations of either nucleocapsids or isolated N protein could be used as immunogen. Such virus-specific antigens would almost certainly be easier to obtain free of host plant antigens than intact virions.

VIII. VIRUS-PLANT RELATIONSHIPS

Those rhabdoviruses which are sap transmissible are the most amenable to comprehensive studies on the nature of virus infection, replication and pathogenesis. Even so, they provide an unusually difficult system to work with because of their instability and complexity. Moreover, progress is limited by failure to obtain significant incorporation of radioactive precursors into virus components.[104] Their only advantages are their size

TABLE 6

Events During the Multiplication of LNYV in *Nicotiana glutinosa*

Days after inoculation	Physiological events	Cytological events in systemically infected leaf	Phase	Ref.
	Inoculated leaf		Nuclear phase	
0.1	Ultraviolet sensitive phase in inoculated leaf			105
3—4	Local symptoms appear; virus translocated from inoculated leaf, invades plant systemically			51
	Systemically infected leaf			
5—6	Latent phase; infection specific polyribosomes detected	Blistering of outer nuclear membrane; spherical vesicles produced in perinuclear space		106, 107
7	Infectious virus detected	Virions sometimes detected in perinuclear space		51, 107
8	Symptoms appear, leaf growth slows	Virions and "viroplasm" detected in cytoplasm	Cytoplasmic phase	51, 107
9—12	Virus reaches maximum and then declines; symptom severity increases; chloroplast ribosomes, fraction I protein, protein synthesis decrease	Virions accumulate in cytoplasm associated with endoplasmic reticulum; occasional virions seen budding from endoplasmic reticulum		51, 107

and easily recognizable structure in the electron microscope making them popular in studies involving the examination of virus *in situ*, a subject which has been recently reviewed by Martelli and Russo[12] and thus here will only be treated superficially.

N. glutinosa infected with LNYV is the most thoroughy studied rhabdovirus-host plant system. Both inoculated and systemically infected leaves have been studied. The time of infection of the first systemically infected leaf can be predicted with reasonable accuracy[51] and has been used to study the early events of virus replication without the complications of phenomena associated with mechanical wounding of tissue. Using this system, biochemical studies supplemented by microscopic observations by a number of workers allow us to reconstruct some of the events leading to the synthesis and accumulation of LNYV in plant cells; these are summarized in Table 6.

Following inoculation of the leaf epidermis, there is an ultraviolet sensitive phase which lasts 2 to 3 hr followed by an increase in the UV resistance of the infected centres reaching a maximum after 4 to 5 hr.[105] This suggests that the viral nucleic acid is unprotected for 2 to 3 hr, after which it gains protection, possibly by association with host cell components. Virus synthesis and cell to cell infection proceeds, and at 25°C under continuous lighting, local chlorotic lesions appear on inoculated leaves after 3 to 4 days.[51] Virus is translocated from the inoculated leaf at about the time lesions appear, both to the developing leaves and to the roots of the plant, presumably through the phloem. At this time, the first systemically infected leaf is 3 to 4 cm long; it is this leaf which has been studied in detail. Infectious virus is first detected after another 1 to 2 days, during which time putative virus-specific polyribosomes are detected.[106] The initial systemic symptoms, clearing and yellowing of veins, are visible one day later. At this stage, virions are detectable by sucrose density-gradient analysis of leaf extracts. As symptoms develop to general chlorosis and then vein necrosis, virus concentration declines. Leaf growth virtually ceases at about the time first symptoms appear.

The distribution of the virus in the leaves is related to the distribution of symptoms. In the inoculated leaves, virus is confined to the lesions and in the systemically infected ones, to chlorotic and necrotic tissue.[108] Roots are invaded to within 0.5 mm of the tip.[108] Virus particles have been observed in all the types of cells examined.[4]

Cytological changes are seen within 2 days of systemic invasion, when blistering of the outer nuclear membrane occurs and perinuclear spaces appear.[107] At this time actinomycin D-resistant RNA synthesis is observed in the nucleus. Particles can then sometimes be observed in the perinuclear spaces. This nuclear phase is followed by a cytoplasmic phase when actonomycin-resistant RNA synthesis occurs mainly in the cytoplasm.[107] Virions are then observed in the cytoplasm, sometimes associated with viroplasms, and some particles are seen apparently budding from cytoplasmic membranes.[73,107]

Cytopathic effects are first observed after symptoms appear. LNYV infection has a marked effect on chloroplasts[109] although they seem to have no direct involvement in virus replication. Within 1 day of symptom appearance, chloroplast ribosomes decline in concentration, and are completely undetectable 1 to 3 days later. There are parallel reductions in the ribosomal RNA synthesis, concentration of Fraction I protein, and size of the chloroplasts.[51] Effects on cytoplasmic ribosome and protein concentration are much less pronounced.[51] Electron microscopy of chloroplasts shows displacement and disorganization of the lamellar membranes, the appearance of vacuoles, osmiophilic granules, and other inclusions; and at a late stage, the chloroplast membrane and starch grains disappear.[109] One of the first signs of degeneration in the nucleus is the appearance of clear areas around the nucleolus, and possible clumping of chromatin. Chromatin disappears and the nucleolus loses its granular texture and becomes fibrous; the degenerate nucleus finally resembles an empty, membrane-bound vesicle.[109] Less severe effects on mitochondria have been observed.[109] Reduced protein synthesis in leaves accompanies symptom development, together with a general increase in the content of soluble amino acids, amines, and peptides.[106]

Little is known of the cellular events leading to the synthesis of intact LNYV particles. The structure or role of viroplasms is obscure. It would be expected from the evidence presented, and also by analogy with other viruses, that the nucleus[107] and polyribosomes[106] are implicated in the synthesis of virus-specific nucleic acids and proteins, respectively. However, questions on aspects such as the role of the virion-associated RNA transcriptase, sites and mode of synthesis of the virus message and genome, enzyme, structural protein, and glycoprotein, have yet to be answered.

With other rhabdoviruses, numerous papers report the localization of nucleocapsids and virions in infected cells, and attempt to deduce the sites of assembly and acquisition of the viral envelope upon maturation. The general conclusions that can be drawn are that particles normally appear as membrane-bound groups in several different cellular locations.[12] Membranes appear to be involved in morphogenesis, and are thought to represent the site of particle assembly. Plant rhabdoviruses can be divided into three groups depending on the mode and location of their assembly in the infected plant cell:

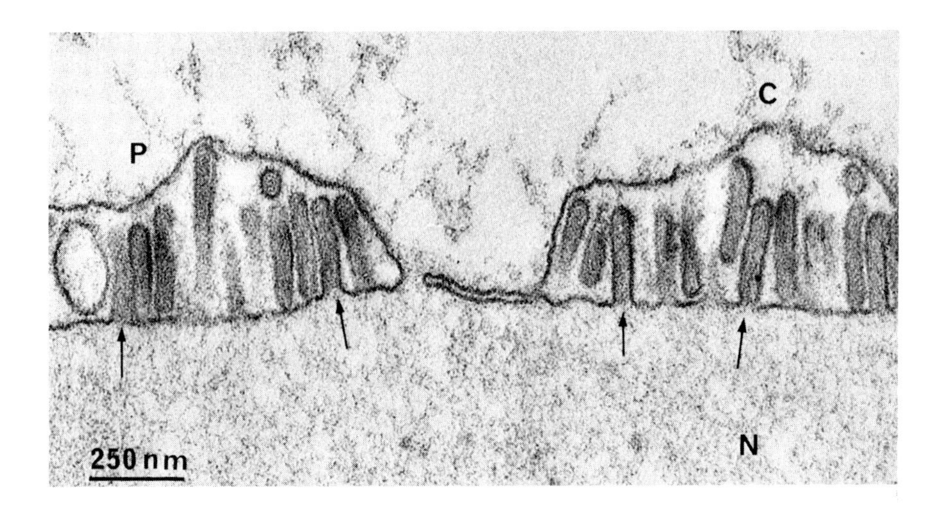

FIGURE 9. Assembly of EMDV in the perinuclear space of an infected plant cell showing the budding of the particles (arrowed) from the inner membrane of the nucleus (N) with the perinuclear space (P). No particles are seen in the cytoplasm (C). Micrographs supplied by G. P. Martelli.

1. Viruses whose maturation is associated with the endoplasmic reticulum as exemplified by LNYV which matures mainly in association with its membranes and which accumulates almost exclusively in vesicles formed from the endoplasmic reticulum. It would appear that BNYV[110] (Figure 6) and viruses infecting *Laelia*,[111] *Iris germanica*,[112] *Zea mays*,[113] and *Cucumis melo*[114] behave in a similar way.
2. Viruses whose maturation is associated with the inner nuclear membrane and which accumulate largely in the perinuclear spaces. Numerous viruses such as EMDV (Figure 9) appear to belong to this group.[12]
3. Viruses whose maturation is associated with viroplasms as exemplified by BYSMV[29] and NCMV.[28] Cells infected by these viruses develop extensive, membrane-bound viroplasms in the cytoplasm from which virus particles bud and accumulate in vacuole-like spaces (Figure 10).

IX. VIRUS-VECTOR RELATIONSHIPS

Plant rhabdoviruses have a specific association with one or a few vector species, they appear generally to be propagative in their vectors, and they can possibly be considered viruses of their aphid or leafhopper vectors as well as pathogens of plants.[4] Similar principles apply to the virus-vector relationships of aphid-borne and leaf hopper-borne rhabdoviruses, and where it is appropriate they are considered together in this section.

Some leafhopper-borne plant rhabdoviruses have been acquired by their vectors during feeds as brief as 1 min.[114] The aphid-borne SYVV was acquired by *H. lactucae* in 2 hr,[115] the shortest period tested, and the threshold could be well below this. The efficiency of acquisition increases with longer access times, but it appears to be independent of temperature.[114] Little attention has been paid to sites of virus acquisition from the plant. Boakye and Randles[116] have estimated the depth of penetration of the stylets of *H. lactucae* with time, and conclude that even after feeding for 1 hr, vascular tissue would not have been reached. Determination of acquisition threshold periods could help determine whether virus is acquired from epidermal, mesophyll, or vascular tissue.

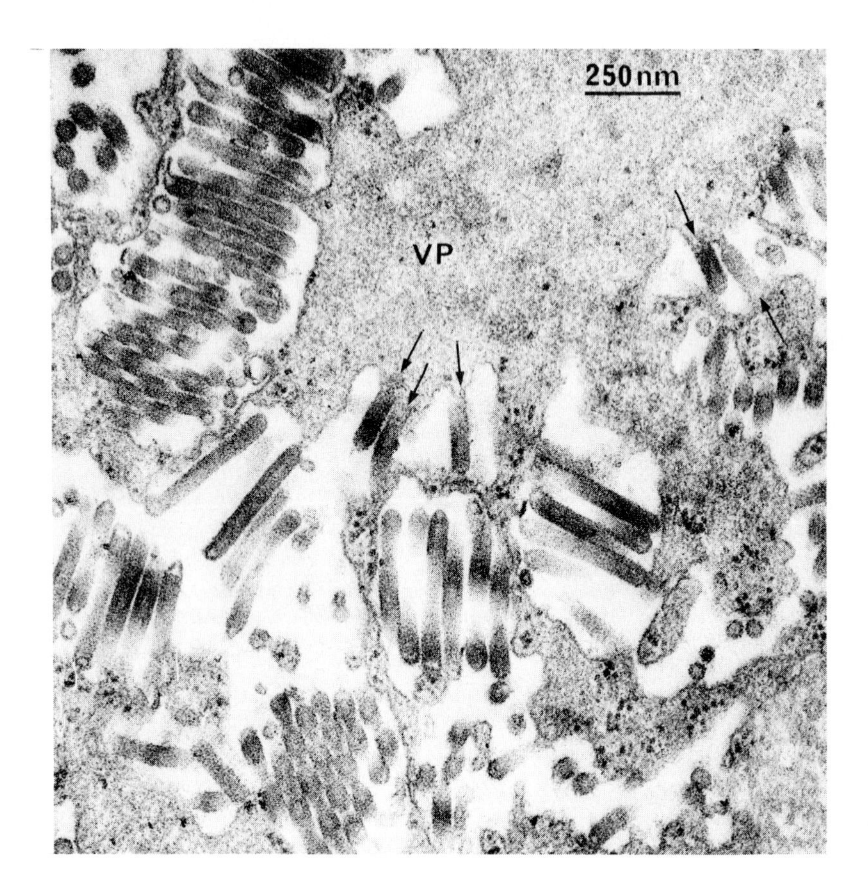

FIGURE 10. Virus-induced viroplasm (VP) in the cytoplasm of a plant cell infected with BYSMV showing particles budding (arrowed) from the membrane enclosing the viroplasm. Micrograph supplied by A. Appiano and M. Conti.

The viruses undergo a temperature dependent incubation period in the vector, the minima ranging from 4 days for WSMV[114] to 5 to 8 days for SYVV[115] and LNYV[116] when virus was acquired by feeding, or 4 days for SCV[117] when virus was injected into the hemocoele. The incubation time is longer at sub- or supra-optimal temperatures, or when the amount of virus acquired is very small.[118] Inoculativity is generally retained almost until the death of the insect. With SYVV and LNYV, approximately half of the viruliferous *H. lactucae* cease to transmit 1 to 2 days before they die.[115,116] Efficiency of transmission declines with age[117] but this is possibly related to a change in feeding behavior rather than to reduced virus concentration in the salivary glands, as virus can be recovered from insect extracts after they have ceased transmitting.[4,11]

Some rhabdoviruses have been shown to be propagative in their vectors, and it seems probable that all may be found to fall into this category. Direct evidence, based on serial transfer of inoculum by injection through a number of generations of insect, has been obtained for SYVV, SCV, PYDV, WSMV, rice transitory yellowing virus (RTYV), and NCMV.[4,11] Other viruses for example, LNYV, BNYV, MMV, SYNV, BYSMV, (Russian) winter wheat mosaic virus (WWMV), Digitaria striate mosaic virus, and cereal chlorotic mottle virus[4,11] are circulative, and virus particles have been seen in thin sections of vector tissues. This is strong circumstantial evidence that they are propagative. A low percentage of transovarial transmission has been demonstrated for SYVV[119] and LNYV[116] in the aphid *H. lactucae*. Of the leafhopper-borne viruses, only BYSMV and WWMV have been reported as infecting through the egg.[11]

Following acquisition of WSMV by *E. inimica,* there appears to be sequential infection of vector tissues. The virus can be detected in the gut after 2 days and in the hemolymph, hemocytes, and salivary glands after 4 days.[120] Virus concentration increased rapidly in the gut to a peak after 8 days, then declined slightly. WSMV was recovered from the fat body, brain, and mycetomes, but not Malpighian tubules, ovaries, or testes.[120] LNYV and SYVV particles have been detected in salivary glands, brain, muscle, fat body, mycetomes, ovaries, and esophagus but not in embryos.[121,122] The presence of virions in mycetomes suggests a mechanism of transovarial transission. Infection of embryos may occur by the transportation of virus particles with maternal mycetomes when they migrate into the developing egg.[119,123]

Particles of SYVV are seen in the nucleoplasm and cytoplasm, and Sylvester and Richardson[122] suggest that the nucleocapsids acquire their envelope while passing from the nucleoplasm into the perinuclear space. BNYV particles accumulate in random arrangements within nuclei,[110] and it is suggested that unlike the situation in the plant, where cytoplasmic vesicles are thought to be the site of virus maturation, the nuclei are the main sites of virus development in the aphid.

BYSMV particles are found in the cytoplasm but not the nuclei, mitochondria, or other organelles of salivary gland cells of *L. striatellus.*[124] They occur as membrane-bound aggregates associated with long tubular structures. In contrast, WSMV virions are seen in the nuclei of salivary gland cells of *E. inimica,*[125] but the convoluted nature of the nuclei in these cells complicates the interpretation of the micrographs. As with plants, time course studies are required to determine whether different intracellular sites are important at different stages of virus replication within the insect.

The effect of rhabdoviruses on their vectors indicate that some of them can be considered as insect as well as plant pathogens. Both LNYV and SYVV infection significantly shortens the life span of *H. lactucae,* and one isolate of SYVV also reduces the rate of larviposition and excretion.[116,126] It has been suggested that rhabdoviruses may be considered as "zoophytic"[126] or "bridging"[124] viruses between plant and invertebrate hosts. That the "bridge" may also be extended to vertebrate hosts is suggested by the demonstration that VSV of vertebrates multiplies in *P. maidis,* the vector of MMV.[127]

The requirement for propagative viruses to infect, systemically invade, and be secreted by the salivary glands of their insect vector, probably is the main determinant of their vector specificity. Behncken[128] has shown that *Macrosiphum euphorbiae* can be infected by SYVV, although infected aphids only rarely transmit the virus. It is suggested that in this case the inability to transmit, despite susceptibility of aphid tissue to infection, is due to the salivary glands not becoming infected.

Variation in the ability of different races of leafhoppers to transmit PYDV[129] illustrates the genetic control of the association between virus and vector. Further evidence for the specificity of the virus-vector relationships is provided by variation in the efficiency of transmission of LNYV by different clones of *H. lactucae.*[130] Characteristics of the virus may likewise determine transmissibility by a vector. The transmissibility of SYVV was lost in only seven serial passages by injection through a series of insects.[128]

X. EPIDEMIOLOGY

The study of outbreaks of virus diseases is important for developing control measures, and in the long term may allow accurate prediction of the time and severity of disease outbreaks. At present there is no single case where the epidemiology of a plant rhabdovirus is sufficiently understood to predict outbreaks reliably.[11]

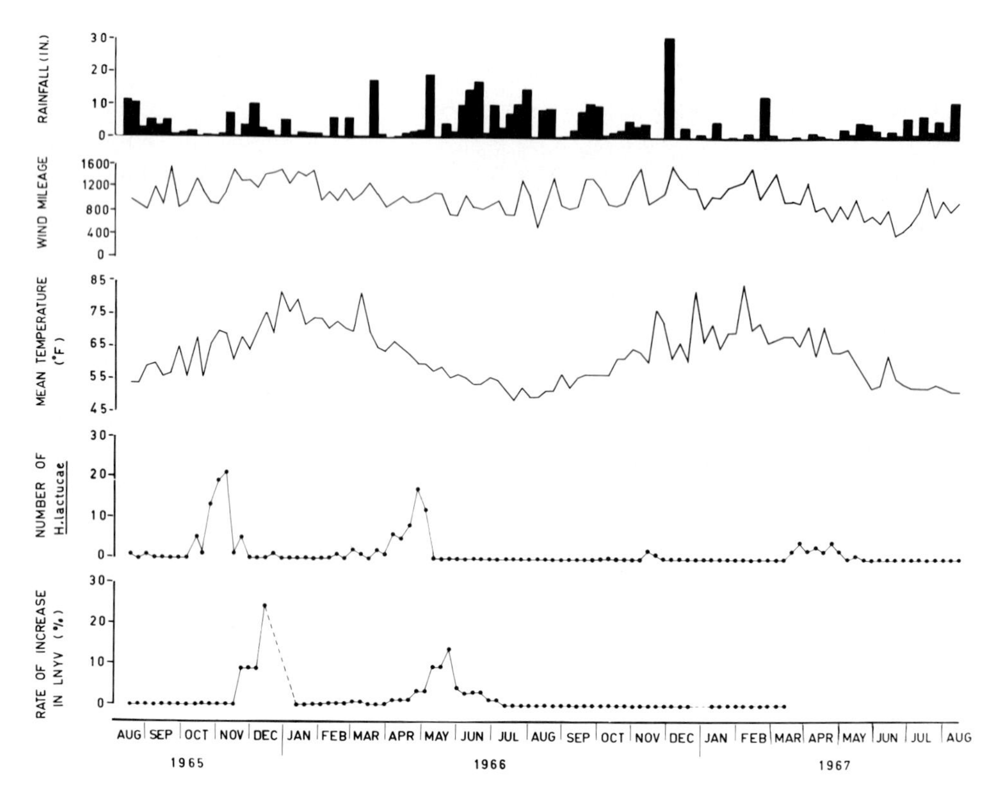

FIGURE 11. The relationship of LNYV outbreaks in lettuce crops with aphid vector flights and environmental conditions. (From Randles, J. W. and Crowley, N. C., *Aust. J. Agric. Res.*, 21, 447, 1970. With permission.)

Rhabdoviruses of plants have a relatively simple ecology because of their vector specificity and limited host range. A useful first approach to understanding disease epidemiology is a survey of disease incidence, vector activity, and climatic factors, like that presented in Figure 11 for LNYV.[131] Relationships between the activity of known and putative vectors, and disease incidence, can be shown. Climatic factors such as mean weekly temperature, extremes of wind, and onset of heavy rain can be implicated, and tested further. In the case of LNYV, correlation between vector flight activity and disease incidence indicates that most of the flying aphids are viruliferous. In the case of RTYV the leafhopper vectors are active throughout the year, and it is necessary to determine fluctuations in the percentage carrying virus to predict times of most active spread.[132]

The epidemiology of LNYV is probably better understood than that of any other rhabdovirus. The possibility that LNYV originated in Australia in one of the endemic perennial hosts, *Sonchus hydrophilus* Boulos or *Embergeria megalocarpa* (Hook) has been discussed by Randles and Carver.[133] However, *S. oleraceus* apears now to be the principal host of both the virus and *H. lactucae*, the important aphid vector.[50,130,133,134] The limited distribution of the other vector and host plant species indicates that their contribution to the spread of LNYV is probably minor. LNYV can therefore be considered as a virus of *S. oleraceus.* It is symptomless in this host, and its economic importance arises from the transmission of virus into lettuce crops by *H. lactucae* where symptoms and crop losses occur. Because *H. lactucae* does not colonize lettuce, the only source of virus for lettuce crops is *S. oleraceus.* Lettuce-to-lettuce spread does not occur.

Assuming that a virtually unlimited reservoir of LNYV and *H. lactucae* exists in *S. oleraceus*, the factors which determine the development and migration of *H. lactucae*, and its ability to feed on and infect lettuce, are clearly important in determining the incidence of disease in lettuce crops.

H. lactucae is probably anholocyclic in most parts of Australia, but sexual forms have been found at two sites. Boakye[130] has shown that the developmental zero for the aphid is 10°C, and the upper limit for development is about 28°C. The reproductive rate is influenced by population density, so that as population density increases, reproduction rate slows. As the population density increases, more alatae (winged forms) are produced, and more aphids leave the plant. Migrant alatae are principally responsible for the colonization of young plants, but apterae (wingless forms) may crawl short distances to new hosts. Field trials in South Australia[130] suggest that long range dispersal of alatae is important, but in Victoria, short-range dispersal was implicated in a field trial;[134] both modes may be important under different conditions.

LNYV is propagative in *H. lactucae*, and nymphs which acquire the virus from the host after birth complete the incubation period approximately 24 hr after reaching the fourth instar (preadult stage).[116,130] Aphids are therefore viruliferous when they migrate. Nymphs infected by transovarial transmission are able to transmit shortly after birth, but this is thought to have no significant effect on disease epidemiology. When a viruliferous aphid alights on a new *S. oleraceus* seedling, inoculation is achieved within the first 5 to 30 min of feeding. Under field conditions LNYV becomes available to feeding aphids approximately 12 days later,[130] and aphids borne after this would be expected to be viruliferous at maturity. Thus, there is a lag of 2 to 3 weeks between a viruliferous aphid inoculating a seedling, and the migration of its progeny as viruliferous adults from that plant.

The period over which a plant can act as a source of viruliferous aphids will depend on the period over which the plant is acceptable to aphids. The factors which determine the acceptability of plants to aphids are not well understood, but as *H. lactucae* feed and breed preferentially on inflorescences (Figure 12), the number of these would be expected to be an important determinant of the rate of reproduction of aphids on the plant.[135]

Of interest to epidemiologists is the observation that peaks in aphid populations occur at mean temperatures between 15 and 17°C, which is below those shown to be optimal for their reproduction in the laboratory (22 to 28°C). This occurs with several aphid species and Maelzer[135] suggests that it is due to the lower temperature (1) coinciding with the season of abundant food which results from rapid plant growth, (2) reducing the rate of senescence of food plants and allowing more time for aphid density to increase on a given plant, and (3) not favoring predator buildup. The second point is supported by the results of D. Martin (cited by Maelzer[135]) who counted the number of aphids produced by a single aphid on a single inflorescence, up to the time it became unacceptable as a food source. At 15°C, production of progeny was 2.5-fold that at 20°C, and 4-fold that at 25°C. Further studies on the ecology of *S. oleraceus* and the biology and behavior of *H. lactucae* are needed to understand the timing of migration, the feeding behavior of migrants, and the duration of their migratory activity.

The transfer of LNYV to lettuce probably occurs fortuitously during host-seeking migration. The fact that lettuce becomes infected at all is of some interest, because *H. lactucae* transferred directly from *S. oleraceus* to lettuce will walk off the plants without feeding. Boakye and Randles[116] have shown that starvation and desiccation of aphids in the light enhances the tendency for them to settle in a feeding position on lettuce, and probe for a period longer than the inoculation threshold which is between 1 and 5 min. During this time probing takes place with little or no imbibition of sap. It seems likely, therefore, that the starvation which occurs during migration lowers the

FIGURE 12. Inflorescence of *S. oleraceus* (sowthistle) the field reservoir plant for both LNYV and its vector *H. lactucae* showing heavy infestation by the aphids. Photograph supplied by G. T. O'Loughlin.

settling or probing threshold, so that even unpalatable plants are probed sufficiently long for them to be inoculated. Furthermore, if migrants alight in a lettuce crop where *S. oleraceus* is rare, it seems likely that probing and migratory behavior would become intensified, at least until continued starvation, or autolysis of flight muscle, prevented further migration.

Although the epidemiology of LNYV is still incompletely understood, the relative simplicity of the model compared with other plant virus diseases makes it a useful system for studying the host-virus-vector interaction. For any disease, the modeling of disease outbreaks is a complex undertaking. The components which require examination are

1 Virus host plants — the relative abundance of different host species, their suitability as vector host species
2. Vectors — number of species, parameters of transmission, population dynamics, migratory and feeding behavior
3. Climate — components of temperature, wind, rainfall, and photoperiod which influence reservoir and economic host plants and vector
4. Cultural practices — modifications to the ecosystem arising from farm operations

With most plant rhabdoviruses, the main effort so far has centered on virus-vector interactions. Vector population dynamics and feeding behavior need more study.

ACKNOWLEDGMENTS

We thank Doctors J. Dale, A. O. Jackson, P. E. Lee, G. Martelli, D. Peters, M. Rubio-Huertos, and P. Signoret and their associates for providing unpublished data and manuscripts prior to publication; Doctors A. Appiano, M. Conti, R. G. Garrett, R. Greber, T. Hatta, A. O. Jackson, D. Lesemann, and G. T. O'Loughlin for illustration material; and Mrs. L. Wichman for preparing the figures. The research on lettuce necrotic yellows virus in the authors' laboratory has been generously supported by the Australian Research Grants Committee.

REFERENCES

1. **Herold, F., Bergold, G. H., and Weibel, J.**, Isolation and electron microscopic demonstration of a virus infecting corn (*Zea mays* L.), *Virology*, 12, 335, 1960.
2. **Harrison, B. D. and Crowley, N. C.**, Properties and structure of lettuce necrotic yellows virus, *Virology*, 26, 297, 1965.
3. **Fenner, F.**, Classification and nomenclature of viruses (Second Report of the International Committee on Taxonomy of Viruses), *Intervirology*, 7, 1, 1976.
4. **Francki, R. I. B.**, Plant rhabdoviruses, *Adv. Virus Res.*, 18, 257, 1973.
5. **Howatson, A. F.**, Vesicular stomatitis and related viruses, *Adv. Virus Res.*, 16, 195, 1970.
6. **Hummeler, K.**, Bullet shaped viruses, in *Comparative Virology*, Maramorosch, K. and Kurstak, E., Eds., Academic Press, New York, 1971, 361.
7. **Knudson, D. L.**, Rhabdoviruses, *J. Gen. Virol.*, 20, 105, 1973.
8. **Russo, M. and Martelli, G. P.**, I rhabdovirus parassiti submicroscopici delle piante dalla inconsueta morphologia, *Ital. Agric.*, 111, 86, 1974.
9. **Wagner, R. R.**, Reproduction of rhabdoviruses, in *Comprehensive Virology*, Vol. 4, Fraenkel-Conrat, H. and Wagner, R. R., Eds., Plenum Press, New York, 1975, 1.
10. **Martelli, G. P. and Russo, M.**, Plant virus inclusion bodies, *Adv. Virus Res.*, 21, 175, 1977.
11. **Jackson, A. O., Milbrath, G. M., and Jedlinski, H.**, Rhabdovirus diseases of the Gramineae, in press.
12. **Martelli, G. P. and Russo, M.**, Rhabdoviruses of plants, in *Insect and Plant Viruses: an Atlas*, Maramorosch, K., Ed., Academic Press, New York, 1977, 181.
13. **Francki, R. I. B. and Randles, J. W.**, Composition of the plant rhabdovirus lettuce necrotic yellows virus in relation to its biological properties, in *Negative Strand Viruses*, Vol. 1, Mahy, B. W. J. and Barry, R. D., Eds., Academic Press, London, 1975, 223.
14. **Greber, R. S.**, Rhabdovirus in Queensland grasses and cereals, Aust. Plant Pathol. Soc. Second National Conference, Abstract, 228, 1976.
15. **Greber, R. S.**, Cereal chlorotic mottle virus (CCMV), a rhabdovirus of *Gramineae* transmitted by the leafhopper *Nesoclutha pallida*, *Aust. Plant Pathol. Soc. Newsletter*, 6, 17, 1977.
16. **Herold, F.**, Maize mosaic virus, *C.M.I./A.A.B. Descriptions of Plant Viruses*, No. 94, 1972.
17. **Martelli, G. P., Russo, M., and Malaguti, G.**, Ultrastructural aspects of maize mosaic virus in host cells, *Phytopathol. Mediterr.*, 14, 140, 1975.
18. **Francki, R. I. B. and Randles, J. W.**, Lettuce necrotic yellows virus, *C.M.I./A.A.B. Descriptions of Plant Viruses*, No. 26, 1970.
19. **Campbell, R. N. and Lin, M. T.**, Broccoli necrotic yellows virus, *C.M.I./A.A.B. Descriptions of Plant Viruses*, No. 85, 1972.
20. **Jackson, A. O. and Christie, R. G.**, Purification and some physicochemical properties of *Sonchus* yellow net virus, *Virology*, 77, 344, 1977.
21. **Peters, D.**, Sow thistle yellow vein virus, *C.M.I./A.A.B. Descriptions of Plant Viruses*, No. 62, 1971.
22. **Sylvester, E. S., Frazier, N. W., and Richardson, J.**, Strawberry crinkle virus, *C.M.I./A.A.B. Descriptions of Plant Viruses*, No. 163, 1976.
23. **Jones, A. T., Murant, A. F., and Stace-Smith, R.**, Raspberry vein chlorosis virus, *C.M.I./A.A.B. Description of Plant Viruses*, No. 174, 1977.
24. **Leclant, F., Alliot, B., and Signoret, P. A.**, Transmission et épidémiologie de la maladie a énations de la luzerne (LEV), premiers résultats, *Ann. Phytopathol.*, 5, 441, 1973.

25. **Black, L. M.**, Potato yellow dwarf virus, *C.M.I./A.A.B. Descriptions of Plant Viruses,* No. 35, 1970.
26. **Sinha, R. C. and Behki, R. M.**, American wheat striate mosaic virus, *C.M.I./A.A.B. Descriptions of Plant Viruses,* No. 99, 1972.
27. **Shikata, E.**, Rice transitory yellowing virus, *C.M.I./A.A.B. Descriptions of Plant Viruses,* No. 100, 1972.
28. **Toriyama, S.**, Electron microscopy of developmental stages of northern cereal mosaic virus in wheat plant cells, *Ann. Phytopathol., Soc. Japan,* 42, 563, 1976.
29. **Conti, M. and Appiano, A.**, Barley yellow striate mosaic virus associated viroplasms in barley cells, *J. Gen. Virol.,* 21, 315, 1973.
30. **Razvjaskina, G. M. and Poljakova, G. P.**, Electron microscopic study of the winter wheat mosaic virus, *Plant Virology Proc. 6th Conf. Czek. Plant Virol.,* Publ. House Czech. Acad. Sci. Prague, 1967, 129.
31. **Jedlinski, H.**, Oat striate, a new virus disease in Illinois spread by the leafhopper, *Graminella nigrifrons* (Forbes.) *Proc. Am. Phytopathol. Soc.,* 3, 19, 1976.
32. **Signoret, P. A., Conti, M., Leclant, F., Alliot, B., and Giannotti, J.**, Données nouvelles sur la maladie des stries chlorotiques du Blé ("Wheat chlorotic streak mosaic virus" = "WCSMV"), *Ann. Phytopathol.,* 9, 3, 1978.
33. **Gollifer, D. E., Jackson, G. V. H., Dabek, A. J., Plumb, R. T., and May, Y. Y.**, The occurrence and transmission of viruses of edible aroids in the Solomon Islands and the South West Pacific, *Pest Articles and News Summaries,* 23, 171, 1977.
34. **Martelli, G. P. and Russo, M.**, Eggplant mottled dwarf virus, *C.M.I./A.A.B. Descriptions of Plant Viruses,* No. 115, 1973.
35. **Kitajima, E. W. and Costa, A. S.**, Morphology and developmental stages of *Gomphrena* virus, *Virology,* 29, 523, 1966.
36. **Vega, J., Gracia, O., Rubio-Huertos, M., and Feldman, J. M.**, Transmission of a bacilliform virus of sow thistle: Mitochondria modifications in the infected cells, *Phytopathol. Z.,* 85, 7 1976.
37. **Peña-Iglesias, A., Rubio-Huertos, M., and Moreno San Martin, R.**, Un virus de tipo baciliforme en alcachofa (*Cynara scolymus* L.), *An. Inst. Nac. Invest. Agrar. (Spain) Ser. Prot. Veg.* 2, 123, 1972.
38. **Doi, Y., Chang, M. U., and Yora, K.**, Orchid fleck virus, *C.M.I./A.A.B. Descriptions of Plant Viruses,* No. 183, 1977.
39. **Stace-Smith, R. and Leung, E.**, Small bacilliform particles associated with *Rubus* yellow net virus, *Proc. Am. Phytopathol. Soc.,* 3, 320, 1976.
40. **Kitajima, E. W. and Costa, A. S.**, Particulas bacilliformes associadas à mancha anular do cafeeiro, *Cienc. Cult. (Sao Paulo),* 24, 542, 1972.
41. **Kitajima, E. W., Müller, G. W., Costa, A. S., and Yuki, W.**, Short, rod-like particles associated with *Citrus* leprosis, *Virology,* 50, 254, 1972.
42. **Jackson, A. O.**, Partial characterization of the structural proteins of Sonchus yellow net virus, *Virology,* 87, 172, 1978.
43. **Putz, C. and Meignoz, R.**, Electron microscopy of virus-like particles found in mosaic-diseased raspberries in France, *Phytopathology,* 62, 1477, 1972.
44. **Stace-Smith, R. and Lo, E.**, Morphology of bacilliform particles associated with raspberry vein chlorosis virus, *Can. J. Bot.,* 51, 1343, 1973.
45. **Lesemann, D.**, Nachweis eines bazilliformen virus in *Atropa belladonna, Phytopathol. Z.,* 73, 83, 1972.
46. **Maramorosch, K., Hirumi, H., Kimura, M., Bird, J., and Vakili, N. G.**, Diseases of pigeon pea in the Caribbean area: an electron microscopy study, *FAO Plant Prot. Bull.,* 22, 32, 1974.
47. **Black, L. M.**, Properties of the potato yellow-dwarf virus, *Phytopathology,* 28, 863, 1938.
48. **Crowley, N. C.**, Factors affecting the local lesion response of *Nicotiana glutinosa* to lettuce necrotic yellows virus, *Virology,* 31, 107, 1967.
49. **Black, L. M.**, Strains of potato yellow-dwarf virus, *Am. J. Bot.,* 27, 386, 1940.
50. **Stubbs, L. L. and Grogan, R. G.**, Necrotic yellows: a newly recognized virus disease of lettuce, *Aust. J. Agric. Res.,* 14, 439, 1963.
51. **Randles, J. W. and Coleman, D. F.**, Loss of ribosomes in *Nicotiana glutinosa* L. infected with lettuce necrotic yellows virus, *Virology,* 41, 459, 1970.
52. **Hsu, H. T. and Black, L. M.**, Inoculation of vector cell monolayers with potato yellow dwarf virus, *Virology,* 52, 187, 1973.
53. **Lee, P. and Peters, D.**, Electron microscopy of sowthistle yellow vein virus in cells of sowthistle plants, *Virology,* 48, 739, 1972.
54. **Lee, P. E.**, Developmental stages of wheat striate mosaic virus, *J. Ultrastruct. Res.,* 31, 282, 1970.
55. **Maramorosch, K.**, Semiautomatic equipment for injecting insects with measured amounts of liquids containing viruses or toxic substances, *Phytopathology,* 46, 188, 1956.

56. **Rochow, W. F,** Transmission of barley yellow dwarf virus acquired from liquid extracts by aphids feeding through membranes, *Virology,* 12, 223, 1960.
57. **Brakke, M. K.,** Systemic infections for the assay of plant viruses, *Annu. Rev. Phytopathol.,* 8, 61, 1970.
58. **Reddy, D. V. R.,** Techniques of invertebrate tissue culture for the study of plant viruses, in *Methods of Virology,* Vol. 6, Maramorosch, K. and Koprowski, H., Eds., Academic Press, New York, 1977, 393.
59. **Peters, D. and Black, L. M.,** Infection of primary cultures of aphid cells with a plant virus, *Virology,* 40, 847, 1970.
60. **Matisova, J. and Valenta, V.,** Aphid cell cultures, in *Aphids as Virus Vectors,* Harris, K. F. and Maramorosch, K., Eds., Academic Press, New York, 1977, 339.
61. **Hsu, H. T. and Black, L. M.,** Multiplication of potato yellow dwarf virus on vector cell monolayers, *Virology,* 59, 331, 1974.
62. **Hsu, H. T. and Black, L. M.,** Comparative efficiencies of assays of a plant virus by lesions on leaves and on vector cell monolayers, *Virology,* 52, 284, 1973.
63. **Francki, R. I. B.,** Purification of viruses, in *Principles and Techniques in Plant Virology,* Kado, C. I. and Agrawal, H. O., Eds., Van Nostrand Reinhold, New York, 1972, 295.
64. **Mossop, D. W., Francki, R. I. B., and Grivell, C. J,** Comparative studies on tomato aspermy and cucumber mosaic viruses. V. Purification and properties of a cucumber mosaic virus inducing severe chlorosis, *Virology,* 74, 544, 1976.
65. **Hsu, H. T. and Black, L. M.,** Polyethylene glycol for purification of potato yellow dwarf virus, *Phytopathology,* 63, 692, 1973.
66. **Lin, M. T. and Campbell, R. N.,** Characterization of broccoli necrotic yellows virus, *Virology,* 48, 30, 1972.
67. **McLean, G. D. and Francki, R. I. B.,** Purification of lettuce necrotic yellows virus by column chromatography on calcium-phosphate gel, *Virology,* 31, 585, 1967.
68. **Ovenstein, J., Johnson, L, Shelton, E., and Lazzarini, R. A.,** The shape of vesicular stomatitis virus, *Virology,* 71, 291, 1976.
69. **MacLeod, R.,** An interpretation of the observed polymorphism of potato yellow dwarf virus, *Virology,* 34, 771, 1968.
70. **Wolanski, B. S., Francki, R. I. B., and Chambers, T. C.,** Structure of lettuce necrotic yellows virus. I. Electron microscopy of negatively stained preparations, *Virology,* 33, 287, 1967.
71. **Wolanski, B. S. and Francki,R. I. B,** Structure of lettuce necrotic yellows virus. II. Electron microscopic studies on the effect of pH of phosphotungstic acid stain on the morphology of the virus, *Virology,* 37, 437, 1969.
72. **Wagner, R. R., Prevec, L., Brown, F., Summers, D. F., Sokol, F., and MacLeod, R.,** Classification of rhabdovirus proteins: a proposal, *J. Virol.,* 10, 1228, 1972.
73. **Chambers, T. C., Crowley, N. C., and Francki, R. I. B.,** Localization of lettuce necrotic yellows virus in host leaf tissue, *Virology,* 27, 320, 1965.
74. **Randles, J. W. and Francki, R. I. B.,** Infectious nucleocapsid particles of lettuce necrotic yellows virus with RNA-dependent RNA polymerase activity, *Virology,* 50, 297, 1972.
75. **Francki, R. I. B. and Randles, J. W.,** Some properties of lettuce necrotic yellows virus RNA and its in vitro transcription by virion-associated transcriptase, *Virology,* 54, 359, 1973.
76. **Reeder, G. S., Knudson, D. L., and MacLeod, R.,** The ribonucleic acid of potato yellow dwarf virus, *Virology,* 50, 301, 1972.
77. **Sinha, R. C., Sehgal, O. P., and Thottappilly, G.,** Effect of temperature on infectivity and some physico-chemical properties of purified wheat striate mosaic virus, *Phytopathol. Z.,* 84, 300, 1975.
78. **Sinha, R. C., Harwalkar, V. R., and Behki, R. M.,** Chemical composition and some properties of wheat striate mosaic virus, *Phytopathol. Z.,* 87, 314, 1976.
79. **Knudson, D. L. and MacLeod, R.,** The proteins of potato yellow dwarf virus, *Virology,* 47, 285, 1972.
80. **Dale, J. and Peters, D.,** private communication.
81. **Ziemiecki, A. and Peters, D.,** The proteins of sowthistle yellow vein virus: characterization and location, *J. Gen. Virol.,* 32, 369, 1976.
82. **Trefzger-Stevens, J. and Lee, P. E.,** The structural proteins of wheat striate mosaic virus, plant rhabdovirus, *Virology,* 78, 144, 1977.
83. **Ahmed, M. E., Black, L. M., Perkins, E. G., Walker, B. L., and Kummerow, F. A.,** Lipid in potato yellow dwarf virus, *Biochem. Biophys. Res. Commun.,* 17, 103, 1964.
84. **Toriyama, S.,** Sterol composition of northern cereal mosaic virus, *Ann. Phytopathol. Soc. Japan,* 42, 494, 1976.
85. **Huang, A. S., Baltimore, D., and Stampfer, M.,** Ribonucleic acid synthesis of vesicular stomatitis virus. III. Multiple complementary messenger RNA molecules, *Virology,* 42, 946, 1970.

86. **Baltimore, D., Huang, A. S., and Stampfer, M.,** Ribonucleic acid synthesis of vesicular stomatitis virus. II. An RNA polymerase in the virion, *Proc. Natl. Acad. Sci. U.S.A.*, 66, 572, 1970.
87. **Bishop, D. H. L. and Flamand, A.,** Transcription processes of animal RNA viruses, in *Control Processes in Virus Multiplication,* Burke, D. C. and Russell, W. C., Eds., Cambridge University Press, London, 1975, 95.
88. **Francki, R. I. B. and Randles, J. W.,** RNA-dependent RNA polymerase associated with particles of lettuce necrotic yellows virus, *Virology,* 47, 270, 1972.
89. **MacLeod, R., Peters, D., and Jackson, A. O.,** private communications.
90. **Duda, C. T.,** Plant RNA polymerases, *Annu. Rev. Plant Physiol.*, 27, 119, 1976.
91. **Ziemiecki, A. and Peters, D.,** Selective proteolytic activity associated with purified sowthistle yellow vein virus preparations, *J. Gen Virol.*, 31, 451, 1976.
92. **Francki, R. I. B. and Peters, D.,** The interference of cytoplasmic membrane-bound material from plant cells with the detection of a plant rhabdovirus transcriptase, *J. Gen. Virol.*, 41, 467, 1978.
93. **Peters, D., McLean, G. D. and Francki, R. I. B.,** unpublished data.
94. **Wolcyrz, S. and Black, L. M.,** Serology of potato yellow-dwarf virus, *Phytopathology,* 46, 32, 1956.
95. **Wolcyrz, S. and Black, L. M.,** Origins of vectorless strains of potato yellow dwarf virus, *Phytopathology,* 47, 38, 1957.
96. **Chiu, R-J., Liu, H-Y, MacLeod, R., and Black, L. M.,** Potato yellow dwarf virus in leafhopper cell culture, *Virology,* 40, 387, 1970.
97. **McLean, G. D., Wolanski, B. S., and Francki, R. I. B.,** Serological analysis of lettuce necrotic yellows virus preparations by immunodiffusion, *Virology,* 43, 480, 1971.
98. **Sinha, R. C.,** Serological detection of wheat striate mosaic virus in extracts of wheat plants and vector leafhoppers, *Phytopathology,* 58, 452, 1968.
99. **Thottappilly, G. and Sinha, R. C.,** Serological analysis of wheat striate mosaic virus and its soluble antigen, *Virology,* 53, 312, 1973.
100. **Thottappilly, G. and Sinha, R. C.,** Serological analysis of antigens related to wheat striate mosaic virus in *Endria inimica, Acta Virol. Engl. Ed.*, 18, 358, 1974.
101. **Hackett, A. J., Sylvester, E. S., Richardson, J., and Wood, P.,** Comparative electron micrographs of sowthistle yellow vein and vesicular stomatitis viruses, *Virology,* 36, 693, 1968.
102. **Toriyama, S.,** Purification and some properties of northern cereal mosaic virus, *Virus,* 23, 114, 1972.
103. **Lastra, R. J.,** Maize mosaic and other maize virus and virus-like diseases in Venezuela, in *Proc. Int. Maize Virus Disease Colloquium and Workshop,* Williams, L. E., Gordon, D. T., and Nault, L. R., Eds., Ohio Agr. Res. Devel. Center, 1976, 30.
104. **Randles, J. W.,** unpublished data.
105. **McLean, G. D. and Crowley, N. C.,** Inactivation of lettuce necrotic yellows virus by ultraviolet irradiation, *Virology,* 37, 209, 1969.
106. **Randles, J. W. and Coleman, D. F.,** Changes in polysomes in *Nicotiana glutinosa* L. leaves infected with lettuce necrotic yellows virus, *Physiol. Plant Pathol.*, 2, 247, 1972.
107. **Wolanski, B. S. and Chambers, T. C.,** The multiplication of lettuce necrotic yellows virus, *Virology,* 44, 582, 1971.
108. **McLean, G. D.,** *Studies on Lettuce Necrotic Yellows Virus,* Ph.D. thesis, University of Adelaide Library, 1969.
109. **Wolanski, B. S.,** Electron Microscopy of Lettuce Necrotic Yellows Virus, Ph.D. thesis, University of Melbourne Library.
110. **Garrett, R. G. and O'Loughlin, G. T.,** Broccoli necrotic yellows virus in cauliflower and in the aphid *Brevicoryne brassicae, Virology,* 76, 653, 1977.
111. **Peters, K-R.,** Orchid viruses: a new rhabdovirus in *Laelia* red leafspots, *J. Ultrastruct. Res.*, 58, 166, 1977.
112. **Rubio-Huertos, M.,** A rhabdovirus in *Iris germanica, Phytopathol. Z.*, 92, 294, 1978.
113. **Rubio-Huertos, M.,** A rhabdovirus in *Zea mays L.* in Spain, *Phytopathol. Z.*, 93, 1, 1978.
114. **Slykhuis, J. T. and Sherwood, P. L.,** Temperature in relation to the transmission and pathogenicity of wheat striate mosaic virus, *Can. J. Bot.*, 42, 1123, 1964.
115. **Duffus, J. E.,** Possible multiplication in the aphid vector of sowthistle yellow vein virus, a virus with an extremely long insect latent period, *Virology,* 21, 194, 1963.
116. **Boakye, D. and Randles, J. W.,** Epidemiology of lettuce necrotic yellows virus in South Australia. III. Virus transmission parameters, and vector feeding behaviour on host and non-host plants, *Aust. J. Agric. Res.*, 25, 791, 1974.
117. **Sylvester, E. S., Richardson, J., and Frazier, N. W.,** Serial passage of strawberry crinkle virus in the aphid *Chaetosiphon jacobi, Virology,* 59, 301, 1974.
118. **Sylvester, E. S., Richardson, J., and Behncken, G. M.,** Effect of dosage on the incubation period of sowthistle yellow vein virus in the aphid *Hyperomyzus lactucae, Virology,* 40, 590, 1970.

119. **Sylvester, E. S.**, Evidence of transovarial passage of the sowthistle yellow vein virus in the aphid *Hyperomyzus lactucae*, *Virology*, 38, 440, 1969.
120. **Sinha, R. C. and Chiykowski, L. N.**, Synthesis, distribution and some multiplication sites of wheat striate mosaic virus in a leafhopper vector, *Virology*, 38, 679, 1969.
121. **O'Loughlin, G. T. and Chambers, T. C.**, The systemic infection of an aphid by a plant virus, *Virology*, 33, 262, 1967.
122. **Sylvester, E. S and Richardson, J.**, Infection of *Hyperomyzus lactucae* by sowthistle yellow vein virus, *Virology*, 42, 1023, 1970.
123. **Nasu, S.**, Electron microscopic studies on transovarial passage of rice dwarf virus, *Jpn. J. Appl. Entomol. Zool.*, 9, 225, 1965.
124. **Conti, M. and Plumb, R. T.**, Barley yellow striate mosaic virus in the salivary glands of its planthopper vector *Laodelphax striatellus* Fallen, *J. Gen. Virol.*, 34, 107, 1977.
125. **Bell, C. D., Omar, S. A., and Lee, P. E.**, Electron microscopic localization of wheat striate mosaic virus in its leafhopper vector, *Endria inimica*, *Virology*, 86, 1, 1978.
126. **Sylvester, E. S.**, Reduction of excretion, reproduction and survival in *Hyperomyzus lactucae* fed on plants infected with isolates of sowthistle yellow vein virus, *Virology*, 56, 632, 1973.
127. **Lastra, J. R. and Esparza, J.**, Multiplication of vesicular stomatitis virus in the leafhopper *Peregrinus maidis* (Ashm), a vector of a plant rhabdovirus, *J. Gen. Virol.*, 32, 139, 1976.
128. **Bechncken, G. M.**, Evidence of multiplicaion of sowthistle yellow vein virus in an inefficient aphid vector, *Macrosiphum euphorbiae*, *Virology*, 53, 405, 1973.
129. **Nagaraj, A. N. and Black, L. M.**, Hereditary variation in the ability of a leafhopper to transmit two unrelated plant viruses, *Virology*, 16, 152, 1962.
130. **Boakye, D. B.**, Transmission of Lettuce Necrotic Yellows Virus by *Hyperomyzus lactucae* (L.) (Homoptera: Aphididae): With Special Reference to Aphid Behaviour, Ph.D. thesis, University of Adelaide.
131. **Randles, J. W. and Crowley, N. C.**, Epidemiology of lettuce necrotic yellows virus in South Australia. I. Relationship between disease incidence and activity of *Hyperomyzus lactuae* (L.), *Aust. J. Agric. Res.*, 21, 447, 1970.
132. **Hsieh, S.**, An analysis of field collections of the rice green leafhoppers for infection with transitory yellowing virus, *Plant Prot. Bull. (Taiwan)*, 11, 171, 1969.
133. **Randles, J. W. and Carver, M.**, Epidemiology of lettuce necrotic yellows virus in South Australia. II. Distribution of virus, host plants and vectors, *Aust. J. Agric. Res.*, 22, 231 1977.
134. **Stubbs, L. L., Guy, A. D., and Stubbs, K. J.**, Control of lettuce necrotic yellows virus disease by the destruction of common sowthistle *(Sonchus oleraceus)*, *Aust. J. Exp. Agric. Anim. Husb.*, 3, 215, 1963.
135. **Maelzer, D.**, The ecology of aphid pests in Australia, in *Ecology of Insect Pests in Australia*, Kitching, R. C. and Jones, R. E., Eds., in preparation.

Chapter 8

BOVINE EPHEMERAL FEVER VIRUS

A. J. Della-Porta and W. A. Snowdon

TABLE OF CONTENTS

I. INTRODUCTION

Bovine ephemeral fever (BEF) is a febrile disease of cattle characterized by high morbidity in herds affected, and by low mortality in those animals infected with the disease. The virus associated with the disease is more often cone shaped rather than bullet shaped, and it has only been provisionally classified as a rhabdovirus.[1] Further, an unconfirmed report that the virion contained double-stranded 12S ribonucleic acid as its genetic material suggested that bovine ephemeral fever virus (BEFV) should not be classified as a rhabdovirus.[2]

In this chapter we shall review the natural history, epidemiology, and what is known of the disease producd by BEFV. Further, we shall draw on new findings on the physicochemical structure of this virus which suggest that it is indeed a rhabdovirus.

II. EPIZOOTIOLOGY

BEF is widespread throughout Asia, Africa, Australia, and the Middle East (see review of Burgess).[3] "Three-day-sickness", "stiffsickness", or ephemeral fever has been described in South Africa,[4] Rhodesia,[5] Sudan,[6] Nigeria,[7] Kenya,[8] Iran,[9] Japan,[10] Philippines,[11] India,[12] Indonesia,[13,14] and in Australia.[15-17] However, it is possible that the clinical picture of the disease has been confused with other diseases, such as Ibaraki disease in Japan and Kotonkan virus infection of cattle in Africa.[10,18]

The disease appears to be widespread throughout Africa.[3] BEFV has been isolated in South Africa,[19] Nigeria,[7] and Kenya.[20] In Kenya there were major outbreaks in 1968 and 1972, and a wide distribution of antibody-positive animals over all regions of the country.[8] Serological evidence indicates that BEFV infection occurred between the two epizootics, probably in a subclinical form. In Nigeria, limited studies indicate that the peak of BEF infection occurs at the beginning of the wet season,[21] between May and June. Also in Nigeria, the rabies serogroup virus Kotonkan produces a similar febrile disease in cattle, but it can be easily distinguished on serology[18] and is often found to infect the same cattle as BEFV and at a similar, although not always the same, time of year[21]

In the Middle East, the disease is believed to be widespread,[3] but positive identification has only been confirmed in Iran where in 1974 the disease was seen in the south and eastern regions of the country, and BEFV was isolated from blood samples collected at Meshed.[9]

The distribution of BEF in Japan has been described in a review by Inaba.[10] A major outbeak occurred in 1949 to 1950, and then only sporadic reports of the disease were made until 1956 to 1958. In particular, there were reports of repeated small outbreaks of BEF preceding major epizootics. In 1968, the disease was confirmed by virus isolation.[22,23] Inaba[10] points out that some of the reports of clinical disease may have been confused with Ibaraki disease, until serological confirmation became available in 1968. In Japan, the disease was originally known as bovine epizootic fever until it was shown that the Japanese isolate was serologically indistinguishable from South African and Australian isolates of BEFV.[24,25] The disease has only been seen in central and western Japan, with the main incidence from late summer till late autumn.

It would appear that BEFV extends throughout Southeast Asia, from Japan in the north to Australia in the south. Major epizootics have been reported in Australia for 1936 to 1937,[26,27] 1955 to 1956,[28-30] 1967 to 1968,[31-34] 1970 to 1971,[17] and a minor outbreak in 1972 to 1973.[17,35] Seddon[26] concluded that BEFV was probably insect-transmitted; the southward extension from northern Australia to the Kimberley Ranges in the west and northern Victoria in the east, possibly being associated with the prevailing winds distributing the virus-infected insects.

The next outbreak studied in detail was the 1967 to 1968 epizootic. Similar in extent to the 1936 to 1937 epizootic, the disease moved very much more rapidly; in six weeks it spread a distance of 1200 miles along an eventual front of 500 miles.[33] Analysis of the spread of the disease and the prevailing wind patterns lead to the same conclusion as Seddon[26] that virus-infected insects, carried by the prevailing wind, transmitted the disease.[32,33] The virus was isolated a number of times from cattle bloods collected during the epizootic.[29,36,37] It is now known that BEF is found in Australia between the major epizootics,[17,30,32] and that separate enzootic focuses are sometimes found, such as in the Hunter Valley area in New South Wales.[17]

It would appear that BEF is seen more frequently in Queensland and New South Wales than previously, especially as small outbreaks. However this may be due to the greater use of serological testing, more veterinary practitioner awareness of the disease, and the use of sentinel herds of cattle to monitor its spread.[17]

Some common features are now emerging out of studies of the epizootiology of bovine ephemeral fever. First, it is now apparent that the disease is present between major epizootics and does not require a fresh introduction of the disease. It may be that between major epizootics there are less virulent strains by BEFV in circulation,[8,30] these strains producing weaker neutralizing antibody responses, as suggested by Snowdon.[30] Secondly, it would appear that the disease is almost certainly spread by insects, and the prevailing wind patterns may play a part in the dissemination of the disease.

III. NATURAL HISTORY

A. Host Range

Cattle are generally accepted as being the principal hosts for BEFV although other hosts have been suspected for some time.[3,5,10,16,17,28] In particular, sheep have been suspected as hosts,[5] but until recently neither virus multiplication nor antibody response had been demonstrated.[29] Recently, Hall et al.[38] were able to demonstrate that BEFV could replicate in merino sheep and produce an antibody response. No antibodies have been detected in sera collected from sheep in the field,[39,40] even in the midst of an outbreak of BEF in cattle.[41] Buffalo may also become infected with BEFV; serum neutralizing antibodies to the virus have been detected in 54% of buffaloes tested in Kenya and 16% of buffaloes tested from northern Austalia,[8,11,17] with there being a 60% positive rate for bulls over 4 years.

Studies in Kenya[8] have further extended the host range to include a number of game animals: waterbuck, wildebeest, and hartebeest. There is no evidence for infection of impala, Grant's gazelle, Thompson's gazelle, eland, and oryx, although very few sera were tested from elands and oryxes. Pigs and rabbits failed to give both clinical signs and antibody responses when infeced with virulent BEFV.[29,42] From serological surveys in Australia[16,39,40] the following species do not appear to be hosts for BEFV: horse, kangaroo, wallaby, rat, bandicoot, domestic fowl, dog and man.

B. The Disease in Cattle

1. Clinical Picture

In cattle, the disease is characterized by a fever (41 to 42°C) which usually lasts for 24 hr but may extend to 2, or more rarely, up to 4 days.[10,27,29,43] The temperature may show a biphasic response within a 2-day period.[27,44,45] The temperature rise is usually accompanied by lameness or stiffness which may last up to a week but is usually present for 1 to 2 days.[5,10,12,32,43,46] The respiration shows considerable variation, but is often slightly increased around the time of fever.[10,27,32,43] A nasal discharge and increased salivation is usually seen.[10,43,47] Although the morbidity may be high in an infected herd, mortality is usually less than 0.5% with the heavier animals being most seriously affected.[10,16,32] A few animals may abort.[16,32]

Major loss is probably associated with the reduction in milk production in BEFV-infected cows. During the 1967 to 1968 epizootic in northern Queensland, the milk production in one herd declined by 30 to 70% during the height of the disease and only recovered to 85% of its previous production levels 2 to 3 weeks after the disease.[34] In Victoria, during the same epizootic, cows affected in late lactation often failed to come back into production.[32] A more controlled study by Theodoridis et al.[48] revealed that cows affected during early lactation dropped their daily milk production by 44 ± 13% at the height of fever, those at middle lactation by 73 ± 27%, and those at late lactation by 59 ± 25%. They estimated that total losses are greater than this because cows at different stages of lactation, particularly those at late stages of lactation, had different abilities to recover their milk production.

Studies on virus excretion in the semen of bulls, attempts to infect cows by inseminating with virus-contaminated semen, and studies of congenital infections have produced mainly negative results. Burgess[49] was unable to infect cattle by inoculation of virus into the cervix. Tzipori and Spradbrow[50] found no evidence of fetal infection as a result of viremia, and indeed produced little evidence of fetal infection after direct inoculation. Burgess and Chenoweith[51] showed a sperm abnormality following BEFV infection, but this abnormality was probably not related to virus infection.[52] Parsonson and Snowdon[52] showed that excretion of virus in bull semen can occur, but as a rare event, following experimental infection, and could show no evidence of infection of heifers or cows following the introduction of virus and semen into the uterus at estrus. They were also unable to produce ill effects on the fetus following experimental infection of pregnant cattle.[53]

The disease is usually studied under experimental conditions by intravenous inoculation of cattle with virus-infected blood or crude white cell suspensions and observing their clinical response.[29] However, this may be considerably different to the natural infection involving an arthropod vector.

Recent, unconfirmed findings of St. George et al.[54] suggest that the incubation period for the disease may be of considerable duration. They maintained a group of 12 calves at a sentinel herd site near Brisbane, Queensland. The calves, aged 10 to 24 months, were bled daily for 60 days in the expectation that they would become naturally infected in an epizootic of BEF. Seven animals of the group developed clinical disease. Five showed no clinical signs or significant temperature responses even though they developed specific neutralizing antibodies during the period, indicating that infection had occurred. The animals with clinical disease developed low levels of neutralizing antibody 2 to 10 weeks before becoming clinically ill with typical BEF. The antibody disappeared during the clinical illness and then rose in 3 to 4 days to a plateau level suggesting an anamnestic response. The virus was isolated in mice from three animals and hematological changes were characteristic of BEFV infection. These results suggested that virus transmission was occurring a considerable period before clinical signs were recorded.

Until controlled experiments using insect transmission with the vector or vectors are performed, the actual significance of these findings is difficult to evaluate. However, they do stress the importance of not placing too much reliance on the results from studies using an experimental method of infection which is obviously different to that which occurs in nature.

2. Pathology

There are a number of pathological findings described for BEFV infection that vary considerably between groups. It may be that a number of descriptions in the literature relate to other diseases, such as those caused by Ibaraki virus or Kotonkan virus.[10,18] Brief summaries of the major findings have been provided by Burgess,[3] French,[16] and Inaba.[10]

The hematology reveals that most animals rapidly develop a leucocytosis or lymphopemia and also less marked neutrophilia.[10,27,43] Mackerras et al.[27] also noted a drop in erythrocyte numbers in the early stage of infection but this has not been confirmed by other workers.[43]

The macroscopic lesions described include edematous lymph nodes,[5,27,55] congestion and hemorrhages in the trachea, nasal cavity and pharynx,[10,16,27] and hydropericardium.[5,27,34] Edema of the limbs and the submandibular space has been described in some cases.[32] However, except for a small number of fatal cases, the pathological changes caused by this disease are often mild and almost inapparent.

The histopathology of the disease is far from defined. What tissues and cells the virus grows in and what changes the virus brings about to cause disease have yet to be defined. Burgess and Spradbrow[43] described fluorescent antibody studies in which they found fluorescence in individual cells, resembling alveolar macrophages, in the lungs; individual cells in the spleen and lymph nodes as well as neutrophils; and they could not exclude the possibility that immunocytes fluoresced. Theodoridis[56] observed specific fluorescence in leucocytes from infected cattle. Lesions have been described in the venules and capillaries in synovial membranes, muscle, and skin. These include hypoplasia of the epithelium, perivascular neutrophilic infiltration and edema followed by proliferation of pericytes with a predominance of round cells, focal or complete necrosis of vessel walls, thrombosis, and perivascular fibrosis.[27,55] Mackerras et al.[27] also reported hemosiderosis of the spleen and lymph nodes.

C. Insect Transmission

The spread of virus in Australia in the 1936 to 1937 and in the 1967 to 1968 epizootics led Seddon[26] and Murray[33] to propose that the disease was transmitted by insects and that the prevailing wind patterns played a large part in the distribution of the disease. The disease is not transmitted by contact but can be transmitted by intravenous injection of blood from infected animals.[5,10,19,27,29] So far, insect vectors have not been conclusively demonstrated for BEFV, but there is increasing evidence that they are likely to be involved.

BEFV was isolated in Kenya,[20] during an outbreak of disease in 1972 to 1973, from a mixed pool of *Culicoides* spp. (biting midges). The pool contained 4000 parous females and the species composition was *C. kingi* (67%), *C. nivosus* (24%), *C. bedfordi* (8%), *C. pallidipennis* (1%), and *C. cornutus* (1%). All except *C. nivosus* are known to take blood meals from cattle, but *C. pallidipennis* is also known to feed on sheep as it is the African vector of bluetongue virus.[57] In Australia it has been widely speculated that *Culicoides* are the likely vector(s) of BEFV.[15,26] Similar speculation has also continued in Kenya, *C. shultzei* being suggested as a likely candidate.[58]

Standfast et al.[19] recently made two isolations of BEFV from mosquitoes caught in northern Australia. From Etna Creek, Queensland, they isolated a virus from a mixed pool of mosquitoes including four *Culex* (Lophoceraomyia) species, four *Uranotaenia nivipes*, one *Uranotaenia albescens* and one *Aedes* (Verralina) *carmenti*. From Beatrice Hill, Northern Territory, they isolated another virus from a pool of 77 *Anopheles* (Anopheles) *bancroftii* mosquitoes. These authors concluded that the distribution of *A. bancroftii* and its seasonal abundance indicated that it was unlikely to be an important vector in the epizootic spread of BEF. One point that these isolations emphasize is that mosquitoes, or some other insect, rather than *Culicoides*, may be involved in the transmission of the disease.

Early attempts to transmit BEFV using the arthropods *Stomoxys calcitrans, Aedes vigliax, Aedes alternans,* and *Culex annulirostris,* that had been allowed to feed on infected cattle, were unsuccessful.[27] The virus has been shown to multiply in the mosquitoes *Aedes aegypti* and *Culex fatigans* following intrathoracic inoculation.[36] It has now been shown that *Aedes aegypti* and *Aedes vigilax* will not support virus multiplication when fed a mixture of blood and mouse-adapted BEFV, but that *Culex annulirostris* will.[15,20] Standfast et al.[15] recovered virus from 17% of *C. annulirostris* 8 days after feeding but could not recover virus on days 2, 4, or 6. Similarly, virus could be recovered from *Culicoides brevitarsis* and *Culicoides marksi* 8 days after feeding a mixture of mouse brain virus and sugar, but not at 2, 4, or 6 days.

IV. ISOLATION AND GROWTH

A. Isolation Systems

Primary virus isolation from infected animals was initially done by injecting blood from infected cattle intravenously into susceptible cattle.[27,46] In 1967, Van der Westhuizen[19] reported adaption of BEFV to 1 to 3-day-old suckling mice. Intracerebral inoculation of mice with infected buffy coat cells from cattle resulted in death of 17% of the mice in an average time of 12 to 14 days post infection. In subsequent passages he was able to increase the mortality to 62% and then 100% and reduce the mean survival time to 8.3, 5.0, 4.8, 4.6, 4.1, 3.6, and 2.8 days. Similar results have since been reported by others.[22,23,36,37]

Virus has also been isolated from cattle blood by intracerebral inoculation into 1-day-old suckling hamsters and subsequently adapted to grow in 1-day-old rats and suckling mice.[23]

Since isolation of virus in experimental animals is not always practical or possible, a number of groups have investigated the use of cell-culture systems for the primary isolation of BEFV. Snowdon[29] reported the isolation of BEFV from a crude buffy coat suspension using BHK-21 cells. On the initial passage, 8 days after inoculation, a small number of scattered cells in one tube showed rounding, the cells and fluid from this tube were inoculated into fresh tubes of BHK-21 cells, and focal areas of cytopathic effect (CPE) developed after 4 days. On subsequent passages, complete cell degeneration occurred within 4 days. Tzipori[61] subsequently isolated BEFV from cattle blood using Vero cells and noted that freezing and thawing the blood reduced the chances of virus isolation and that dilution of the blood resulted in more rapid development of CPE. Perhaps the use of insect cells, such as Singh's *Aedes albopictus* cells, may aid in the adaption of BEFV to vertebrate cell culture, as has been described for the rabies serogroup viruses, Obodhiang and Kotonkan.[62]

B. Virus Growth

The most universally used cell line for the growth of BEFV has been BHK-21 cells,[23,29,64] followed by Vero cells.[61,63-65] Other cell lines used for the growth of BEFV include HmLu-1 (a stable line of suckling hamster lung), Hmt (hamster kidney cells), SVP (Super-Vero-PS cells), MS (monkey kidney cells).[23,42,64,66] Lack of growth has been reported in the following cells: mouse kidney, mouse embryo, chicken kidney, calf leukocytes, calf thymus, calf thyroid, calf bone marrow, calf adrenal, PS (porcine stable), PS-EK, and LLC-MK2 (rhesus monkey kidney cell line).[23,29,42,66]

Growth cycle studies indicate that BEFV has a latent period of 8 to 14 hr in HmLu-1, MS, Vero, and SVP cells and that the amount of intracellular virus formed is similar to the amount of extracellular virus.[23,61,64,66] Inaba et al.[23] reported that yields obtained at 34°C or 30°C were higher than at either 38°C or 25°C and that the CPE is more

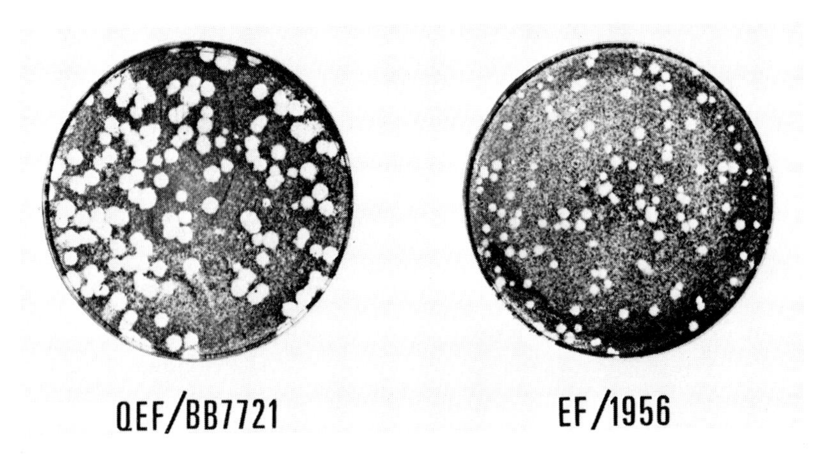

FIGURE 1. Plaque assay of BEFV using SVP cells and a nutrient agar overlay containing 0.1 μg actinomycin D/mℓ of overlay.[66] Virus strains are as described by Doherty et al.[36] and Snowdon.[29]

distinct. Snowdon recommended rolling BHK-21 cells in order to obtain better results.[29] Yields are usually below 10^6 $TCID_{50}$/mℓ, although when multiplicities of infection around 0.01$TCID_{50}$/cell are used, yields up to 10^7TCID_{50}/mℓ can be obtained.[66,67]

Virus growth has also been reported in weaned mice,[10,68] chicken embryos when inoculated intravenously after 10 to 11 days incubation,[68] 3-day-old kittens,[68] and suspected in mature rats because of the development of serum neutralizing antibodies.[10] BEFV has also been found to grow well in Singh's *Aedes albopictus* cells.[69]

C. Virus Assay

Quantal assays in suckling mice or in BHK-21 cells have been the most commonly used methods for titrating virus infectivity.[10,19,29,36]

Plaque assays in MS and Vero cells were first described by Heuschele.[64] Serum-free medium was found to give optimum results, with plaques being visible in 8 to 10 days postinfection, with a diameter of 1 to 1.5 mm. Theodoridis[70] found considerable variation in plaque size, from 1 mm to 5 mm, and a great insensitivity of the plaque assay in Vero cells ($10^{2.8}$ PFU/mℓ) when compared to BHK-21 roller tubes (10^6TCID_{50}/mℓ). Tzipori[65] was able to plaque various strains of BEFV in Vero cells. He found that 2-day-old monolayers gave the best results with plaques from 1.2 ± 0.4 mm to 0.4 ± 0.4 mm after 4 days or 4-day-old monolayers giving plaques of 2.3 ± 0.8 mm to 1.2 ± 0.2 mm after 7 days. However, no comparison of the sensitivity of this assay and the quantal assay was made.

Della-Porta and Snowdon[66] also had difficulty in producing reproducible plaquing results with BEFV. They found that the inclusion of 0.1 μg to 0.05 μg Actinomycin D (ACD) per milliliter of overlay enabled plaques to be readily visualized in SVP cells after 4 to 5 days. The plaques were up to 4 mm in diameter and seldom less than 1 mm. Figure 1 illustrates plaques seen for two Australian isolates of BEFV (strain EF/1956 and QEF/BB7721). Note the appearance of more than one type of plaque population in each isolate. The amount of ACD incorporated in the overlay was about one sixth the level (0.6 μg/mℓ) required to inhibit the incorporation of 90% of ^{3}H-Uridine into SVP cellular RNA. Plaques could also be obtained in Vero cells using 0.5 to 0.1 μg ACD/mℓ of overlay, but they were not as clear as those obtained in SVP cells. The plaque assay was found to be slightly more sensitive than assay in BHK-21 roller tubes (0.4 to 0.5 log) and very reproducible. It has been suggested.[66,71] that the use of ACD may produce better plaquing either by inhibition of interferon production or, more likely, by inhibiting cell growth and making the cells more susceptible to destruction and hence plaque formation, which is also the probable basis for the plaquing method of Heuschele.[64] The use of ACD may also be of use for plaquing other slow-growing rhabdoviruses.

V. VIRUS STRUCTURE AND COMPOSITION

A. Electron Microscopy

There appears to be general agreement concerning the intracellular development of BEFV when thin sections of infected cells are examined in the electron micro scope.[72-76] BEFV budding takes place from the cytoplasmic membrane in mouse neurones, sometimes from extended cytoplasmic processes, and in Vero cells. In its morphogenesis, BEFV closely resembles other rhabdoviruses, such as vesicular stomatitis, Flanders-Hart Park and Kern Canyon viruses.[76-78] The punctate or diffuse cytoplasmic fluorescence seen in BEFV-infected cells is also characteristic of the antigen distribution in vesicular stomatitis virus-infected cells.[56,76,79]

Holmes and Doherty[74] noted that the budding and extracellular virus particles, seen in infected mouse brain, had a diameter of 70 nm (range 60 to 80 nm) and a mean length of 145 nm (range 120 to 170 nm) although they sometimes saw particles over 200 nm long. The particles were slightly tapered. Sometimes, giant budding particles 100 nm in diameter were seen. No fine detail of the inner shell could be seen, but it appeared to have an outer diameter 10% smaller than the enveloped particle. These authors also observed that in some neurones, the most noticeable feature was a striking increase in the electron density of the ergastoplasm.

Murphy et al.[76] confirmed that viral morphogenesis took place primarily upon plasma membranes in association with small accumulations of cytoplasmic matrix. Fusion of viral envelope was observed and early syncytium formation occurred in infected cell cultures. They also noted, early in the growth of BEFV in Vero cells, 185 × 73 nm cone-shaped particles with nearly parallel sides. Later in infection the particles became more cone-shaped with a wider base. The presence of truncated or T-particles was noted at all stages of the infection. Studies by Bauer and Murphy[80] of Obodhiang and Kotonkan viruses, two rabies serogroup members, showed such similar morphogenesis and virus structure to BEFV that they suggested that BEFV may be related to these viruses.

Lecatsas et al.[72,73] examined South African, Japanese, and Australian isolates of BEFV and found the South African isolate more conical, with a particularly sharp point. A reexamination by Theodoridis and Lectasas[81] of the South African and Japanese isolates showed that both bullet- and cone-shaped forms were present. Ito et al.[75] reported that only bullet-shaped forms were present in the Japanese isolate grown at 33 to 34°C, whereas Theodoridis and Lectasas grew the virus at 37°C and suggested that temperature may have an effect on the pleomorphism observed. Murphy et al.[76] came to a similar conclusion that variation in morphology was not necessarily related to "strain variation" but may be due to varying growth rates and T-particle interference.

Ito et al.[75] examined BEFV and found a mean length of 140 nm and diameter of 80 nm. They found degraded particles in sucrose density gradients and in CsCl a peak of virus infectivity at 1.196 g/cc. Bullet-shaped particles were associated with the peak of infectivity. No cross-striations could be found. In contrast, Murphy et al.[76] found that BEFV possessed a precisely coiled helical nucleocapsid with 35 cross-striations at a 4 to 8 nm interval.

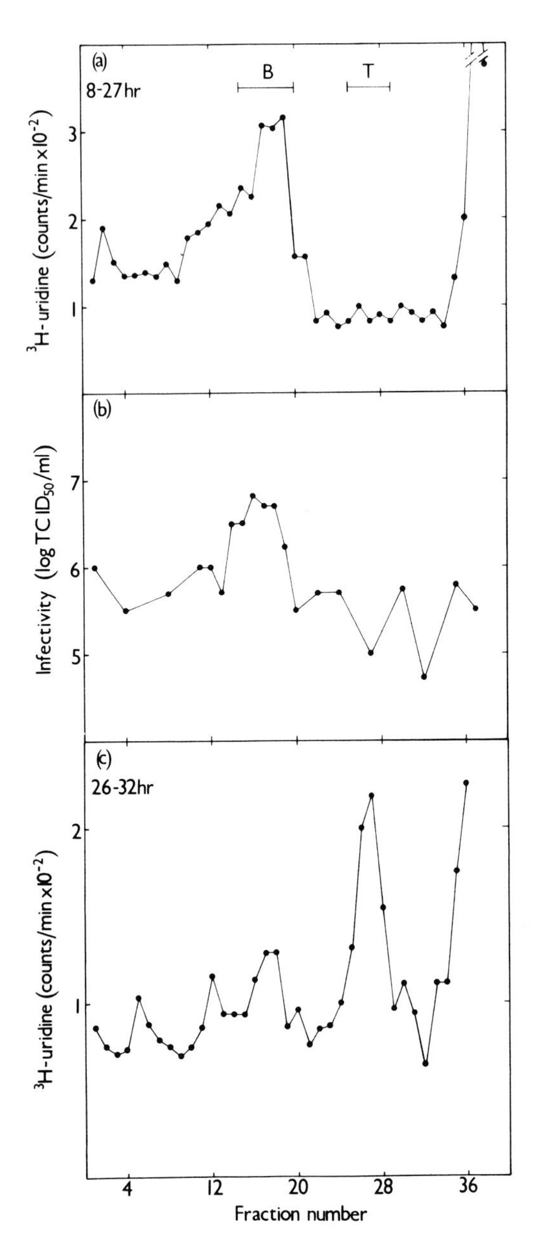

FIGURE 2. Purification of BEFV (strain EF/1956) labeled with ^{3}H-uridine, in the presence of 0.03 μg ACD/mℓ and grown in BHK 21 cells, after sedimentation, at 20,000 rpm for 1 hr in an MSE-SW30 rotor, from clarified culture fluid. The resuspended virus was sedimented through a 15 to 45% sucrose density gradient in NTE buffer (pH 7.6) including 0.1% BSA, in an MSE-SW30 rotor at 20,000 rpm for 2 hr and then 0.5 mℓ fractions collected dropwise from the bottom of the centrifuge tube. Sedimentation is from right to left. Total ^{3}H-uridine radioactivity in 25 $\mu\ell$ samples was counted, and infectivity was assayed usng BHK-21 roller tubes. (A) Virus labeled between 8 and 27 hr postinfection. (B) Infectivity assay for purification gradient (A). (C) Virus labeled between 26 and 32 hr postinfection. Infectivity assay for gradient (C) similar to results shown in (B).

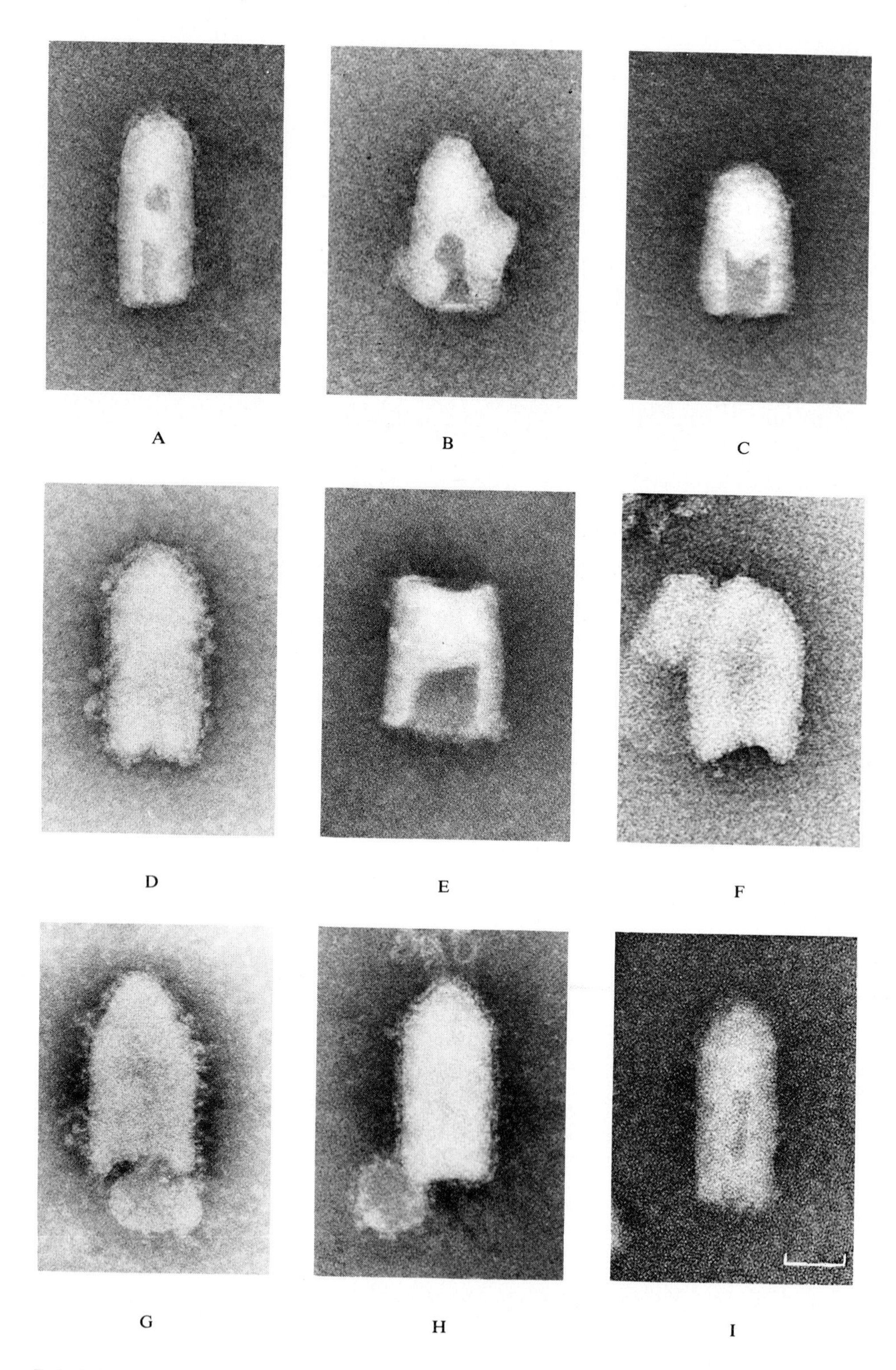

FIGURE 3. The appearance of sucrose density gradient banded infectious BEFV (strain EF/1956) particles stained with 4% ammonium molybdate (A to F) or 2% PTA (pH 6.5) (G) and VSV (strain Indiana) particles stained with 2% PTA (pH 6.0) (H) or 4% ammonium molybdate (I) and examined in a Phillips EM301 electron microscope.[82] The bar represents 50 nm.

Examination of sucrose density gradient purified BEFV (Figure 2) showed that the lower radioactivity peak, associated with virus-infectivity, consisted of bullet-shaped or sometimes slightly conical particles (Figure 3A to G).[82] BEFV particles stained particularly well with ammonium molybdate (Figure 3A to F) but less well with phosphotungstic acid (Figure 3G), sodium silicotungstate, and uranyl acetate. Also, there was significant variation in forms of BEFV visualized, from typical bullet shapes to cone shapes. In contrast, vesicular stomatitis virus (VSV) stained better with phosphotungstic acid (Figure 3H) than with ammonium molybdate (Figure 3I) and was uniformly bullet-shaped. The higher peak of radioactivity in the purification gradient (Figure 2) was associated with small cap-like or T-particles (Figure 4A and B). These T-particles were not associated with virus infectivity and are defective-interfering (DI) particles (see discussion in section on DI particles). The BEFV T-particles had a similar morphology to VSV T-particles prepared at the same time (Figure 4C and D). The size and shape of the virion, the T-particles, and the cross-striations of the coiled nucleocapsid, suggest that BEFV is a rhabdovirus, but the differences of staining and shape suggest that it is not closely related to VSV.

B. Physicochemical Properties

BEFV was shown to be an enveloped virus because it was inactivated with 20% ethyl ether, 5% chloroform, and 0.1% sodium deoxycholate.[19,23,36,83] Further, the virus was inactivated with 1.0% or 0.5%, trypsin (Difco 1:250).[83] Filtration experiments indicated that the virus size was in the range of 100 to 220 nm.[64,83]

Attempts to purify the virus using sucrose density gradients proved unsuccessful because of low recovery of infectivity (0.08%).[83] However, Della-Porta and Snowdon[66] showed that the inclusion of a protein stabilizer in the gradient (0.1% bovine serum albumin) led to almost 100% of the infectivity being recovered. The main peak of incorporated ^{3}H-uridine and virus infectivity coincided in the purification gradient (Figure 2). Comparison of the sedimentation behavior of BEFV and VSV in sucrose density gradients (Figures 2, 6, and 8) showed that both viruses have similar sedimentation coefficients (around 625S).[84] Equilibrium centrifugation in cesium chloride revealed that BEFV had a density of 1.19 g/cc which is similar to the reported values of 1.18 g/cc for VSV and 1.17 g/cc for rabies.[2,83-85]

The virus is considerably more stable in protein-containing solutions than in buffers free of protein.[64,66] The infectivity is rapidly lost at pH 2.5 and pH 12.0 (4.5 log lost in less than 10 min), fairly rapidly at pH 5.1 (4.5 log in 60 min) or pH 9.1 (4.5 log in 90 min), and less rapidly between pH 7.0 and pH 8.0[23,64,83] The effect of temperature has been described by Tanaka et al.[83] and Heuschele.[64] The virus was rapidly inactivated at 56°C (about 0.5 log/min), and less rapidly at 37°C (about 0.2 log/hr) and 30°C (about 0.7 log/day). Storage at 4°C for up to 30 to 40 days resulted in a loss of 1 to 2 log in titre. BEFV was stable at −80°C (little loss after 134 days and less than 2 log at 278 days) but at −20°C a 200-fold drop in titer occurred within 73 days.

C. Chemical Structure

1. Nucleic Acid

The failure of the inhibitors of DNA synthesis, 5-iodo-2′-deoxyuridine and 5-bromodeoxyuridine, to inhibit the growth of BEFV strongly indicated that the virus contained RNA.[70,83] Staining of BHK-21 cells infected with BEFV using acridine orange showed that the cytoplasm fluoresced with a reddish to intense flame color, which was homogeneous rather than granular, while the nuclei of cells remained normal.[70] These results are characteristic of an RNA virus replicating in the cytoplasm of the cell. The lack of inhibition of BEFV replication by ACD suggests that the DNA-dependent RNA

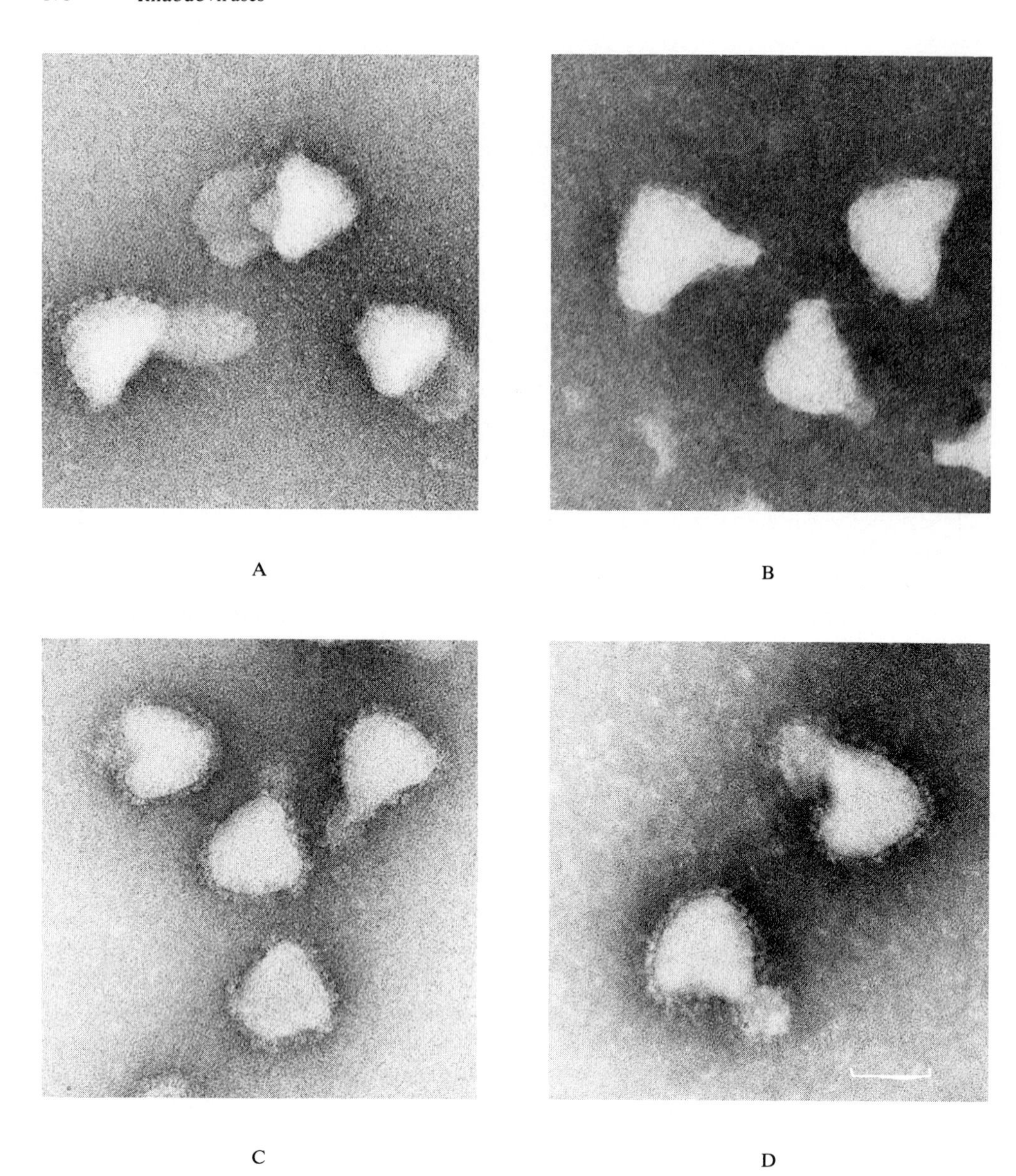

FIGURE 4. The electron microscopic appearance of sucrose density gradient banded truncated (T) particles of BEFV (strain EF/1956) stained with 4% ammonium molybdate (A) or 2% PTA (pH 6.5) (B) and VSV (strain Indiana) particles stained with 4% ammonium molybdate (C) or 2% PTA (pH 6.0) (D).[82] The bar represents 50 nm.

synthesis step was not involved in the virus replication.[61,66,70] The demonstration that ^{3}H-uridine was incorporated into the virus was further evidence that BEFV contains RNA (Figure 2).[2,83]

However, the unconfirmed report that BEFV contained a 12S piece of double-stranded RNA has led to the virus only being provisionally classified as a rhabdovirus.[1,2] Tanaka et al.[2] purified the virus by equilibrium centrifugation in cesium chloride, a method long known not to separate the virion from T-particles of the rhabdoviruses.[86] Furthermore, the low titer of the virus $10^{5.5}$ $TCID_{50}/m\ell$, in the purification peak, would be expected to reveal very few particles in the electron microscope. However, their Figure 3 shows many short, cone-shaped particles, similar to the T-particles shown in Figure 4. Further, the use of RNase, at 25 $\mu g/m\ell$, to clean up the virion, may have led to breakdown of any single-stranded RNA when extraction was carried out. Della-Porta and Brown[87] found that after such RNase treatment, followed by virus purification, some RNase was still attached to the virions and led to breakdown as the extraction was carried out.

A reexamination of the nucleic acid of BEFV led to the conclusion that the virus did indeed contain RNA but that this RNA had all the characteristics of RNAs found in members of the rhabdovirus family.[87] The viral nucleic acid was labeled with ^{3}H-uridine, in the presence of 0.03 μg ACD/mℓ and growing the virus in BHK-21 cells. The virus labeled under these conditions was pelleted, at 20,000 rpm (in an MSE SW30 rotor) for 1 hr, from clarified infectious culture fluid. The pellet was resuspended in 0.12 M NaCl, 0.012 M Tris (pH 7.6), 0.0012 *M* disodium ethylenediaminetetra-acetate (NTE) buffer, and further purified on a 15 to 45% (v/v) sucrose density gradient in NTE buffer containing 0.1% BSA using an MSE SW30 rotor at 20,000 rpm for 2 hr.[88] Two peaks of incorporated ^{3}H-uridine could be seen, the virus infectivity corresponding to the lower peak (Figure 2). The RNA was extracted from the purified virus using 0.1 *M* acetate and 0.1% sodium dodecyl sulphate (SDS) buffer at pH 5.0, and precipitated in the presence of BHK-21 ribosomal RNA carrier using two volumes of absolute ethanol at −20°C overnight. The precipitated RNA was analyzed after treatment with 0.01 μg of pancreatic RNase/mℓ or without treatment. Sedimentation was for 16 hr at 16,000 rpm in a Beckman-Spinco SW41 rotor at 20°C in 5 to 25% sucrose density gradients in 0.1 M acetate/0.1% SDS, pH 5.0, buffer.[89] The optical density was read at 260 nm for the BHK-21 cellular 28S and 18S ribosomal RNA markers and then the ^{3}H-uridine labeled RNA counted after precipitation, in the presence of bovine serum albumin carrier, with 10% trichloroacetic acid (Figure 5). The RNA had a sedimentation coefficient in the range 42 to 43S and was single-stranded, as judged by its digestion by RNase.

Further, the analysis of ^{32}P-labeled RNA showed that BEFV-RNA had a sedimentation coefficient of 42S and that it was similar to VSV-RNA sedimented in a parallel gradient (Figure 7). Hence, it can be concluded that BEFV possessed 42S single-stranded RNA, similar to the prototype of the rhabdovirus family, VSV-Indiana.

2. Proteins

There are no published reports describing the protein composition of BEFV. An analysis has been made of the viral structural proteins that suggested that BEFV had similar proteins to other members of the rhabdovirus family.[87] The virus was labeled with ^{35}S-methionine, in the presence of ACD (0.03 μg/mℓ), and the labeled virus concentrated from clarified culture fluid by centrifugation at 20,000 rpm for 1 hr in an MSE-SW30 rotor. The virus was resuspended in NTE buffer and further purified by sedimentation in a 15 to 45% (w/v) sucrose density gradient in NTE buffer containing 0.1% bovine serum albumin (BSA) and centrifugation at 20,000 rpm for 2 hr in an MSE-SW30 rotor (Figure 8). A parallel preparation of ^{35}S-methionine labeled VSV-Indiana was also purified at the same time (Figure 8). The purified viruses were dissociated in 0.5 *M* urea, 0.1% SDS, and 0.1% 2-mercaptoethanol in 0.01 *M* phosphate buffer, pH 7.2, and heated at 100°C for 2 min. The virus proteins were analyzed on 7.5% SDS-phosphate polyacrylamide gels run at 4 mA/gel overnight, until the dye front had migrated 90 mm.[90] The gels were frozen and sliced into 1 mm slices using a gel slicer (Mickel Laboratory Engineering Co., Gomshall, Surrey), the slices incubated with 1 mℓ of NCS solubilizer (Amersham/Searle, The Radiochemical Centre, Amersham, Bucks) for 1 hr at 56°C, and then counted in a toluene-based cocktail using a Packard Tricarb liquid-scintillation counter. Figure 9 shows a comparison of the profiles for BEFV and VSV run in separate gels. When the virus proteins for BEFV and

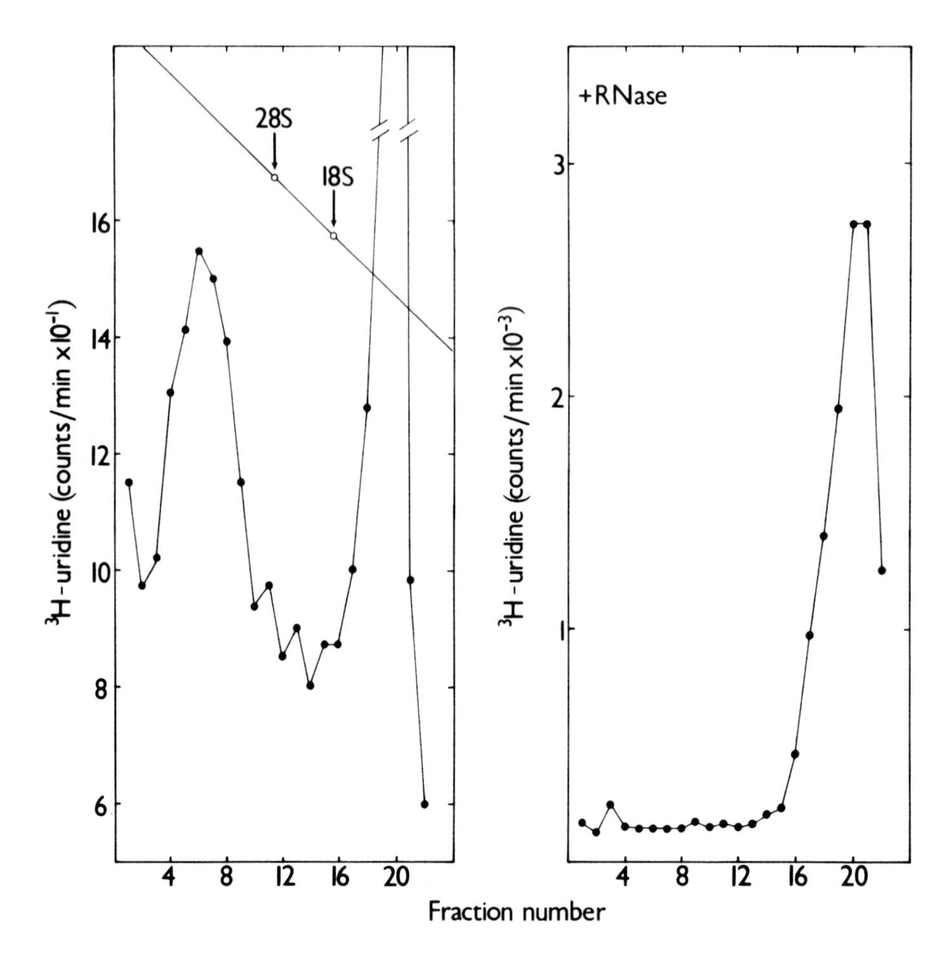

FIGURE 5. Analysis of ^{3}H-uridine labeled RNA extracted from purified BEFV (strain EF/1956) (Figure 4A), Fractions 14 to 18). The RNA was sedimented, in the presence of BHK-21 ribosomal RNA markers (28S and 18S), without or after RNase treatment (0.01 μg RNase/mℓ) through a 5 to 25% sucrose density gradient in 0.1 *M* acetate/0.1% SDS (pH 5.0) buffer in a Spinco-SW41 rotor at 16,000 rpm for 16 hr at 20°C 0.5 mℓ fractions were collected from the bottom of the gradient. The position of the ribosomal RNA markers was determined by optical density at 260 nm, and total radioactivity in each fraction was determined after precipitation with trichloroacetic acid.

VSV were coelectrophoresed, the N protein of VSV migrated 4 to 5 fractions further than BEFV "N" protein, and thus the BEFV "N" protein would appear to have a higher molecular weight.

It would appear that BEFV possesses a typical rhabdovirus protein profile but differs both from VSV (Figure 9) and from rabies.[90-92] The actual identity of the proteins of BEFV has yet to be established. It would be of particular interest to compare the BEFV proteins with the two rabies serogroup viruses, Obodhiang and Kotonkan, which appear structurally similar in the electron microscope to BEFV.[80]

D. Defective-Interfering (DI) Particles

As is common with most rhabdoviruses, BEFV exhibits interference when used to infect cells at a high multiplicity of infection (MOI).[61,66,67] At MOIs approaching 1 PFU per cell, the yield of virus from both BHK and SVP cells is reduced to about 10^3PFU/mℓ of culture fluid, whereas when the MOI is reduced to around 0.01 PFU per cell the yield can be increased up to 10^7PFU/mℓ.[66]

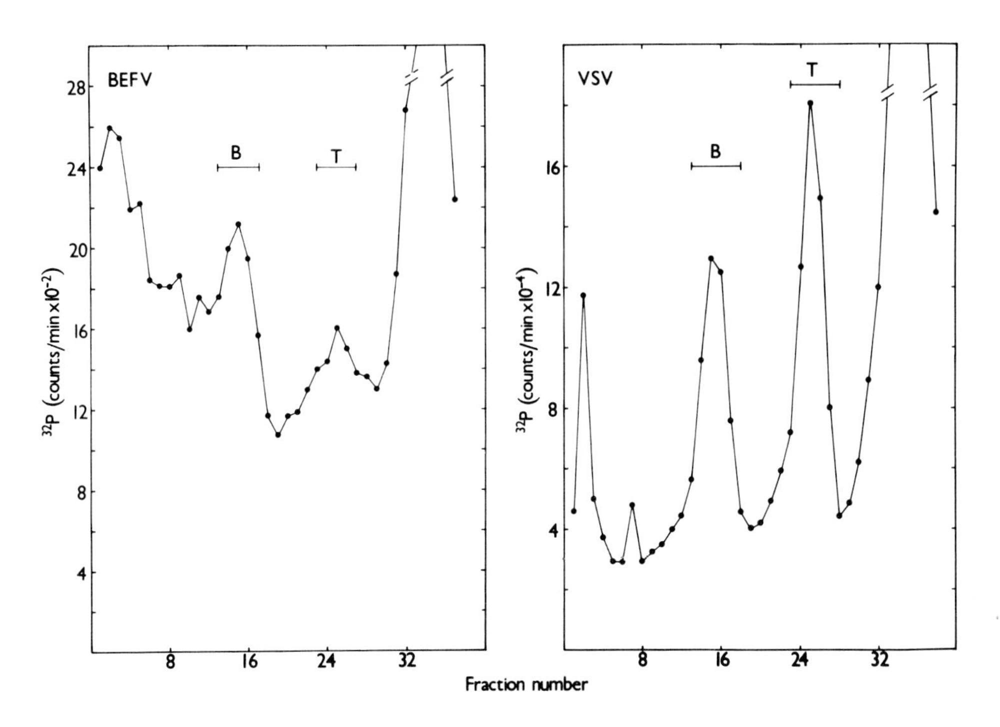

FIGURE 6. Purification of ^{32}P-orthophosphate labeled BEFV (strain EF/1956) (grown in the presence of 0.03 μg ACD/mℓ and labeled between 16 and 40 hr postinfection) and VSV (strain Indiana). Virus concentration and purification procedures as for Figure 4. Total ^{32}P counted by Chernekov radiation.

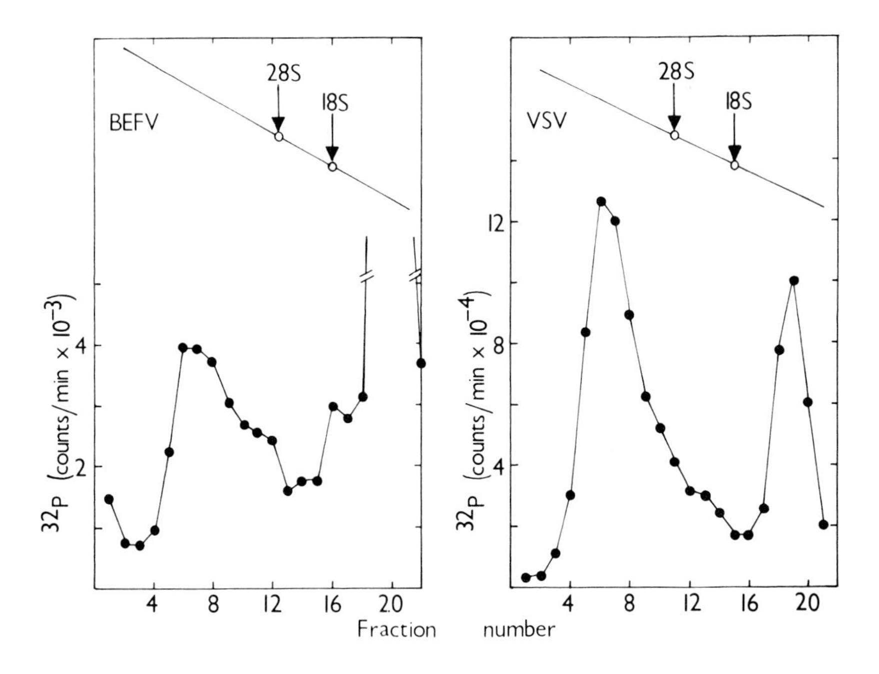

FIGURE 7. Comparison of ^{32}P-labeled RNA extracted from BEFV-B and VSV-B particles (Figure 6) and analyzed under the conditions described for Figure 5, excluding the RNase treatment.

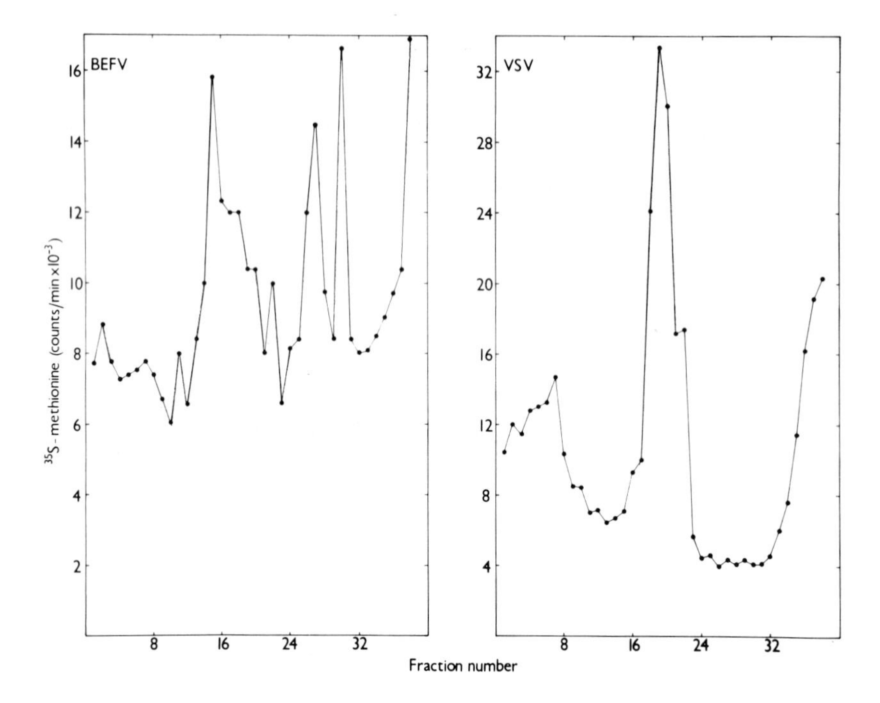

FIGURE 8. Purification of ^{35}S-methionine labeled BEFV (strain (EF/1956) (grown in the presence of 0.03 μg ACD/mℓ and labeled between 16 and 40 hr postinfection) and VSV (strain Indiana) under the condition described for Figure 4. Total ^{35}S-methionine in 25 μℓ samples of each fraction was counted.

Murphy and Fields[76] observed a number of cone-shaped T-particles in samples taken from any stage of the infection cycle. These particles may represent the classical rhabdovirus DI particle. Such particles are shown in Figure 3 and represent particles recovered from the upper peak of virus material in a BEFV purification gradient (Figure 4)[82] and does not correspond to the major peak of virus infectivity. They are similar both in sedimentation behavior (Figure 6 and 8) and in the electron microscope (Figure 3) to VSV T-particles and rabies DI particles.[89] It is also of interest to note that the T-particles shown in Figure 3 are similar to the short, cone-shaped particles shown in Figure 3 of Tanaka et al.,[2] from which it was claimed that 12S double-stranded RNA was obtained.

The BEFV T-particles appear to contain single-stranded RNA of approximately 18 to 20S (Figure 10), similar to VSV T-particle RNA.[87] The T-particles are preferentially labeled later in the virus growth cycle (Figure 4), suggesting that they are formed later in the growth cycle than B-particles of BEFV, a similar process to that described for VSV.[93] The BEFV T-particles have been shown to cause interference with the growth of BEFV but not with VSV.[87]

VI. SEROLOGY

Until the development of methods to grow the virus in mice and in cell culture, the only method available to identify isolates of BEFV was the cross-protection test in cattle.[27,94,95] This test is still the most reliable indicator that isolates of virus are in fact BEFV.[29,45] By the cross-protection test Takematsu et al.[94] showed that three Japanese isolates of BEFV were indistinguishable, as were the Australian isolates from the 1956 and 1967 to 1968 epizootics.[29]

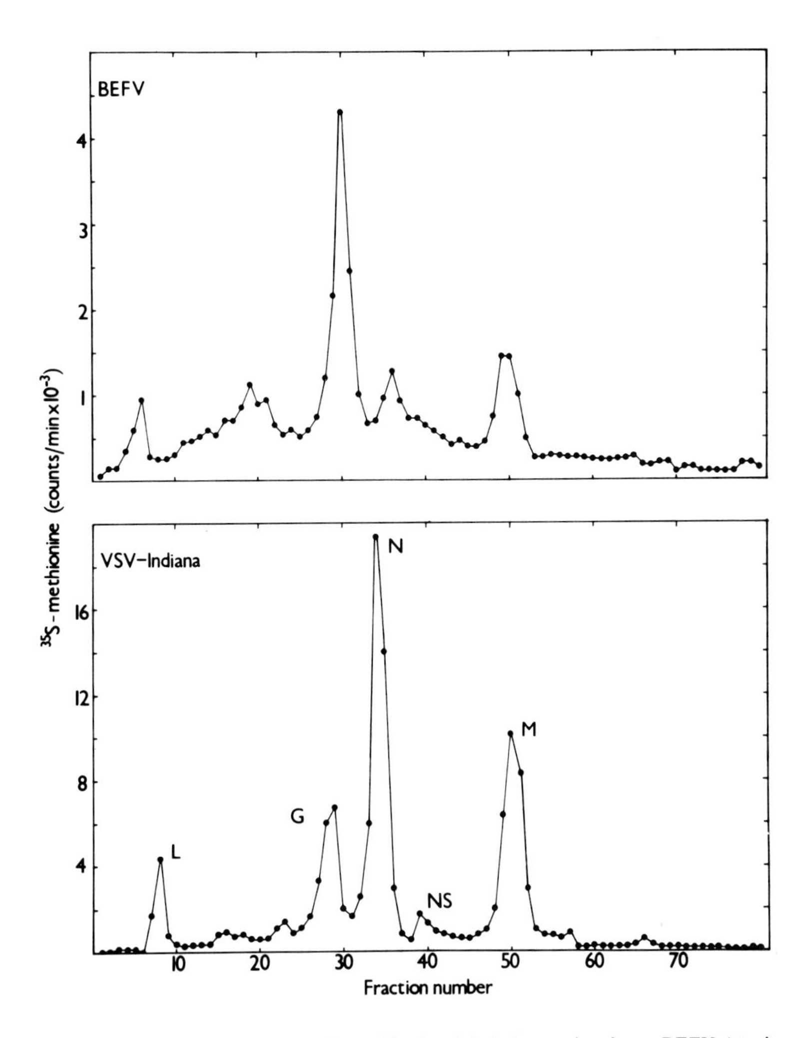

FIGURE 9. Comparison of ^{35}S-methionine labeled proteins from BEFV (strain EF/1956) and VSV (strain Indiana) from purification gradients (Figure 9) after dissociation and running on a 7.5% SDS-phosphate polyacrylamide gel.[90] Analyses of both the proteins were done at the same time and a coelectrophoresis of both preparations in the same gel indicated that the migrations of the N proteins differed by 4 to 5 fractions.

The serum neutralization test has virtually replaced the expensive and time-consuming cross-protection test. This is performed using either suckling mice[19,36] or cell cultures.[22,29] BHK-21 cells have become the standard detection system and the maintenance of the cells by rolling has resulted in better reproducibility of this test.[29] Serological comparison of a Japanese strain (YHK), a South African strain (EFI), and an Australian strain (EF/1956) by cross-neutralization tests has shown that these isolates cannot be differentiated.[24] In South Africa, another Australian isolate (QEF/BB7721) appeared very similar to the Japanese (YHK) and South African (EFI) isolates, although there were slight differences.[72] Further, an isolate from Nigeria was shown to be indistinguishable from an Australian and a South African isolate;[7] an isolate from Kenya was indistinguishable from an Australian isolate;[20] and an isolate from Iran was indistinguishable from a Japanese isolate.[9] Thus by cross-neutralization tests it appears that all of the BEFV isolates are very closely related, if not identical.

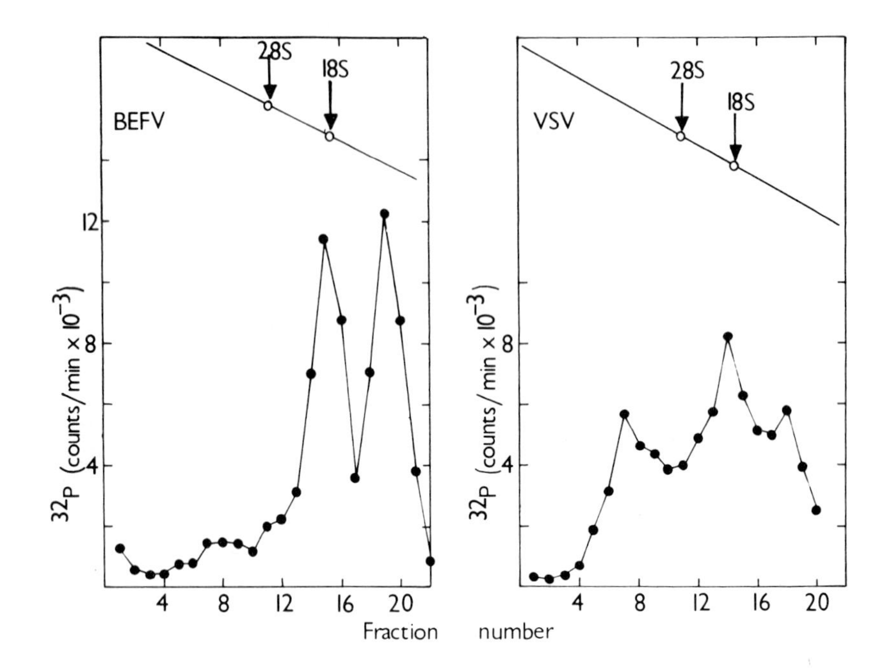

FIGURE 10. Comparison of ^{32}P-labeled RNA extracted from BEFV and VSV T-particles (Figure 6) and analyzed under the same conditions described for Figure 7.

However, recently two BEFV isolates from mosquitoes caught in Northern Australia[59] were shown to differ to some extent when compared in cross-neutralization tests with an Australian reference strain QEF/BB7721 (Table 1)[96] Cross-protection studies with these two isolates showed that the sixth mouse brain passage, when injected intravenously into cattle, did not produce clinical signs of disease but did produce serum neutralizing antibodies. Challenge with strain EF/1956,[29] a strain indistinguishable in cross-neutralization tests from QEF/BB7721,[29,45] showed that the insect isolates produced cross-protection and hence were strains of BEFV. However, the antibody responses following challenge were not typical.[97] The significance of these results is not apparent but does suggest that strains could differ slightly. Perhaps these isolates could represent milder strains of BEFV, which produce weaker serological responses, as has been postulated for interepizootic periods.[30]

The neutralizing antibody response in cattle can first be detected 7 to 14 days after infection and rises slightly until about 35 days.[29] Antibodies persist in cattle in excess of 320 to 422 days, when animals can still resist challenge. The first antibodies detected are of the 19S class, and these remain fairly constant till about 30 days. By day 36 after infection, the 19S activity has almost completely disappeared. The 7S class of antibody was first detected 22 days after infection and continued to rise until 36 days.[30,97]

Three other serological tests which have been reported are complement-fixation (CF), immunofluorescence, and immunodiffusion. CF tests have mainly been used to demonstrate serological relationships, or lack of them, with BEFV.[36,76] Thus, Murphy et al.[76] were unable to demonstrate any CF relationships between BEFV and 15 rhabdoviruses. Inaba et al.[10] found that both hyperimmune and convalescent bovine sera gave highest CF titers when heated at 45°C for 30 min rather than 56°C for 30 min. The addition of the heat labile factor (using 1:20 normal bovine serum) to the heat-inactivated sera, may restore CF activity as seen for other virus CF systems involving bovine sera.[98] CF antibodies in cattle were first detected 3 weeks after infection, which

TABLE 1

Comparison of a Cattle Isolate With Two Insect Isolates of BEFV in the Neutralization Test

	Neutralization Index[a] for Antiserum[b]		
Virus	BB7721	Etna Creek	BH0698
BB7721[c]	2.6	2.6	2.1
Etna Creek[d]	1.5	5.6	2.2
BH0698[d]	2.3	2.2	4.0

[a] Neutralization index assayed in suckling mice.
[b] Immune mouse ascites fluid.
[c] QEF/BB7721 virus isolate of Doherty et al.[36] isolated in mice from blood of an infected cow.
[d] Etna Creek and BH0698 virus isolates of Standfast et al.[96] isolated in mice from mosquito pools. Indistinguishable by CF from QEF/BB7721.

is 1 to 2 weeks later than the detection of neutralizing antibodies.[10] An immunodiffusion test has been described by Heuschele and Johnson.[47] The fluorescent antibody test has been used mainly for experimental studies.[43,56,76] The technique would appear to have limited value in the diagnosis of the disease but may be of use in studying the pathogenesis of infection. An antibody-blocking immunofluorescent assay may be of use for serological assays, but methods based upon the enzyme-linked immunoadsorbent assay (ELISA) or radioimmunoassay (RIA) are more likely to be of use for widespread serological surveys.

VII. VACCINES AND IMMUNITY

It has long been known that animals that have recovered from infection with BEFV can resist challenge with virulent virus for long periods after the original infection.[5,27,46] Van der Westhuizen[19] was the first to describe studies in cattle where the serum neutralizing antibody response was related to resistance to intravenous challenge with virulent BEFV. He also observed that virus grown in suckling mouse brain, with a low passage level (third passage), would not produce clinical signs of disease, an antibody response, or immunity. However, repeated vaccinations with the passaged virus in Freund's complete adjuvant produced a neutralizing antibody response in five out of five cattle, four of which resisted challenge.

Further studies using live vaccines showed that such vaccines could not only elicit a neutralizing antibody response in cattle but could also produce immunity.[29,47,99-102] Again, the lack of clinical signs of disease and the poor antibody response were noted after various isolates of BEFV (with up to 10^6TCID_{50}) had undergone a few passages in mice or cell culture.[29,47,68,103] An interesting observation by Snowdon[29] was that even when the neutralizing antibody levels were low, cattle immunized with low passage virus could resist challenge. For example, one of the cattle shown in Table 3 of Snowdon's paper had no detectable BEFV neutralizing antibodies before challenge and yet resisted the challenge.

TABLE 2

Comparison of Neutralizing Antibody Response and Resistance to Challenge Produced by Vaccinating Cattle With Inactivated and Live BEFV Vaccines[a]

	Vaccine[b]					Serum neutralizing[c] antibody titres		
Animal number	Virus[d] strain	Number of passages	Titre (log PFU)	Inactivated	Live	Prechallenge	Postchallenge	Clinical response to challenge
F60	BB7721	29	8	+	−	280	7,000	+
U27	BB7721	29	8	+	−	74	3,125	+
A41	Control	N.A.[e]	N.A.	N.A.	N.A.	<1	1,845	+
A20	EF/1956	7	7.2	+	−	125	625	+
A22	EF/1956	7	7.2	+	−	25	15,625	+
A25	EF/1956	7	7.2	+	−	1	15,625	+
A31	EF/1956	7	7.2	+	−	125	125	+
A2	Control	N.A.	N.A.	N.A.	N.A.	<1	125	+
A10	Control	N.A.	N.A.	N.A.	N.A.	<1	25	+
S1	EF/1956	7	4.7	−	+	25	25	−
X2	EF/1956	7	4.7	−	+	1	625	+
D76	EF/1956	7	4.7	−	+	1	125	−
T66	EF/1956	7	4.7	−	+	25	25	−
X21	Control	N.A.	N.A.	N.A.	N.A.	<1	25	+
X54	Control	N.A.	N.A.	N.A.	N.A.	<1	25	+

[a] Based on data of Della-Porta and Snowdon.[45]

[b] Virus vaccines made up in an equal volume of Freund's incomplete adjuvant, following inactivation with 0.05%, β-propiolactone for the killed virus, and given intramuscularly at 4-week intervals. Challenge was with virulent EF/1956 (approximately 100 cattle infective doses) and temperature responses and clinical signs were monitored for 14 days after challenge. A nonimmune control was used for every two test animals, in order to check the challenge virus.

[c] The reciprocal of the dilution of serum that neutralized 100 $TCID_{50}$ units of QEF/BB7721.[29] Prechallenge sera taken immediately before challenge and postchallenge sera taken 2 weeks after challenge.

[d] QEF/BB7721 and EF/1956 were indistinguishable in a cross-neutralization test.[29,36,45]

[e] N.A. — not applicable.

BEFV is probably an insect-transmitted virus but, because the vector has not been identified, back passage of live "attenuated" virus cannot be done to test the safety of live vaccines. In nonendemic areas, such as southeastern Australia, the spread of live vaccine virus could cause both safety problems and problems meeting the export requirements of cattle from Australia. It is problems such as these that have stimulated an extensive debate on the use of live "attenuated" and inactivated vaccines against BEFV.[104-108] In Japan, studies using a formalin-inactivated, aluminum phosphate gel adsorbed vaccine indicated that reasonable neutralizing antibody levels could be attained after two doses, and that cattle would resist challenge.[109] However, the challenge virus was virulent BEFV that had four passages in cell culture, and as the virus rapidly loses its ability to produce clinical disease on passage in cell culture, it may not have been a challenge as severe as virulent, cattle-passage virus.[19,29]

In a similar study in Australia,[45,66] using a vaccine of β-propiolactone inactivated virus in Freund's incomplete adjuvant, it was shown that good neutralizing antibody responses could be produced but that they bore little direct relationship to protection against challenge with virulent BEFV (Table 2). Passage level did not affect the antibody response, as virus passaged as many times as 27 (10^8 PFU) produced a very high antibody level (Table 2). Low passage level virus (seven passages, Table 2), when inactivated ($10^{7.2}$ PFU) or live ($10^{4.7}$ PFU), produced very similar neutralizing antibody responses after two vaccinations. However, cattle that received the inactivated vaccine failed to resist challenge (zero out of four) whereas three out of four of those that received the live vaccine were resistant.

The assumption that the serum neutralizing antibody level is the indicator of resistance to infection by BEFV needs thorough reappraisal. Recent studies with rabies virus suggest that neutralizing antibodies are not necessarily related directly to resistance to rabies and that T-cell dependence and cell-mediated immunity may also be involved in resistance.[110-114] The studies reported by Della-Porta and Snowdon[45] would suggest that resistance to BEF-like rabies, may be more complex than previously thought. Adequate challenge studies, in the natural host (or a satisfactory model), which approximate the natural infective process as closely as possible are required for any vaccine trials. Furthermore, more detailed studies of the pathogenesis of, and resistance to, BEF are needed, although such studies may require the development of new techniques for handling BEFV and for assessing immunity in the natural host.

VIII. CONCLUSIONS

BEFV can now be definitely classified as a rhabdovirus. Its biophysical behavior is very similar to the prototype of the family VSV-Indiana. BEFV contains 42S single-stranded RNA and a protein composition typical of members of the rhabdovirus family. The virus is found as infectious full-size bullet- and cone-shaped ("B") particles. Also, a truncated, cone-shaped T-particle is found, typical of the rhabdovirus DI particles. However, BEFV cannot, as yet, be grouped with any similar rhabdoviruses. The morphologically similar rabies serogroup viruses, Obodhiang and Kotonkan, have yet to be fully compared with BEFV.

The search for an insect vector(s) of BEFV is still under way. Once an insect vector has been found, transmission studies may yield insights into many unanswered questions such as the actual length of time from infection till clinical signs of disease and the factors involved in disease. More sensitive and quantitative methods of virus isolation from the tissues of cattle would certainly aid pathogenesis studies. Then, with an understanding of transmission and pathogenesis of BEF, a more thorough assessment of the factors involved in resistance to infection can be made. Dissection of the immune response in cattle will be difficult, but the rewards for successfully accomplishing this shall be immense.

IX. ADDENDUM

Further characterization of the proteins of BEFV has shown that there are probably six virus proteins,[87] with a similar molecular weight distribution to those for the rabies serogroup viruses, Kotonkan and Obodhiang. The preliminary designation (and molecular weight $\times 10^{-3}$) is as follows: L (164), G (101), N (64), NS (53), M1 (43), and M2 (29). P 101 is a trypsin sensitive protein on the outside of the virion and can be labeled with glucosamine. P 64 is present in the highest molar concentrations of any of the virus proteins and is phosphorylated.

ACKNOWLEDGMENTS

This work was supported in part by a research grant from the Australian Meat Research Committee. Some of the biochemical comparisons were done while one of us (A.J.D.P.) was at the Animal Virus Research Institute, Pirbright, England, supported by CSIRO. We would like to thank Dr. Fred Brown for his collaboration in the biochemical studies and for helpful discussions. We would also like to thank Doctors Inaba, St. George, Spradbrow, Standfast, Theodoridis, and Tzipori for supplying information which was helpful in writing this review. Finally we would like to thank members of the staff of CSIRO and AVRI who helped in the production of this chapter.

REFERENCES

1. **Fenner, F.**, Classification and nomenclature of viruses. Second report of the International Committee on Taxonomy of Viruses, *Intervirology*, 7, 1, 1976.
2. **Tanaka, Y., Inaba, Y., Ito, Y., Sato, K., Omori, T., and Matumoto, M.**, Double-strandedness of ribonucleic acid of bovine ephemeral fever virus, *Jpn. J. Microbiol.*, 16, 95, 1972.
3. **Burgess, G. W.** Bovine ephemeral fever: a review, *Vet. Bull. (London)*, 41, 887, 1971.
4. **Theiler, A.**, Stiffsickness or three-day sickness, *Transvaal Gov. Vet. Bacteriol. Rep.*, 22, 1906—1907.
5. **Bevan, L. E. W.**, Preliminary report on the so-called "stiffsickness" or "three-day sickness" of cattle in Rhodesia, *J. Comp. Pathol. Ther.*, 20, 104, 1907.
6. **Wakeem, A. A.**, Cases of ephemeral fever, "Three day sickness", *Sudan J. Vet. Sci. Anim. Husb.*, 2, 192, 1961.
7. **Kemp, G. E., Mann, E. D., Tomori, O., Fabiyi, A., and O'Connor, E.**, Isolation of bovine ephemeral fever virus in Nigeria, *Vet. Rec.*, 93, 107, 1973.
8. **Davies, F. G., Shaw, T., and Ochieng P.**, Observations on the epidemiology of ephemeral fever in Kenya, *J. Hyg.*, 75, 231, 1975.
9. **Hazrati, A., Hessami, M., Roustai, M., and Dayhim, F.**, Isolation of bovine ephemeral fever virus in Iran, *Arch. Inst. Razi*, 27, 81, 1975.
10. **Inaba, Y.**, Bovine ephemeral fever (three-day sickness). Stiff sickness, *Off. Int. Epizoot. Bull.*, 79, 627, 1973.
11. **Topacio, T., Farinas, E. C., Yutuc, L. M., de Jesus, Z., and Munoz, M.**, Report of the Committee on Diseases. III. Three-day fever of caraboa (influenza or ephemeral fever?), *Philipp. J. Anim. Ind.*, 4, 113, 1937.
12. **Meadows, D.**, Notes on an ephemeral fever of Indian cattle resembling South African "three days sickness", *Vet. J.*, 75, 138, 1919.
13. **Merkens, J.**, Een Ziekte Onder Melkkoeien, *Ned.-Indische Bladen Diergeneestkd.*, 31, 48, 1919.
14. **Burggraaf, H.**, "Dreitage-krankheit" op de Oostkust van Sumatra, *Tijdschr. Diergeneeskd.*, 59, 234, 1932.
15. **Standfast, H. A., Murray, M. D., Dyce, A. L., and St. George, T. D.**, Report on ephemeral fever in Australia, *Off. Int. Epizoot. Bull.*, 79, 615, 1973.
16. **French, E. L.**, A review of arthropod-borne virus infections affecting man and animals in Australia, *Aust. J. Exp. Biol. Med. Sci.*, 51, 131, 1973.
17. **St. George, T. D., Standfast, H. A., Christie, D. G., Knott, S. G., and Morgan, I. R.**, The epidemiology of bovine ephemeral fever in Australia and Papua-New Guinea, *Aust. Vet. J.*, 53, 17, 1977.
18. **Kemp, G. E., Lee, V. H., Moore, D. L., Shope, R. E., Causey, O. R., and Murphy, F. A.**, Kotonkan, a new rhabdovirus related to Mokola virus of the rabies serogrup, *Am. J. Epidemiol.*, 98, 43, 1973.
19. **Van der Westhuizen, B.**, Studies on bovine ephemeral fever. I. Isolation and preliminary characterization of a virus from natural and experimentally produced cases of bovine ephemeral fever, *Onderstepoort J. Vet. Res.*, 34, 29, 1967.
20. **Davies, F. G. and Walker, A. R.**, The isolation of ephemeral fever virus from cattle and Culicoides midges in Kenya, *Vet. Rec.*, 95, 63, 1974.
21. **Tomori, O., Fagbami, A., and Fabiyi, A.**, Serum antibodies to two rhabdoviruses (bovine ephemeral fever and Kotonkan) in calves on the University of Ibadan agricultural farm, *Bull. Anim. Health. Prod. Afr.*, 23, 39, 1975.
22. **Inaba, Y., Tanaka, Y., Sato, K., Ito, H., Omori, T., and Matumoto, M.**, Propagation in laboratory animals and cell cultures of a virus from cattle with bovine epizootic fever, *Jpn. J. Microbiol.*, 12, 253, 1968.
23. **Inaba, Y., Tanaka, Y., Sato, K., Ito, H., Omori, T., and Matumoto, M.**, Bovine epizootic fever. I. Propagation of the virus in suckling hamster, mouse and rat, and hamster kidney BHK21-W12 cell, *Jpn. J. Microbiol.*, 12, 457, 1968.

24. **Inaba, Y., Tanaka, Y., Omori, T., and Matumoto, M.**, Serological relation between bovine epizootic fever and ephemeral fever, *Jpn. J. Microbiol.*, 13, 129, 1969.
25. **Inaba, Y., Sato, K., Tanaka, Y., Ito, H., Omori, T., and Matumoto, M.**, Serological identification of bovine epizootic fever virus as ephemeral fever virus, *Jpn. J. Microbiol.*, 13, 388, 1969.
26. **Seddon, H. R.**, The spread of ephemeral fever (three-day sickness) in Australia in 1936—37, *Aust. Vet. J.*, 14, 90, 1938.
27. **Mackerras, I. M., Mackerras, M. J., and Burnet, F. M.**, Experimental studies of ephemeral fever in Australian cattle, *Aust. C.S.I.R. Bull.*, 136, 1, 1940.
28. **Seddon, H. R.**, Diseases of Domestic Animals in Australia. IV. Protozoan and Virus Diseases, 2nd ed., Rev. by H. E. Albiston, Dept. Health, Canberra, *Aust. Dep. Health Div. Vet. Hyg. Serv. Publ.*, No. 8, 1966.
29. **Snowdon, W. A.**, Bovine ephemeral fever: the reaction of cattle to different strains of ephemeral fever virus and the antigenic comparison of two strains of virus, *Aust. Vet. J.*, 46, 258, 1970.
30. **Snowdon, W. A.**, Some aspects of the epizootiology of bovine ephemeral fever in Australia, *Aust. Vet. J.*, 47, 312, 1971.
31. **Gee, R. W., Hall, W. T. K., Littlejohns, I., and Snowdon, W. A.**, The 1967—68 outbreak of ephemeral fever in cattle, *Aust. Vet. J.*, 45, 132, 1969.
32. **Morgan, I. and Murray, M. D.**, The occurrence of ephemeral fever of cattle in Victoria in 1968, *Aust. Vet. J.*, 45, 271, 1969.
33. **Murray, M. D.**, The spread of ephemeral fever of cattle during the 1967—68 epizootic in Australia, *Aust. Vet. J.*, 46, 77, 1970.
34. **Newton, L. G. and Wheatley, C. H.**, The occurrence and spread of ephemeral fever of cattle in Queensland, *Aust. Vet. J.*, 46, 561, 1970.
35. **St. George, T. D., Standfast, H. A., Armstrong, J. M., Christie, D. G., Irving, M. R., Knott, S. G., and Rideout, B. L.**, A report on the progress of the 1972/73 epizootic of ephemeral fever — 1 December 1972 to 30 April 1973, *Aust. Vet. J.*, 49, 441, 1973.
36. **Doherty, R. L., Standfast, H. A., and Clark, I. A.**, Adaptation to mice of the causative virus of ephemeral fever of cattle from an epizootic in Queensland, 1968, *Aust. J. Sci.*, 31, 365, 1969.
37. **Spradbrow, P. B. and Francis, J.**, Observations on bovine ephemeral fever and isolation of virus, *Aust. Vet. J.*, 45, 525, 1969.
38. **Hall, W. T., Daddow, K. N., Dimmock, C. K., St. George, T. D., and Standfast, H. A.**, The infection of merino sheep with bovine ephemeral fever virus, *Aust. Vet. J.*, 51, 344, 1975.
39. **Doherty, R. L.**, Arboviruses of Australia, *Aust. Vet. J.*, 48, 172, 1972.
40. **Doherty, R. L., Carley, J. G., Standfast, H. A., Dyce, A. L., and Snowdon, W. A.**, Virus strains isolated from arthropods during an epizootic of bovine ephemeral fever in Queensland, *Aust. Vet. J.*, 48, 81, 1972.
41. **Della-Porta, A. J. and Murray, M. D.**, unpublished data, 1976.
42. **Matumoto, M., Inaba, Y., Tanaka, Y., Ito, H., and Omori, T.**, Behaviour of bovine ephemeral fever virus in laboratory animals and cell cultures, *Jpn. J. Microbiol.*, 14, 413, 1970.
43. **Burgess, G. W. and Spradbrow, P. B.**, Studies on the pathogenesis of bovine ephemeral fever, *Aust. Vet. J.*, 53, 363, 1977.
44. **Mulhearn, C. R.**, Ephemeral or three-day-fever in Northern Queensland, its diagnosis and some preliminary investigations, *Aust. Vet. J.*, 13, 186, 1937.
45. **Della-Porta, A. J. and Snowdon, W. A.**, An experimental inactivated virus vaccine against bovine ephemeral fever. II. Do neutralizing antibodies protect?, *Vet. Microbiol.*, in press.
46. **Bevan, L. E. W.**, Ephemeral fever, or three days sickness of cattle, *Vet. J.*, 68, 458, 1912.
47. **Heuschele, W. P. and Johnson, D. C.**, Bovine ephemeral fever. II. Responses of cattle to attenuated and virulent virus, *Proc. Annu. Meet. U.S. Anim. Health Assoc.*, 73, 185, 1969.
48. **Theodoridis, A., Giesecke, W. H., and Du Toit, I. J.**, Effects of ephemeral fever on milk production and reproduction of dairy cattle, *Onderstepoort J. Vet. Res.*, 40, 83 1973.
49. **Burgess, G. W.**, Attempts to infect cattle with bovine ephemeral fever by inoculation of virus into the cervix, *Aust. Vet. J.*, 49, 341, 1973.
50. **Tzipori, S. and Spradbrow, P. B.**, The effect of bovine ephemeral fever virus on the bovine foetus, *Aust. Vet. J.*, 51, 64, 1975.
51. **Burgess, G. W. and Chenoweth, P. J.**, Mid-piece abnormalities in bovine semen following experimental and natural cases of bovine ephemeral fever, *Br. Vet. J.*, 131, 536, 1975.
52. **Parsonon, I. M. and Snowdon, W. A.**, Ephemeral fever virus: excretion in the semen of infected bulls and attempts to infect female cattle by the intrauterine inoculation of virus, *Aust. Vet. J.*, 50, 329, 1974.
53. **Parsonson, I. M. and Snowdon, W. A.**, Experimental infection of pregnant cattle with ephemeral fever virus, *Aust. Vet. J.*, 50, 335, 1974.

54. **St. George, T. D., Cybinski, D. H., Dimmock, G. K., and Murphy, G.,** Aust. C.S.I.R.O. Div. Anim. *Health Annu. Rep.*, Melbourne, 24, 1976.
55. **Basson, P. A., Pienaar, J. G., and Van der Westhuizen, B.,** The pathology of ephemeral fever: a study of the experimental disease in cattle, *J. S. Afr. Vet. Med. Assoc.*, 40, 385, 1969.
56. **Theodoridis, A.,** Fluorescent antibody studies on ephemeral fever virus, *Onderstepoort J. Vet. Res.*, 36, 187, 1969.
57. **Howell, P. G. and Verwoerd, D. W.,** Bluetongue virus, *Virol. Monogr.*, 9, 35, 1971.
58. **Walker, A. R. and Boreham, P. F. L.,** Blood feeding of *Culicoides* (Diptera, Ceratopogonidae) in Kenya in relation to the epidemiology of bluetongue and ephemeral fever, *Bull. Entomol. Res.*, 66, 181, 1976.
59. **Standfast, H. A., St. George, T. D., and Dyce, A. L.,** The isolation of ephemeral fever virus from mosquitoes in Australia, *Aust. Vet. J.*, 52, 242, 1976.
60. **Kay, B. H., Carley, J. G., and Filippich, C.,** The multiplication of Queensland and New Guinean arboviruses in *Culex annulirostris* Skuse and *Aedes vigilax* (Skuse) (Diptera: Culicidae), *J. Med. Entomol.*, 12, 279, 1975.
61. **Tzipori, S.,** The isolation of bovine ephemeral fever virus in cell cultures and evidence for autointerference, *Aust. J. Exp. Biol. Med. Sci.*, 53, 273, 1975.
62. **Buckley, S. M.,** Singh's *Aedes albopictus* cell cultures as helper cells for the adaption of Obodhiang and Kotonkan viruses of the rabies serogroup to some vertebrate cell cultures, *Appl. Microbiol.*, 25, 695, 1973.
63. **Burgess, G. W.,** A microtitre serum neutralization test for bovine ephemeral fever virus, *Aust. J. Exp. Biol. Med. Sci.*, 52, 851, 1974.
64. **Heuschele, W. P.,** Bovine epehemeral fever. I. Characteristics of the causative virus, *Arch. Gesamte Virusforsch.*, 30, 195, 1970.
65. **Tzipori, S.,** Plaque assay and characteristics of strains of bovine ephemeral fever virus in Vero cells, *Br. Vet. J.*, 131, 720, 1975.
66. **Della-Porta, A. J. and Snowdon, W. A.,** An experimental inactivated virus vaccine against bovine ephemeral fever. I. Studies of the virus, *Vet. Microbiol.*, in press.
67. **Sato, K., Inaba, Y., Kurogi, H., Omori, T., and Yamashiro, T.,** Rolling round bottle culture of HmLu-1 cells and the production of bovine ephemeral fever virus, *Natl. Inst. Anim. Health Q.*, 15, 109, 1975.
68. **Tzipori, S. and Spradbrow, P. B.,** Development and behaviour of a strain of bovine ephemeral fever virus with unusual host range, *J. Comp. Pathol.*, 84, 1, 1974.
69. **St. George, T. D.,** personal communication, 1977.
70. **Theodoridis, A.,** *Characterization of Bovine Ephemeral Fever Virus,* Master of Medicine, Veterinary (Virol.), thesis, University of Pretoria, South Africa, 1973.
71. **Westaway, E. G.,** Assessment and application of a cell line from pig kidney for plaque assay and neutralization tests with twelve group B arboviruses, *Am. J. Epidemiol.*, 84, 439, 1966.
72. **Lecatsas, G., Theodoridis, A., and Els, H. J.,** Morphological variation in ephemeral fever virus strains, *Onderstepoort J. Vet. Res.*, 36, 325, 1969.
73. **Lecatsas, G., Theodoridis, A., and Erasmus, B. J.,** Electron microscopic studies on bovine ephemeral fever virus, *Arch. Gesamte Virusforsch.*, 28, 390, 1969.
74. **Holmes, I. H. and Doherty, R. L.,** Morphology and development of bovine ephemeral fever virus, *J. Virol.* 5, 91, 1970.
75. **Ito, Y., Tanaka, Y., Inaba, Y., and Omori, T.,** Electron microscopic observations of bovine epizootic fever virus, *Natl. Inst. Anim. Health Q.*, 9, 35, 1969.
76. **Murphy, F. A., Taylor, W. P., Mims, C. A., and Whitfield, S. G.,** Bovine ephemeral fever virus in cell culture and mice, *Arch. Gesamte Virusforsch.*, 38, 234, 1972.
77. **Murphy, F. A. and Fields, B. N.,** Kern Canyon virus: electron microscopic and immunological studies, *Virology*, 33, 625, 1967.
78. **Murphy, F. A., Coleman, P. H., and Whitfield, S. G.,** Electron microscopic observations of Flanders virus, *Virology*, 30, 314, 1966.
79. **Paucker, K., Schechmeister, I. L., and Birch-Andersen, A.,** Studies in the multiplication of vesicular stomatitis virus with fluorescein and ferritin conjugated antibodies, *Acta Pathol. Microbiol. Scand. Sect. B.*, 78, 317, 1970.
80. **Bauer, S. P. and Murphy, F. A.,** Relationship of two arthropod-borne rhabdoviruses (Kotonkan and Obodhiang) to the rabies serogroup, *Infect. Immun.*, 12, 1157, 1975.
81. **Theodoridis, A. and Lecatsas, G.,** Variation in morphology of ephemeral fever virus, *Onderstepoort J. Vet. Res.*, 40, 139, 1973.
82. **Della-Porta, A. J., Smale, C. J., and Brown, F.,** unpublished data, 1978.
83. **Tanaka, Y., Inaba, Y., Sato, K., Ito, H., Omori, T., and Matumoto, M.,** Bovine epizootic fever. II. Physicochemical properties of the virus, *Jpn. J. Microbiol.*, 13, 169, 1969.

84. **Bradish, C. J., Brooksby, J. B., and Dillon, J. F., Jr.,** Biophysical studies of the virus system of vesicular stomatitis, *J. Gen. Microbiol.,* 14, 290, 1956.
85. **Sokol, F., Kuwert, E., Wiktor, T. J., Hummeler, K., and Koprowski, H.,** Purification of rabies virus grown in tissue culture, *J. Virol.,* 2, 836, 1968.
86. **Knudson, D. L.,** Rhabdoviruses, *J. Gen. Virol.,* 20(Suppl), 105, 1973.
87. **Della-Porta, A. J. and Brown, F.,** The physio-chemical characterization of bovine ephemeral fever virus as a member of the family *Rhabdoviridae, J. Gen. Virol.,* in press.
88. **Brown, F., Cartwright, B., and Smale, C. J.,** The antigens of vesicular stomatitis virus. III. Structure and immunogenicity of antigens derived from the virion by treatment with tween and ether, *J. Immunol.,* 99, 171, 1967.
89. **Crick, J. and Brown, F.,** An interfering component of rabies virus which contains RNA, *J. Gen. Virol.,* 22, 147, 1974.
90. **Cartwright, B., Talbot, P., and Brown, F.,** The proteins of biologically active sub-units of vesicular stomatitis virus, *J. Gen. Virol.,* 7, 267, 1970.
91. **Sokol, F., Stancek, D., and Koprowski, H.,** Structural proteins of rabies virus, *J. Virol.,* 7, 241, 1971.
92. **Tignor, G. H., Murphy, F. A., Clark, H F., Shope, R. E., Madore, P., Bauer, S. P., Buckley, S. M., and Meredith, C. D.,** Duvenhage virus: morphological, biochemical, histopathological and antigenic relationships to the rabies serogroup, *J. Gen. Virol.,* 37, 595, 1977.
93. **Crick, J., Cartwright, B., and Brown, F.,** A study of the interference phenomenon in vesicular stomatitis virus replication, *Arch. Gesamte Virusforsch.,* 27, 221,1969.
94. **Takematsu, M., Sasahara, J., Chikatsune, M., and Okazaki, K.,** Immunological experiments on bovine epizootic fever, *Bull Natl. Inst. Anim. Health,* 31, 25, 1956.
95. **Inaba, Y., Morimoto, T., and Omori, T.,** Transmission of bovine epizootic fever in cattle, *Bull. Natl. Inst. Anim. Health,* 46, 1, 1963.
96. **Standfast, H. A., St. George, T. D., Doherty, R. L., Della-Porta, A. J., and Snowdon, W. A.,** *Aust. C.S.I.R.O. Div. Anim. Health Annu. Rep.,* 24, 1976.
97. **Kurogi, H., Inaba, Y., Takahashi, E., Sato, K., Fusato, S., Taniguchi, S., Satoda, K., and Omori, T.,** Neutralizing antibody sensitive to 2-mercaptoethanol in cattle infected with bovine ephemeral fever virus, *Natl. Inst. Anim. Health Q.,* 17, 126, 1977.
98. **Boulanger, P.,** Technique of a modified direct complement-fixation test for viral antibodies in heat inactivated cattle serum, *Can. J. Comp. Med. Vet. Sci.,* 24, 262, 1960.
99. **Theodoridis, A., Boshoff, S. E. T., and Botha, M. J.,** Studies on the development of a vaccine against bovine ephemeral fever, *Onderstepoort J. Vet. Res.,* 40, 77, 1973.
100. **Tzipori, S. and Spradbrow, P. B.,** Studies on vaccines against bovine ephemeral fever, *Aust. Vet. J.,* 49, 183, 1973.
101. **Inaba, Y., Kurogi, H., Takahashi, A., Sato, K., Omori, T., Goto, Y., Hanaki, T., Yamamoto, M., Kishi, S., Kodama, K., and Harada, K.,** Vaccination of cattle against bovine ephemeral fever with live attenuated virus followed by killed virus, *Arch. Gesamte Virusforsch.,* 44, 121, 1974.
102. **Spradbrow, P. B.,** Attenuated vaccines against bovine ephemeral fever, *Aust. Vet. J.,* 51, 464, 1975.
103. **Inaba, Y., Tanaka, Y., Sato, K. Ito, H., Omari, T., and Matumoto, M.,** Bovine ephemeral fever. III. Loss of virus pathogenicity and immunogenicity for the calf during serial passage in various host systems, *Jpn. J. Microbiol.,* 13, 181, 1969.
104. **Lascelles, A. K.,** Editorial: ephemeral fever vaccination, *Aust. Vet. J.,* 52, 381, 1976.
105. **Francis, J.,** An attenuated vaccine against bovine ephemeral fever, *Aust. Vet. J.,* 52, 537, 1976.
106. **Della-Porta, A. J. and Snowdon, W. A.,** Vaccines against bovine ephemeral fever, *Aust. Vet. J.,* 53, 50, 1977.
107. **Francis, J.,** Vaccines against bovine ephemeral fever, *Aust. Vet. J.,* 53, 198, 1977.
108. **Spradbrow, P. B.,** Vaccines against bovine ephemeral fever, *Aust. Vet. J.,* 53, 351, 1977.
109. **Inaba, Y., Kurogi, H., Sato, K. Goto, Y., Omori, T., and Matumoto, M.,** Formalin-inactivated, aluminum phosphate gel-adsorbed vaccine of bovine ephemeral fever virus, *Arch. Gesamte Virusforsch.,* 42, 42, 1973.
110. **Corey, L., Hattwick, M. A. W., Baer, G. M., and Smith, J. S.,** Serum neutralizing antibody after rabies postexposure prophylaxis, *Ann. Intern. Med.,* 85, 170, 1976.
111. **Wiktor, T. J., Koprowski, H., Mitchell, J. R., and Merigan, T. C.,** Role of interferon in prophylaxis of rabies after exposure, *J. Infect. Dis.,* 133(Suppl.), A260, 1976.
112. **Turner, G. S.,** Thymus dependence of rabies vaccine, *J. Gen. Virol.,* 33, 535, 1976.
113. **Kaplan, M. M., Wiktor, T. J., and Koprowski, H.,** Pathogenesis of rabies in immunodeficient mice, *J. Immunol.,* 114, 1761, 1975.
114. **Wiktor, T. J., Doherty, P. C., and Koprowski, H.,** *In vitro* evidence of cell-mediated immunity after exposure of mice to both live and inactivated rabies virus, *Proc. Natl. Acad. Sci. U.S.A.,* 74, 334, 1977.

Chapter 9

FISH RHABDOVIRUSES

P. Roy

TABLE OF CONTENTS

I. INTRODUCTION

Fish virology is relatively recent; the first virus isolation being reported in 1960.[1] Since then, eight different fish viruses have been isolated and characterized.[2-4] Four of these agents belong to the Rhabdoviridae family. Each is responsible for an economically important disease of fish cultivated either for human consumption or for sport purposes.

In a broad sense, the four rhabdoviruses are classified into two major divisions reflecting the origin of the various agents: (1) the salmonid fish rhabdoviruses and (2) the nonsalmonid fish rhabdoviruses. The basic characterization of these rhabdoviruses, in terms of their cellular and molecular biology, is in its infancy.

This chapter will present a synopsis of what is known about the viruses, starting with their isolation and pathogenicities, and then discuss the relatedness of the viruses on the basis of their structural and known antigenic properties.

A. Origins and Isolation

1. Salmonid fish rhabdoviruses

a. Egtved virus or virus hemorrhagic septicemia (VHS) virus

In 1950, in the eastern part of Jutland (Denmark), a new disease was reported among rainbow trout (*Salmo gairdnerii*). It was called the Egtved disease after the locality where it was first observed. Schäperclaus[5] subsequently showed that this disease was the same as one he had described in 1938 under the designation "Nierenschwellung" and later in 1954 as "Infektiose Nierenschwellung und Leberdegeneration" (INUL).[6] In Europe the disease has been described by many other names: "Bauchwassersucht der Forellen",[7,8] "Die neue Forellenkrankheit",[9,10] "Forellenseuche",[11] "Egtvedsygen",[12] "Anémie infectieuse",[13] "L'anémie pernicieuse des truites",[14] "Syndrôme entérohépatorenal",[15] and "La Lipoidosi Epatica".[16] In 1962, at the first symposium of fish diseases held in Turin (Italy), it was proposed, and accepted, that the disease should be named "viral hemorrhagic septicemia" (VHS).[17] The etiologic agent of this disease has been shown to be a virus, which is therefore termed viral hemorrhagic septicemia virus. Zwillenberg and associates showed the Egtved virus is morphologically a typical rhabdovirus.[18]

VHS is considered a disease primarily of rainbow trout. Brown trout (*S. trutta*) and brook trout (*Salvelinus fontinalis*) are considered to be immune to it,[19] although experimental infections can be induced in grayling (*Thymallus thymallus*), whitefish (*Coregonus* sp.), and *S. trutta.*[20]

b. Infectious hematopoietic necrosis (IHN), Oregon sockeye salmon disease (OSD), and Sacramento River Chinook disease (SRCD)

A number of investigators have described behavioral symptoms and histopathologic lesions of salmon and trout which have been ascribed to a virus infection termed infectious hematopoietic necrosis virus (IHNV). Salmon diseases such as the Oregon sockeye disease, and the Sacramento River Chinook disease, documented since 1941 on the West Coast of the U.S.,[21-24] are very similar to a rainbow trout (*Salmo gairdnerii*) disease caused by a virus isolated by Amend and associates.[25] All three virus isolates are antigenically related to each other.[26] However, both the Oregon sockeye virus isolate and the Sacramento River Chinook salmon virus have a high degree of host specificity and cause epizootics among sockeye salmon (*Oncorhynchus nerka*) and Chinook salmon (*Oncorhynchus tshaycytscha*), respectively.[27]

2. Nonsalmonid fish rhabdoviruses

Two nonsalmonid fish rhabdoviruses have been isolated in Europe. One of these was isolated from carp and the other from pike. The carp virus isolates have come from diseased fish exhibiting either a condition known as infectious dropsy,[28] or a condition involving swim bladder inflammation.[29] The two diseases are quite different in their pathogenicity, seasonal variation, and geographical distribution; however, despite these differences, it has been proposed that both diseases are caused by the rhabdovirus — spring viremia of carp virus (SVCV).[30]

The second and most recent nonsalmonid fish rhabdovirus isolate is pike fry rhabdovirus which causes an acute disease of young pike fry — a disease also known as "red disease of pike".[4]

II. DISEASE SYMPTOMS

A. VHS Virus

Infections of rainbow trout by viral hemorrhagic septicemia (VHS) or Egtved viruses are seasonal in occurrence. Ordinarily the disease culminates during the spring when the water temperature is between 6 and 12°C. The disease usually disappears spontaneously during the early summer when the water temperature rises above 15°C. However, infections can occur sporadically during summer as shown by Scholari who, in 1954, described a characteristic disease that broke out among young fish in June and ended 4 months later.[16]

VHS is a contagious disease. Its outbreak in a hatchery can often be traced to a surivival trout from an infected hatchery. Once infected, a hatchery seldom becomes free from yearly recurrences of the disese. Infected fish seem to lose their appetite and later on break away from the rest of the stock and remain motionless by the banks or near the surface of the water.

Symptoms associated with VHS include darkening of the skin, which becomes purple or black especially on the head and abdomen.[31] In many cases the eyes protrude giving a "popeye" effect due to hemorrhages in the connective tissues of the eye pit. The fish gills become pale or colorless compared to their normal bright red color and the abdomen is distended. These disease signs may be accompanied by hemorrhages at the base of the pectoral fins, and at the lateral line.[19]

Sometimes fish do not show any of the above symptoms, instead they become anemic, very lean, and a nervous syndrome appears 2 to 3 weeks after the onset of the infection. Such infected fish show motor disorders in the form of sudden vigorous twisting of the body.[32,33]

Internally, the fluids present in the abdomen of an infected fish may be colorless or yellowish in appearance. Histological investigations have shown that muscles and most of the organs of the body become affected.[32-35] The stomach often is smaller and full of a clear fluid, the intestines are usually empty and contain only a yellow mucus. The liver is discolored and hypertrophied. Hemorrhages are often seen in the kidneys, mouth cavity, bladder, muscles, and sex organs.[35-38]

B. IHN, OSD, and SRCD Viruses

In the case of Oregon sockeye disease virus (OSDV), the disease occurs in both the anadramous sockeye salmon and in Kokanee salmon, its landlocked form.

In 1946 an outbreak of disease in fingerling sockeye salmon was documented at two hatcheries in the U.S. Both are located in the drainage region of the Columbia River in the nortwestern part of Washington. About the same time fish at various hatcheries in Oregon also showed the same sort of symptoms.[39]

Diseased fish infected with this virus are often lethargic and darker in color. Some develop erratic swimming patterns and other evidence of hyperactivity.[22] As the disease progresses, the fish develop swollen abdomens, pale gills, hemorrhagic areas in, or at the base, of the fins, and in the throat area, and occasionally have protruded eyes. Internally, the spleen, liver, and kidneys are usually light in color, the stomach filled with a milky fluid, while the intestines become filled with a watery yellowish fluid, reddened by petechiae. Petechiae are also often extensively distributed throughout the visceral fat, bladder, and peritoneum.[40]

The Sacramento River Chinook disease virus causes an acute diseae with high mortalities among feeding fingerlings of Chinook salmon. The disease was initially observed to be limited to one locality on the bank of the Sacramento River in northern California, although similar diseases have been subsequently observed in sockeye and spring Chinook salmon in Washington.[20] Epizootics of the disease occur when the water temperature is below 13°C. The disease appears to subside when the water temperature stays above 13°C.[41]

Moribund fish infected with SRCD virus do not feed. They have pale gills and red blotches appear on the skin due to subcutaneous hemorrhages. The symptoms are very similar to sockeye-diseased fish. Internal lesions associated with the disease include necrosis of the spleen and kidneys. The pancreas becomes vacuolated, and both pancreas and adrenal cortical tissues become necrotic.[42] A massive vascular damage in the head can also be observed. Darlington and associates have shown by electron microscopy that virus particles occur mostly in interstitial spaces of diseased organs and occasionally in cytoplasmic vacuoles.[43]

Infectious hematopoietic necrosis disease initially causes necrosis of the hematopoietic tissue of the spleen and anterior kidney. As the disease progresses other tissues become involved. Necrosis is observed in pancreatic acinar and islet cells, and the granular cells in the stratum compactum of the alimentary tract. Focal necrotic areas develop in the liver.[42] Typically, infected fish become anemic. It is believed that the primary cause of death is due to kidney malfunction.[43-46] Epizootics of the disease occur at low temperatures (around 10°C) and never above 15°C. In cell culture the virus replicates at temperatures up to 18°C.[47]

C. Spring Viremia of Carp Virus

Roegner-Aust and associates in 1950 first suggested that the etiologic agent of the infectious dropsy disease of carp was a virus, on the basis that viruslike particles in organs taken from diseased carp could be observed by electron mcroscopy.[48] The Rumanian researchers Dimuslescu and co-workers, as reported by Goncharov,[49] proposed that meningoencephalitis, as one of the primary symptoms of this disease, indicated that the causative agent was viral rather than bacterial in origin.

The symptoms associated with SVCV infections often vary depending on whether an acute, chronic, or asymptomatic, latent form of the disease is present. In an overt disease the initial infected areas include the central nervous system and peripheral nerves.[50] The infected fish become hyperactive. Their gills are pale, the abdomen distended, and the fish scales sometimes protrude where there are ruptured and ulcerated dermal vesicles.[28] The infected fish kidneys and spleen become enlarged and a hemorrhagic necrosis of the intestine often develops.[51] Some diseased fish have protruding eyes.

While infectious dropsy mainly appears in the early spring, the second carp disease, swim bladder inflammation of carp (aerocystitis), occurs mainly during early summer when the water temperature is around 17°C. The symptoms of infected fish are similar to those of infectious dropsy disease, except that the swim bladder of diseased fish becomes inflamed with extensive petechiae in its wall. Petechiae are also observed in in the brain and pericardium of diseased fish.[30]

Virus has been recovered from the liver and swim bladder of infected fish.[29] Virions have also been observed by electron microscopy in the intestinal, kidney, heart, muscle, and brain tissues.[20]

D. Pike Fry Rhabdovirus

Pike fry rhabdovirus is responsible in the Netherlands for two different diseases of pike fry (*Esox lucius*). "Head disease," as named by Dutch fish farmers, is characterized by a swelling or lump on the head. The other, called "red disease of pike," is characterized by large areas of the body and tail becoming swollen and reddish in color. Both diseases can cause severe mortalities among pike fry.[4,52-55]

The hydrocephalus associated with the "head disease" also causes the fish to lose their equilibrium and swim near the surface of the water. Poor growth is associated with this diseased condition. In addition to abnormal amounts of cerebral liquids, petechial hemorrhages have been observed in the brain, spinal cord, spleen, and pancreas of diseased animals, while necrotic, degenerative changes have been observed in the kidney tubules.[52]

Red disease of pike was first observed in 1956 during an epizootic of young fry.[53] Diseased pike fry often have pale gills and severe hemorrhages in their trunks, occurring as bilateral red swollen areas particularly above the pelvic fins. Severe hemorrhages are also observed in muscle connective tissue.

Virus has been observed by electron microscopy in the hematopoietic tissues of kidneys of diseased pike, but not in the skin, muscle, or nervous system tissue.[53]

III. DEFENSE MECHANISMS AND PERSISTENT INFECTIONS OF FISH

That the various components of the fish specific immune response, the cellular inflammatory response, interferon, and nonspecific humoral factors all possibly play a role in determining the outcome of a viral disease is suggested by the fact that the temperature range supporting disease in fish (low temperature for IHN and VHS viruses) often does not correspond to the in vitro range of temperatures which support virus replication. Those components of the teleost protective responses that have been studied, have been shown to be temperature dependent. It is quite possible that at low temperatures these responses may be too slow or weak to combat virulent virus-induced diseases. This is a unique and an important factor in the survival of nonhomeothermic vertebrates, by comparison to their homeothermic counterparts, and possibly a factor which the viruses have, through evolution, adapted to exploit.

There are only a few reports of the role of specific immunity in piscine rhabdovirus infections. Jørgensen described the development of low titered neutralizing antibody in trout given repeated inoculations of Egtved virus.[56] However, de Kinkelin and Dorson[57] have failed to detect neutralizing antibody in trout surviving natural outbreaks of the disease.

It has been suggested that adult trout are asymptomatic carriers of Egtved virus,[58,59] and this may also be true for IHN, ODS, and SRCD viruses.[40]If so, then how the viruses persist is an interesting question.

Fijan has asserted that carp surviving experimental infections with SVCV are not susceptible to disease by reinfection with the virus.[60] Arshaniza and associates have likewise described a "relative immunity" of carp which survive the swim bladder inflammation syndrome induced by SVCV.[61] Although this has yet to be confirmed, it is commonly held that for natural infections caused by SVCV (as well as each of the other fish rhabdoviruses) the adult infected fish "carriers" of the virus are responsible for the year to year maintenance of the infectious cycle.[57]

TABLE 1

Molecular Weight ($\times 10^{-3}$) Estimate of Fish Rhabdovirus Virion Proteins

VIRUS	L	G	N	NS	M1	M2
Pike fry rhadbovirus (PFR)	160 ± 10	85 ± 10	50 ± 5	40—50(P)	—	23 ± 5
Spring viremia of carp virus (SVCV)	160 ± 10	85± 10	50 ± 6(P)	40 ± 5(P)	—	23 ± 2
Infectious hematopoietic necrosis virus (IHNV)	157 ± 10	72 ± 10	40 ± 4(P)	—	23 ± 3(P)	20 ± 1
Viral hemorrhagic septicemia (VHS) virus	157 ± 10	74 ± 10	41 ± 4(P)	—	22 ± 2(P)	19 ± 1

Note: Molecular weight estimates are the average of the reported values.[62-67,69,81] Since different gel systems give different results for the same protein (particularly for NS protein, see Figure 1), the averages quoted reflect these variations. Whether the N protein of SVCV, IHNV, or VHS virus is phosphorylated (P), is a moot question, see Figure 2. It could represent an alternate form of another phosphorylated virion polypeptide, i.e., NS or M1. The molecular weight estimates of the phosphoproteins may be overestimated if their mobilities are affected by their resident phosphate charges.

IV. MOLECULAR ASPECTS OF FISH RHABDOVIRUSES

A. Structural Proteins

The sizes and location of the major structural proteins of the four fish rhabdoviruses are similar to those of other rhabdoviruses.[62-65] All four viruses have an outer glycoprotein (G), an internal nucleocapsid protein (N) associated with viral RNA, and either one internal membrane protein (M) for SVCV and PFR, or two (M1 and M2) in the case of VHS and IHN viruses. All four viruses have minor amounts of a large protein (L) while SVCV and PRF have minor amounts of phosphosylated NS type protein (Table 1).

Both salmonid viruses, IHN and VHS, resemble rabies virus in that they have the second membrane protein.[64] They and SVCV also appear to have a phosphorylated nucleocapsid protein.[64-66] The M1 protein of IHNV appears to be phosphorylated although the M1 protein of VHS virus is not phosphorylated.[64]

Of all the fish viruses, PFR is the most similar to VSV Indiana (Figure 1) since it has three major structural proteins (G, N, and M) and two minor proteins (L and NS) with NS as the only phosphorylated protein[65] (Figure 1). Whether the NS protein of PFR is a transcriptase component remains to be determined. The functions of the two phosphoproteins found in SVCV virions[66] are also not known and, in fact, there is some evidence that the phosphorylated protein that has an electrophoretic mobility like that of the N protein may be distinguishable from the bulk of the nucleocapsid protein,[67] raising the question of whether it is a phosphorylated N, or an alternative phosphorylated form of the other phosphoprotein, NS (Figure 2).

The single, glycosylated, structural protein found in all four virus types can be readily removed by proteolytic digestion leaving a spikeless particle. In the presence of a nonionic detergent and salt, both the G and M proteins of IHN and VHS viruses are solubilized and can be freed from the nucleocapsid,[64] suggesting that they may be associated with the viral envelope.

B. Viral RNA Genome

The genome of the four fish rhabdoviruses has been shown by ribonuclease digestion and the specific incorporation of ribonucleosides to consist of a single-stranded RNA molecule essentially similar in size to that of VSV Indiana.[82]

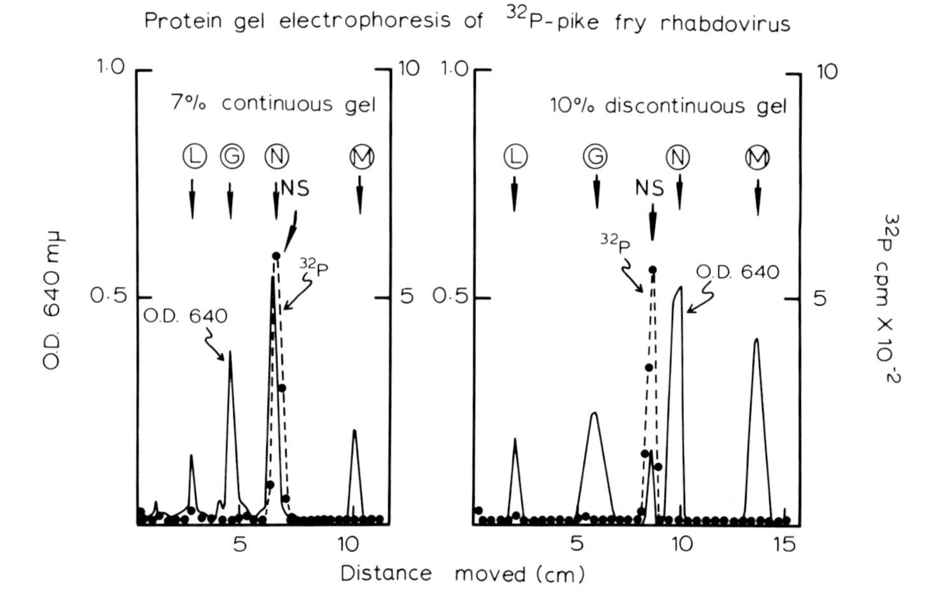

FIGURE 1. Pike fry rhabdovirus proteins. A preparation of ^{32}P-labeled pike fry rhabdovirus was freed from labeled phospholipids, RNA, and oligonucleotides, then dissociated and resolved by polyacrylamide gel electrophoresis by using a continuous or discontinuous gel system.

The 5′ terminal nucleotide SVCV is pppAp like that of VSV Indiana. It is neither capped nor methylated[68,69] (Figure 3). Base ratio analyses of PFR viral RNA indicate that it has an overall composition of 24.9% AMP, 20.2% GMP, 22.5% CMP, and 32.4% UMP.[65] The oligonucleotide fingerprint of the SVCV genome RNA, shown in Figure 4, is quite easily distinguished from those of other rhabdoviruses so far examined.[69]

C. The Viral Transcriptase Activity

The specific activity of the virion RNA-dependent RNA polymerase (transcriptase) varies considerably among the different rhabdovirus serotypes that have been studied.[65,68-73] At least two of the four fish rhabdoviruses (PFR and SVCV) appear to have quite active endogeneous transcriptase activities,[65,68,69] while IHNV and VHS have lower specific activities.[74,82] For all four viruses the transcriptase temperature optimum is significantly lower than that of VSV Indiana — reflecting the lower temperature optimum for growth.

Evidence has been obtained indicating that the transcriptase of PFR and SVCV synthesizes viral complemetary RNA not only in vitro, but also in vivo in cyclohexamide treated or untreated cells, as shown for a variety of other rhabdoviruses[64,68] (Figures 5 and 6). Unlike PFR, the polymerase of SVCV is stimulated in vitro when a methyl donor (S-adenosyl-L-methionine, SAM) is present. The effects of SAM upon the reaction rates of VSV Indiana, PFR, or SVCV virion transcriptases are shown in Figure 7.

For both salmonid viruses, IHN and VHS, the virion polymerase activity is stimulated by the presence of manganese ions which contrast the magnesium requirement of other rhabdoviruses.[74]

The most striking feature of the fish rhabdovirus transcriptases is their temperature optima which is between 15 and 22°C.[65,68,69,74]

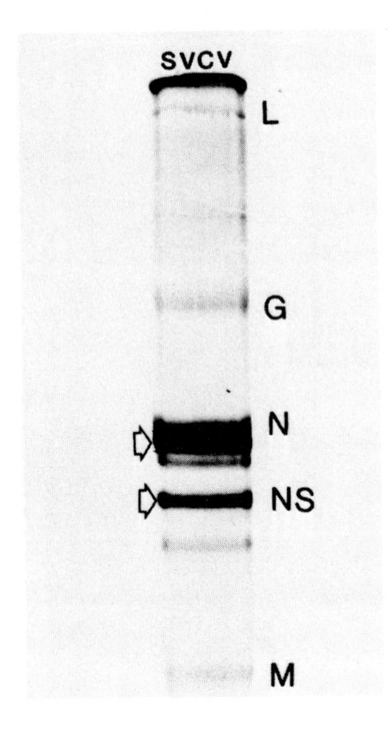

FIGURE 2. High resolution gel electrophoresis of SVCV phosphoproteins: a preparation of ^{35}S-methionine and ^{32}P-phosphate labelled SVCV was extracted for proteins and resolved by electrophoresis on a 30 cm 8% polyacrylamide slab gel at pH 7.0. After autoradiography the positions of the major viral polypeptides (L, G, N, and M) were identified as faint bands (as indicated by the adjacent letters), while the two phospoproteins (arrows) were identified as well pronounced bands. Staining the gel after autoradiography confirmed the positions of the major viral polypeptides (L, G, N, NS, and M). The two unassigned faint bands which migrated respectively further than N phosphoprotein, or NS protein, did not show up as stained polypeptides and presumably are contaminant phosphoproteins of unknown origin. They were not studied further.

D. Protein Kinase Activity of Fish Rhabdoviruses

The presence of a virion protein kinase activity capable of phosphorylating certain viral polypeptides has been demonstrated for VSV and all rhabdoviruses of the vesiculovirus group.[75-77] The origin of this protein kinase enzyme is unknown although specific activity measurments made by Imblum and Wagner[77] for VSV Indiana grown in different cell types have suggested that the enzyme may be a host function which is specifically picked up by budding virus particles. From all the evidence obtained so far it is clear that the preferred substrate for the protein kinase enzyme is the resident viral phosphoprotein. Evidence obtained in the laboratories of Bishop[83] has indicated that the phosphorylation obtained with the protein kinase for VSV Indiana is not a replacement phosphorylation of preexisting phosphate groups on the NS protein, but rather additional phosphorylation.

The proteins kinase activity has been investigated for PFR and SVCV grown in both fathead minnow cells (FHM) and BHK-21 cells. For PFR grown in either cell type, analyses of the phosphate acceptor proteins for endogenously templated reactions indicate that the NS protein is the preferred substrate, although M protein is also apparently phosphorylated (Figure 8). For SVC virus grown in both cell types the preferred substrates of the endogenous protein kinase activity are the two SVCV phophoproteins

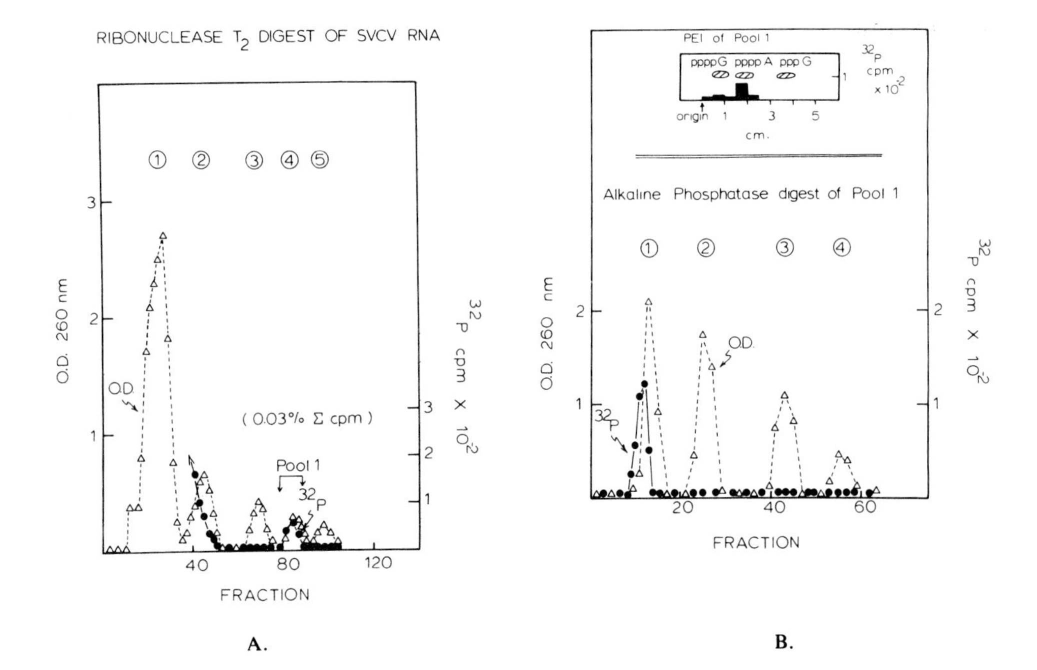

FIGURE 3. Identification of the SVCV RNA 5′-terminal nucleotide by DEAE-cellulose column chromatography. ^{32}P-labeled SVCV viral RNA was digested with RNase T_2, and the resulting nucleotides resolved on DEAE-cellulose, at pH 8, with a pancreatic RNase digest of chicken embryo fibroblast RNA (A). The 5′-terminal nucleotide fragment was recovered, and a portion was treated with alkaline phosphatase and rechromatographed on DEAE-cellulose column (B). A second sample of the termini was analyzed by thin-layer chromatography on PEI cellulose (B insert).

(Figure 9). The protein kinase temperature optimum for SVCV grown in FHM cells is between 25 and 30°C, whereas the temperature optimum for the BHK-21 grown SVCV is between 30 and 35°C. For virus prepared from either cell type, the two phosphoproteins are phosphorylated at either low or high temperatures (Figure 9) with the NS protein being the preferred substrate.

V. INTERFERON

Only very preliminary data is available concerning the nature of fish interferon. Serum interferon induction in trout inoculated with Egtved virus and held at 15°C, has been documented. Titers as high as 2750 units/mℓ of serum were detected 3 days after infection.[78] It has been suggested that interferon production may play a role in rendering trout resistant to virus infection at temperatures above 15°C. Also it has been suggested that the interferon response depends on the body temperature of the host.[41]

VI. PRODUCTION OF DEFECTIVE INTERFERING "T" PARTICLES

The role of "T" particles in fish rhabdovirus infections is unknown. Autointerference associated with the infection of salmon cells with IHNV at high multiplicities of infection and concomitant T particle production has been reported.[79]

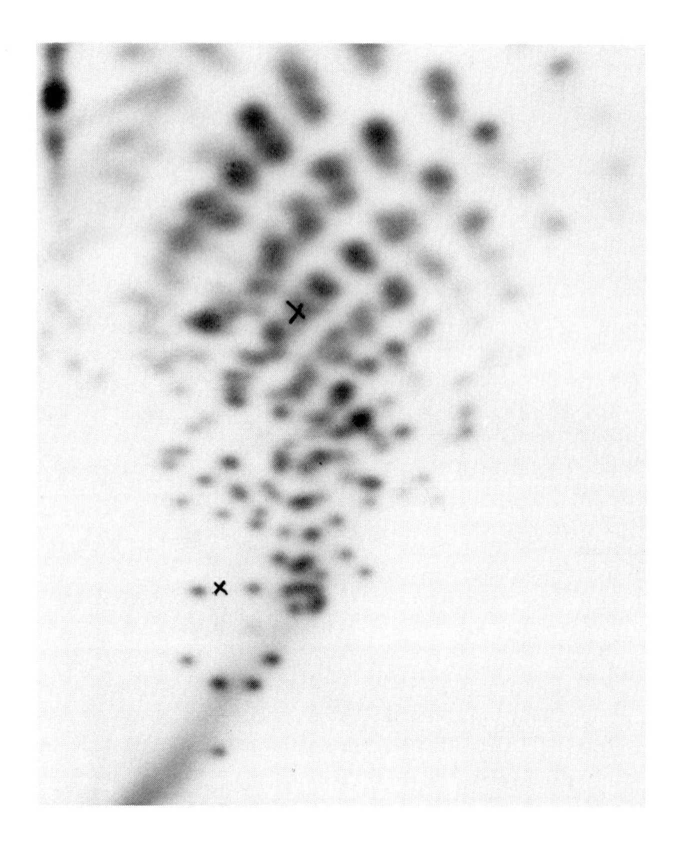

FIGURE 4. Oligonucleotide fingerprint of SVCV B particle RNA. A total of 5 × 10^6 cpm of RNA extracted from purified virions was digested with RNase T1 and resolved by two-dimensional gel electrophoresis. The first dimension was from left to right and the second from bottom to top. The positions of the two dye markers are indicated.

VII. THE SEROLOGICAL RELATIONSHIP OF THE FISH RHABDOVIRUSES

Jørgenson[80] has investigated the antigenic relationship of 76 isolates of VHS (Egtved) from Danish, Norwegian, and Swedish rainbow trout, and one isolate from Italian brown trout. Using neutralization of infectivity tests, 72 of the isolates were found to be essentially identical. However, three Danish virus isolates and one Norwegian isolate differed significantly from the others, suggesting that there may be a second serotype of VHS virus.

Jørgensen has also shown that IHNV can be neutralized to some extent with an anti-Egtved serum,[80] although fluorescent antibody staining of IHNV infected cells by anti-Egtved fluorescein conjugated serum did not confirm this observation.

McCain and associates[26] using plaque reduction tests with the three salmonid rhabdoviruses OSDV, SRCSDV, and IHNV reported that all three virus isolates are antigenically related to each other and that OSDV and IHNV are indistinguishable by this test.

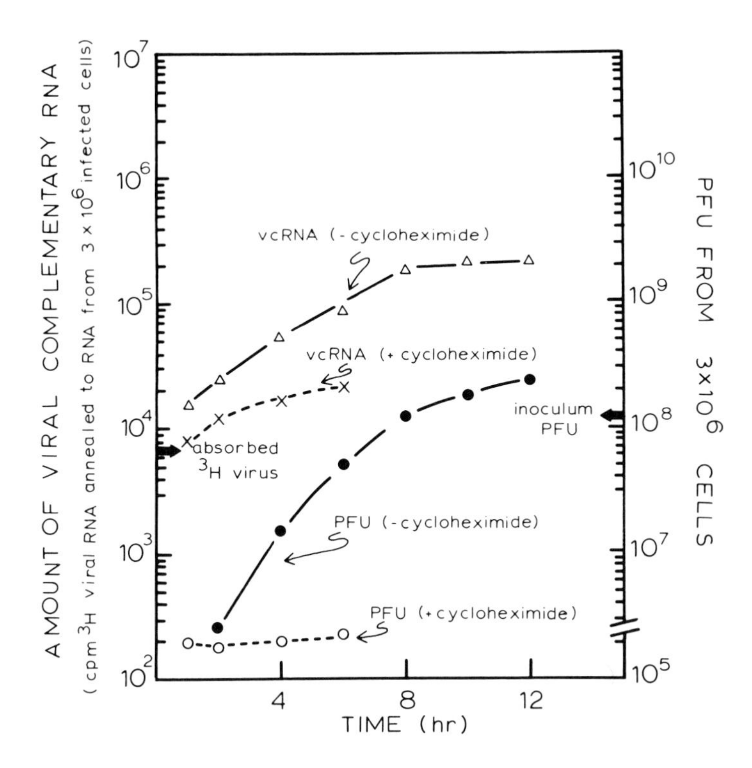

FIGURE 5. In vivo synthesis of viral complementary (vc), RNA by pike fry rhabdovirus in FHM cells. A preparation of [^{3}H]cytidine and [^{3}H]adenosine-labeled virus (specific activity, 4 × 10^8 counts per minute per milligram of RNA; i.e., 1 count/min equalled 4 × 10^5 RNA molecules) was used to infect FHM cell monolayers (3 × 10^6 cells per monolayer), and the RNA was extracted at various times postinfection. The amount of ^{3}H-labeled ribonuclease resistance in the RNA was determined before or after self-annealing. The content of viral-complementary RNA in the extracted nucleic acids was determined. Cell monolayers were incubated in the presence or absence of 100 μg of cycloheximide per milliliter, and the occurrence and release of infectious virus into the supernantant fluids were monitored by plaque assays. The average ^{3}H recovered per monolayer (6.8 × 10^3 counts per min per monolayer) is indicated on the left-hand ordinate, while the number of PFUs originally used to infect each monolayer (1.2 × 10^8 PFU per monolayer) is indicated on the right-hand ordinate. From the RNA-specific activity, the amount of ^{3}H recovered from the monolayers and the PFU applied, it was calculated that the original virus inoculum possessed a particle to PFU ratio of about 20:1.

It has also been shown by cross neutralization tests that SVCV and isolates obtained from carp swim bladder inflammation disease are indistinguishable.[29] Such tests indicate that SVCV is not related to PFR, IHNV, or Egtved virus,[63,65] although in recent studies, Hill and collaborators obtained some cross-neutralization between SVCV, PFR, and IHN viruses, but none with VHS virus.[81]

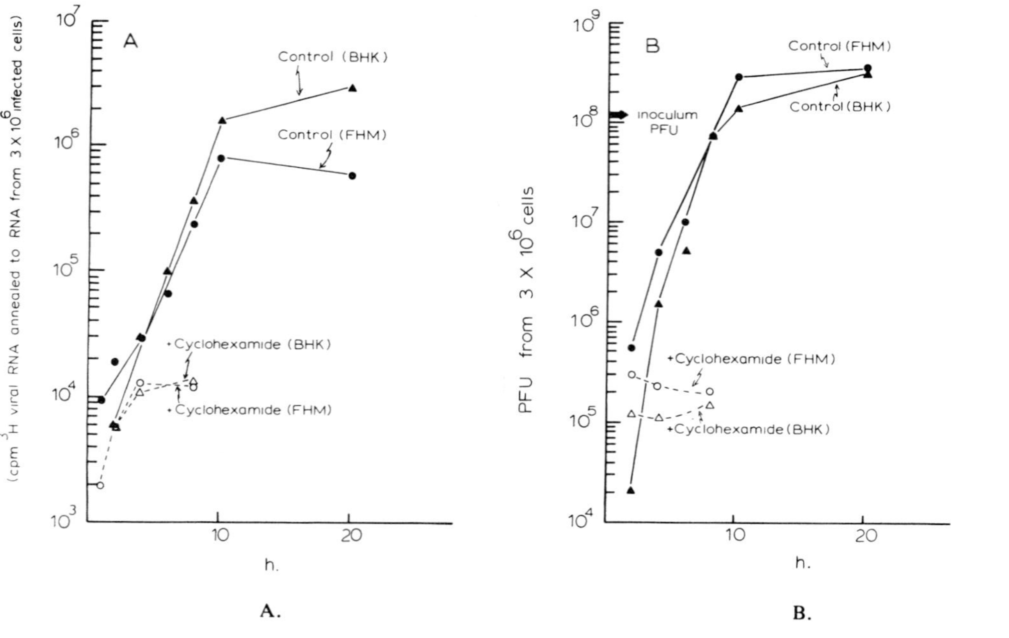

FIGURE 6. The in vivo synthesis of viral complementary RNA by SVCV in BHK and FHM cells. The intracellular viral complementary (vc) RNA produced by SVCV in BHK- or FHM-infected cells was determined at various times postinfection (A). The cell monolayers were incubated at 20°C in the presence or absence of 100μg of cycloheximide per milliliter, and the release of infectious virus into the supernatant fluids (B) was monitored by plaque assays.

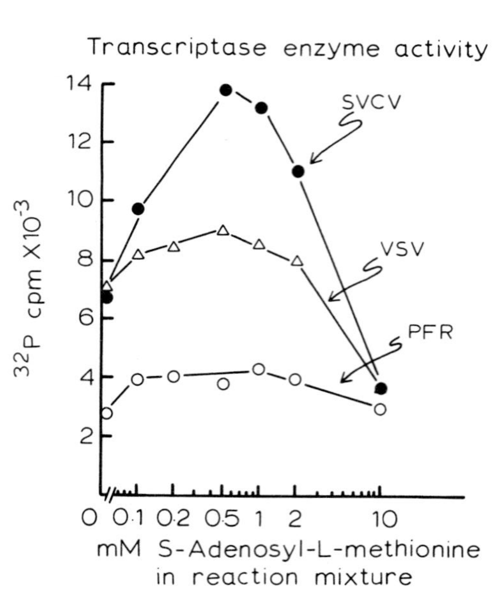

FIGURE 7. Effect of SAM concentration on VSV, PFR, and SVCV virion RNA polymerases. The incorporations of [^{32}P] AMP into product RNA by standard reaction mixtures containing [α-^{32}P]ATP (specific activity, 5mCi/μmol), various concentrations of SAM, and templated by VSV, PFR, or SVCV were determined. The PFR and SVCV reaction mixtures were incubated at 20°C, whereas the VSV reaction mixture was incubated at 28°C. The [^{32}P]AMP incorporations, which were linear for 4 hr of incubation, are plotted.

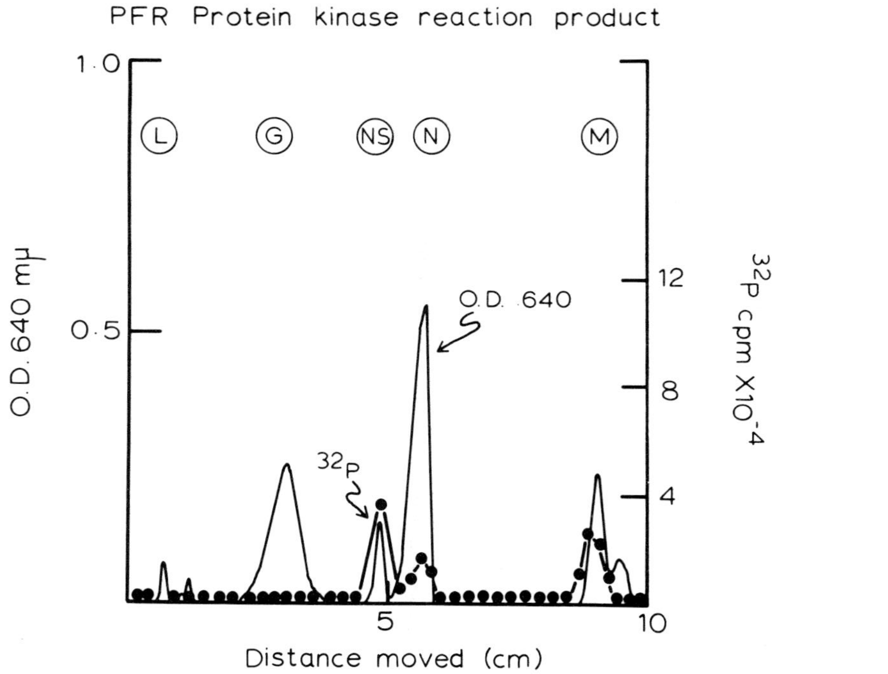

FIGURE 8. In vitro phosporylation of pike fry rhabdovirus proteins. A protein kinase 1-hr reaction product (incubated at 30°C) was purified from [γ-^{32}p]ATP, dissociated, and the distributions of labeled proteins was determined by discontinuous polyacrylamide gel electrophoresis.

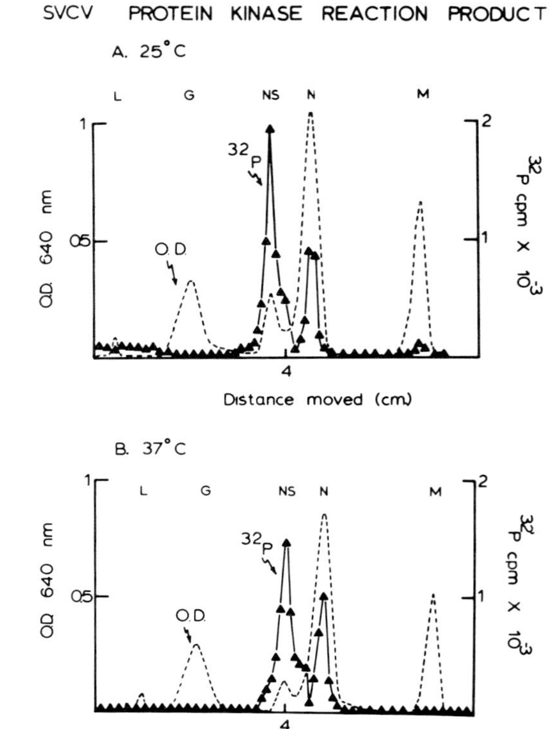

FIGURE 9. Protein kinase reaction products of SVCV. Protein kinase reaction products of SVCV grown in BHK cells, labelled by [γ-^{32}P]ATP, and incubated for 2 hr at (A) 25°C or (B) 37°C, followed by repetitive 5% TCA precipitation, were dissolved in 1% SDS, 1% β-mercaptoethanol, and 0.01 *M* phosphate buffer pH 7.0, and resolved by discontinuous gel electrophoresis at pH 8.9. After staining, the distribution of the ^{32}P was determined.

REFERENCES

1. **Wolf, K., Snieszko, S. F., Dunbar, C. E., and Pyhle, E.,** Virus nature of infectious pancreatic necrosis in trout, *Proc. Soc. Exp. Biol. Med.,* 104, 105, 1960.
2. **Wolf, K.,** Advances in fish virology: a review. 1966—1971. *Symp. Zool. Soc. London,* 30, 305, 1972.
3. **Wolf, K. and Quimby, M. C.,** Fish viruses: buffers and methods for plaquing eight agents under normal atmosphere, *Appl. Microbiol.,* 25, 659, 1973.
4. **de Kinkelin, P., Galimard, P. B., and Bootsma, R.,** Isolation and identification of the causative agent of "Red Disease" of pike (*Esox lucius L.* 1766), *Nature (London),* 241, 465, 1973.
5. **Schaperclaus, W.,** Untersuchungen uber die ansteckende Bauchwassersucht des Karpfens und ihre Bekampfung, *Allg. Fisch. Ztg.,* 37, 1, and 127, 1939.
6. **Schaperclaus, W.,** *Fischkrankheiten,* 3rd ed., Akademie Verlag, Berlin, 1953, 537.
7. **Heuschmann, O.,** Bauchwassersucht bei Regenbogenforellen?, *Allg.Fisch. Ztg.,* 77, 214, 1952.
8. **Liebmann, H.,** Ernarungsstorung und Degeneration als primare Ursache der Bauchwassersucht bei Fischen, *Allg. Fisch. Ztg.,* 81, 68, and 88, 1956.
9. **Klingler, K.,** Die "neue Forellenkrankheit", *Schweiz, Fischereiztg.,* 65, 65, 1957.
10. **Numann, W. and Deufel.,** Vorlaufige Ergebnisse unserer Untersuchungen uber die "neue" Forellenkrankheit,, *Allg. Fisch. Ztg.,* 81, 244, 1956.
11. **Tack, E.,** Ist die neu aufgetretene Forellenkrankheit ansteckend?, *Allg. Fisch. Ztg.,* 82, 61 and 307, 1957.
12. **Rasmussen, C. J.,** Nogle forelobige undersogelser over regnbueorrendens virussygdom (Egtvedsygen), *Med. Forsogsdambruget,* 10, 1, 1959.
13. **Besse, P.,** Recherches sur l'etiologie de l'anemie infectieuse de la truite, *Bull. Acad. Vet. Fr.,* 28, 194, 1955.
14. **Besse, P.,** L'anemie pernicieuse des truites, *Ann. Stn. Cent. Hydrobiol. Appl.,* 6, 441, 1956.
15. **Bellet, R.,** Du syndome entero-hepato-renal chez la truite "arc-en-ciel" de pisciculture, *Bull. Fr. Piscic.,* 189, 113, 1958.
16. **Scholari, C.,** Su di una epizoozia delle trotte iridee d'allevamento "La lipoidosi epatica", *Clin. Vet.,* 77, 102, 1954.
17. Conference in Turin, Italy, October 20—24, 1962, *Off. Int. Epizoot. Bull.,* 59, 1, 1963.
18. **Zwillenberg, L. O., Jensen, M. H., and Zwillenberg, H. H. L.,** Electron microscopy of the virus of viral haemorrhagic septicaemia of rainbow trout (Egtved virus), *Arch. Gesamte Virusforsch.,* 17, 1, 1965.
19. **Bellet, P.,** Viral hemorrhagic septicemia (VHS) of the rainbow trout bred in France, *Ann. N. Y. Acad. Sci.,* 126, 461, 1965.
20. **Wolf, K.,** The fish viruses, *Adv. Virus Res.,* 12, 35, 1966.
21. **Rucker, R. R., Whipple, W. J., Parvin, J. R., and Evans, C. A.,** A contagious disease of salmon possibly of virus origin, *U.S. Fish Wildl. Serv. Fish Bull.,* 76, 35, 1953.
22. **Watson, S. W., Guenther, R. W., and Rucker, R. R.,** A virus disease of sockeye salmon: interim report, *U. S. Fish Wildl. Serv. Spec. Sci. Rep. Fish.,* 138, 35, 1954.
23. **Ross, A. J., Pelnar, J., and Rucker, R. R.,** A virus-like disease of Chinook salmon, *Trans. Am. Fish. Soc.,* 89, 160, 1960.
24. **Wingfield, W. H. and Chan, L. D.,** Studies on the Sacramento River Chinook disease and its causative agent, in A symposium on diseases of fishes and shellfishes, Sniezski, S. F., Ed., *Am. Fish Soc. Spec. Publ.,* 5, 1970, 307.
25. **Amend, D. F., Yasutake, W. T., and Mead, R. W.,** A hematopoietic virus disease of rainbow trout and sockeye salmon, *Trans Am. Fish Soc.,* 98, 796, 1969.
26. **McCain, B. B., Fryer, J. L., and Pilcher, K. S.,** Antigenic relationship in a group of three viruses of salmonid fish by cross neutralization, *Proc. Soc. Exp. Biol. Med.,* 137, 1042, 1971.
27. **Wolf, K. and Quimby, M. C.,** in Progress in Sport Fishery Research. Resource Publication 106, U.S. Department of the Interior, Bureau of Sport Fish and Wildlife, Washington, D.C., 86, 1970.
28. **Fijan, N., Petrinec, Z., Sulimanovic, D., and Zwillenberg, L. O.,** Isolation of the viral causative agent from the acute form of infectious dropsy of carp, *Vet. Arh.,* 41, 125, 1971.
29. **Bachman, P. A. and Ahne, W.,** Isolation and characterization of agent causing swim bladder inflammation in carp, *Nature (London),* 244, 235, 1973.
30. **Bachman, P. A. and Ahne, W.,** Biological properties and identification of the agent causing swim bladder inflammation in carp, *Arch. Gesamte Virusforsch.,* 44, 261, 1974.
31. **Rasmussen, C. J.,** A biological study of the Egtved Disease (INUL), *Ann. N.Y. Acad. Sci.,* 126, 468, 1965.

32. **Ghittino, P.**, Viral hemorrhagic septicemia (VHS) in rainbow trout in Italy, *Ann. N. Y. Acad. Sci.*, 126, 468, 1965.
33. **Klingler, K.**, Forellenfutterung und "neue Krankheit" (infektiöser Nierenschwellung und Leberdegeneration der Regenbogenforellen-INUL), *Allg. Fisch. Ztg.*, 83, 12, 1958.
34. **Lauridsen, O.**, Undersøgelser over den sakaldte ørredwirussygdom hos regnbue ørred, *Nord. Veterinaermed.*, 10, 553, 1958.
35. **Schmidt, B.**, Histologische untersuchungen an der leber von gesunden und an "Infektiöser nierenschwellung und leberdegeneration" erkrankten forellen, *Inaug. Diss. Tierarzt. Fak. Munchen*, 1960.
36. **Stroh, R.**, Histologische untersuchungen an der niere van gesunden und an "Infektiöser neirenschwellung und leberdegeration" erkrankten forellen, *Inaug. Diss. Tierarzt. Fak. Munchen*, 1960.
37. **Ghittino, P.**, "L'ipertrofia renale e degenerazione epatica infettiva" della trota iridea de allevamento *(Salmo gairdnerii), Vet. Ital.*, 13, 457, 1962.
38. **Ghittino, P.**, Differential-Diagnose der bei Regenbogenforellen vorkommenden "Lipoide Leberdegeneration" und "Infektiöser nierenschwellung und leberdegeneration" auf grund histopathologisch-anatomischer untersuchungen, *Allg. Fisch. Ztg.*, 89, 549, 1962.
39. **Parisot, T. J., Lasutake, W. T., and Klantz, G. W.**, Virus diseases of the Salmonidae in Western United States. I. Etiology and epizootiology, *Ann. N. Y. Acad. Sci.*, 126, 502, 1965.
40. **Amend, D. F.**, Infectious hematopoietic necrosis (IHN) virus disease, U.S. Department of Fish and Wildlife Service, FDL-39, 1974.
41. **Amend, D. F.**, Control of infectious hematopoietic necrosis virus disease by elevating the water temperature, *J. Fish. Res. Board Can.*, 27, 1285, 1970.
42. **Amend, D. F.**, Infectious hematopoietic necrosis (IHN) virus disease. U.S. Department of Interior, Fish and Wildlife Service, FDL-39, 1974.
43. **Darlington, R. W., Trafford, R., and Wolf, K.**, Fish rhabdoviruses: morphology and ultrastructure of North American salmonid isolates, *Arch. Gesamte Virusforsch.*, 39, 257, 1972.
44. **Yasutake, W. T. and Amend, D. F.**, Some aspects of the pathogenesis of infectious hematopoietic necrosis, *J. Fish Biol.*, 4, 261, 1972.
45. **Amend, D. F. and Chambers, V. C.**, Morphology of certain viruses of salmonid fishes. I. In vitro studies of some viruses causing hematopoietic necrosis, *J. Fish. Res. Board Can.*, 27, 1285, 1970.
46. **Amend, D. F., Yasutake, W. T., Fryer, J. L., Pilcher, K. S., and Wingfield, W. H.**, Infectious hematopoetic necrosis (IHN), *Eur. Inland Fish. Commis., FAO Tech. Pap.*, 17 (Suppl. 2), 80, 1973.
47. **Clark, H F. and Soriano, E. Z.**, Fish rhabdovirus replication in non-piscine cell culture: new system for the study of rhabdovirus-cell interaction in which the virus and cell have different temperature optima, *Infect. Immun.*, 10, 180, 1974.
48. **Roegner-Aust, S., Brunner, G., and Jaxtheimer, R.**, Elektronenmikroskopische Untersuchungen über den Erreger der inf. Bauchwassersucht der Karpfen-Bakterium? — Virus?, *Allg. Fisch. Ztg.*, 75, 420, 1950.
49. **Goncharov, G. D.**, Tr. *Soveščanij Ichtiologičeskij Komisii ANS-SSSR*, 9, 39, 1959.
50. **Kocylowski, B.**, The role of virus in septicemia of carp (*Cyprinus carpio*) and pox of carp. Influence of environment on infection, *Ann. N.Y. Acad. Sci.*, 126, 616, 1965.
51. **Otte, E.**, Die neutigen ansichten über die ätriologie der "Infektiösen bauchwassersucht" der karpfen, *Wien. Tieraerztl. Wochenschr.*, 11, 995, 1963.
52. **Bootsma, R.**, Hydrocephalus and red-disease in pike fry (*Esox Lucius L.*), *J. Fish Biol.*, 3, 417, 1971.
53. **Bootsma, R. and Van Vorstenbosch, C. J. A. H. V.**, Detection of a bullet-shaped virus in kidney sections of pike fry *(Esox lucius L.), Neth. J. Vet. Sci.*, 98, 86, 1973.
54. **Bootsma, R., de Kinkelin, P., and Le Berre, M.**, Transmission experiments with pike fry (*Esox lucius L.*) rhabdovirus, *J. Fish Biol.*, 7, 269, 1975.
55. **von Drimmelen, D. E.**, Pike Culture, Special Report for the Organization for Improvement of Inland Fisheries (O. V. B.), Utrecht, The Netherlands, 1969.
56. **Jørgenson, P. E. V.**, Egtved virus: demonstration of neutralizing antibodies in serum from artificially infected rainbow trout, *J. Fish. Res. Board Can.*, 28, 875, 1971.
57. **de Kinkelin, P. and Dorson, M.**, Interferon production in rainbow trout (*Salmo gairdnerii*) expermentally infected with Egtved virus, *J. Gen. Virol.*, 19, 125, 1973.
58. **Zwillenberg, L. O. and Zwillenberg, H. H. L.**, Transmission and recurrence problems in viral haemorrhagic septicaemia of rainbow trout, *Bull. Off. Int. Epizool.*, 69, 969, 1968.
59. **Christensen, N. O.**, Some diseases of trout in Denmark, *Symp. Zool. Soc. London*, 30, 83, 1972.
60. **Fijan, N. N.**, Infectious dropsy in carp — a disease complex, *Symp. Zool. Soc. London*, 30, 39, 1972.
61. **Arshaniza, N. M., Bauer, O. N. and Vladimirow, V. L.**, Air bladder disease of carps; its aetiology, epizootiology and control, *Bull. Off. Int. Epizool.*, 69, 999, 1968.
62. **de Kinkelin, P., LeBerre, M., and Lenoir, G.**, Rhabdovirus des poissons. I. Propriétés in vitro du virus de la maladie rouge de l'alevin de brochet, *Ann. Microbiol. Inst. Pasteur*, 125A, 93, 1974.

63. **de Kinkelin, P. and LeBerre, M.**, Rhabdovirus des poissons. II Propriétés in vitro du virus de la virémie printanière de la carpe, *Ann. Microbiol. Inst. Pateur,* 125A, 113, 1974.
64. **McAllister, P. E. and Wagner, R. R.**, Structural proteins of two salmonid rhabdoviruses, *J. Virol.,* 15, 733, 1975.
65. **Roy, P., Clark, H F., Madore, H. P., and Bishop, D. H. L.**, RNA polymerase activity associated with virions of pike fry rhabdovirus, *J. Virol.,* 15, 338, 1975.
66. **Sokol, F., Clark, H F., Wiktor, T. J., McFalls, M. L., Bishop, D. H. L., and Obijeski, J. F.**, Structural phosphoproteins associated with ten rhabdoviruses, *J. Gen. Virol.,* 24, 433, 1974.
67. **Roy, P. and Clewley, J. P.**, Phosphoproteins of spring viremia of carp virus and other rhabdoviruses, in *Negative Strand Viruses,* Barry, R. D., and Mahy, B. W. A., Eds., Academic Press, London, 1978, 117.
68. **Roy, P.**, The transcription process of the RNA polymerase of spring viremia of carp virus, *Fed. Proc. Fed. Am. Soc. Exp. Biol.,* 35, 3346, 1976.
69. **Roy, P. and Clewley, J. P.**, Spring viremia of carp virus ribonucleic acid and virion associated transcriptase activity, *J. Virol.,* 25, 912, 1977.
70. **Knudson, D. L.**, Rhabdoviruses, *J. Gen. Virol.,* 20(Suppl.), 105, 1973.
71. **Wagner, R. R.**, Reproduction of rhabdoviruses, in *Comprehensive Virology,* Vol. 4, Fraenkel-Conrat, H. and Wagner, R. R., Eds., Plenum Press, New York, 1975, 1.
72. **Aaslestad, H. G., Clark, H F., Bishop, D. H. L., and Koprowski, H.**, Comparison of the ribonucleic acid polymerases of two rabdoviruses, Kern Canyon virus and vesicular stomatitis virus, *J. Virol.,* 7, 726, 1971.
73. **Chang, S. H., Hefti, E., Obijeski, J. F., and Bishop, D. H. L.**, RNA transcription by the virion polymerase of five rhabdoviruses, *J. Virol.,* 13, 652, 1974.
74. **McAllister, P. E. and Wagner, R. R.**, Virion RNA polymerases of two Salmonid rhabdoviruses, *J. Virol.,* 22, 839, 1977.
75. **Moyer, S. A. and Summers, D. F.**, Phosphorylation of vesicular stomatitis virus in vivo and in vitro, *J. Virol.,* 13, 455, 1974.
76. **Sokol, F. and Clark, H F.**, Phosphoproteins, structural components of rhabdoviruses, *Virology,* 52, 246, 1973.
77. **Imblum, R. L. and Wagner, R. R.**, Protein kinase and phosphoproteins of vesicular stomatitis virus, *J. Virol.,* 13, 113, 1974.
78. **de Kinkelin, P. and Dorson, M.**, Interferon production in rainbow trout (*Salmo gairdnerii*) experimentally infected with Egtved virus, *J. Gen. Virol.,* 19, 125, 1973.
79. **McAllister, P. E. and Pilcher, K. S.**, Autointerference in infectious hematopoietic necrosis virus of salmonid fish, *Proc. Soc. Exp. Biol. Med.,* 145, 840, 1974.
80. **Jørgensen, P. E. V.**, Egtved virus:antigenic variation in 76 virus isolates examined in neutralization tests and by means of the fluorescent antibody technique, *Symp. Zool. Soc. London,* 30, 333, 1972.
81. **Hill, B. J., Underwood, B. O., Smale, C. J., and Brown, F.**, Physicochemical and serological characterization of five rhabdoviruses infecting fish, *J. Gen. Virol.,* 27, 369, 1975.
82. **Roy, P.**, unpublished observations.
83. **Bishop, D. H. L.**, unpublished observations.

Chapter 10

REPTILE RHABDOVIRUSES

Thomas P. Monath

TABLE OF CONTENTS

I. INTRODUCTION

In the course of field studies on arboviruses in the Amazon basin, Causey et al. isolated three viruses from lizards collected near Belém, Pará State, Brazil, in 1962.[1] These viruses, named Marco, Timbo, and Chaco were found to be antigenically distinct from other known viruses; Timbo and Chaco viruses were serologically related to one another and thereby formed the Timbo serologic group. On the basis of their sensitivity to deoxycholate, pathogenicity for mice, and viral multiplication in the salivary glands of parenterally inoculated mosquitoes,[1,2] the viruses were classified as possible arboviruses.[2] The taxonomic status of Marco, Chaco, and Timbo viruses was only recently elucidated when they were found to be members of the Rhabdoviridae by electron microscopic studies.[3] Addition of these three viruses to the Rhabdoviridae bridges the gap between fish and homeothermic vertebrates and reconfirms the diversity of ecological relationships within the family. A fourth virus, Almpiwar virus, registered in the *International Catalogue of Arboviruses* [2], as been isolated from skinks *(Ablepharus boutonii virgatus)* in Australia; certain biological characteristics of this agent suggest that it, too, may be a rhabdovirus, but morphologic identification has not been made.

II. ECOLOGY

Between 1955 and 1963, 4766 lizards of at least 14 species were examined for viruses at the Belém Virus Laboratory by intracerebral (IC) inoculation of organ suspensions into suckling mice.[1] Most of the specimens (93%) were obtained in 1962 and 1963 during intensive studies on the role of lizards in viral transmission. Fourteen isolations were made, four of Marco, six of Timbo, and four of Chaco virus (Table 1). Thirteen strains were recovered from *Ameiva ameiva ameiva* and a single isolate of Chaco virus from *Kentropyx calcaratus* lizards. Marco, Timbo, and Chaco viruses were not recovered from over 3000 lizards of other species, nor from many thousands of birds, rodents, marsupials, bats, human beings, sentinel mice, and pools of arthropods tested by the Belém laboratory. All of the virus-positive specimens in 1962 to 1963 were captured at the edges of the Utinga and Instituto Agrônomico do Norte forests near Belém, where studies were concentrated. Since 1963, lizards have not been intensively studied in Brazil. A single isolation was made of an agent from *A. a. ameiva* collected in 1976 in the southwest Amazon region (Acre State); this virus is closely related to Timbo Virus.[4]* Neutralizing antibodies to Timbo virus have been found in 5 of 225 *A. a. ameiva* and 5 of 162 other reptiles tested from Pará State, Brazil.[2]

Marco and Chaco viruses, respectively, were carried through multiple serial passages in *Aedes aegypti* mosquitoes by intrathoracic inoculation of salivary gland material.[1,2] Transfer of Marco virus in *Anopheles quadrimaculatus* and of Chaco virus in *A. quadrimaculatus* and *Culex fatigans* failed after two passages.[1,2]

On the basis of these limited observations, few conclusions are possible regarding the ecology of the agents. Their association with lizards, and in particular with the teiid species *Ameiva a. ameiva,* appears well-founded, but it is not known whether this vertebrate develops a viremia sufficient to serve as a source for arthropod infection. Nosquito transmission is suggested by the experimental studies; a number of tropical mosquito species, e.g., of the subgenus *Culex (Melanoconion),* feed extensively on reptiles. The known distribution of the viruses is limited to the Amazon basin, but few attempts to recover viruses from reptiles have been made elsewhere in tropical America. Serological surveys of humans and domestic animals have not been conducted, and the potential public health importance of the viruses remains unknown.

* Mention of this viral isolation herein does not constitute priority description.

TABLE 1

Isolations of Rhabdoviruses (Marco, Timbo, and Chaco) Virus From Lizards Collected Near Belém, Pará State, Brazil[1]

Lizard species	Number tested[a]	Number (%) positive, by virus		
		Marco	Timbo	Chaco
Ameiva ameiva ameiva	1507	4 (0.2)	6 (0.4)	3 (0.2)
Kentropyx calcaratus	4	0	0	1 (25)
Other spp.[b]	3255	0	0	0

[a] Lizards captured at Utinga — Instituto Agrônomico do Norte forest fringe, 1955—1963.

[b] Includes 3226 lizards belonging to 12 species and 29 unidentified specimens.

From Causey, O. R., Shope, R. E., and Bensabath, G., *Am. J. Trop. Med. Hyg.*, 15, 239, 1966. With permission.

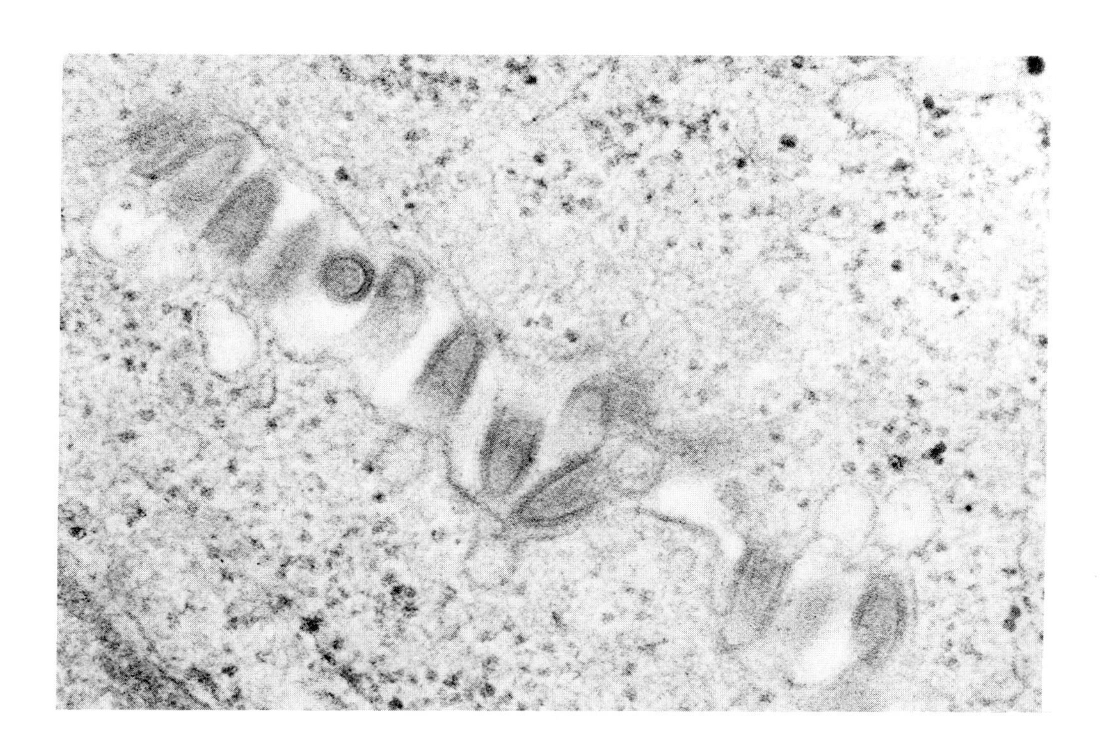

FIGURE 1. (Top of EM figure). Marco virus budding upon endoplasmic reticulum in Vero cells × 84,000. Viral particles are conically-shaped, resembling Kotonkan, Obodhiang, and bovine ephemeral fever viruses.

III. MORPHOLOGY

Marco, Chaco, and Timbo virus particles exhibit all of the structural details of typical rhabdoviruses (surface projections, membranous envelope, coiled nucleocapsid), but differences have been noted between Marco and the other two viruses in shape and size. Marco virus particles are conically-shaped (Figure 1), with a mean particle

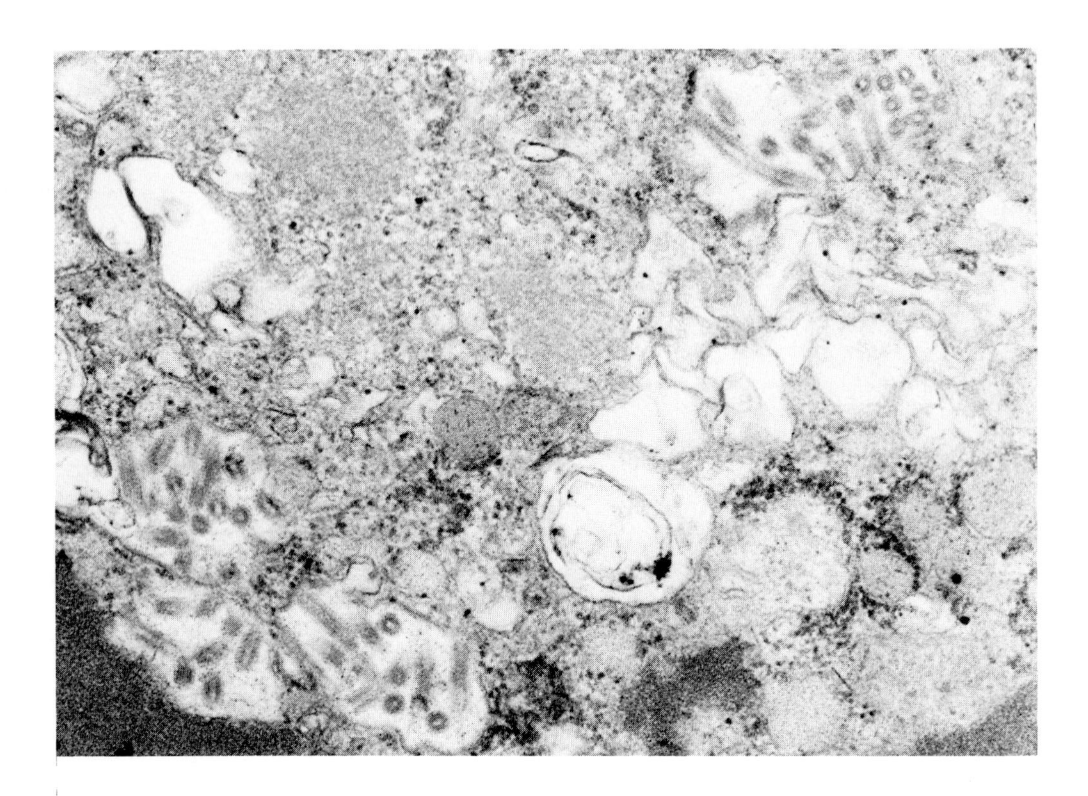

FIGURE 2. (Bottom of EM figure). Chaco virus particles in infected Vero cells × 49,000. Particles are 202 nm in length and cylindrically shaped.

length, as measured by thin section electron microscopy, of 180 nm. Budding occurs primarily from plasma membranes, but also from the endoplasmic reticulum. Marco virus most closely resembles Kotonkan, Obodhiang, and bovine ephemeral fever viruses. The shape of Chaco and Timbo virus particles is cylindrical (Figure 2). Mean particle length is 202 nm, and budding as been seen only upon endoplasmic reticulum. Chaco and Timbo viruses resemble other rhabdoviruses with particle lengths longer than the prototype VSV.

IV. PHYSICOCHEMICAL CHARACTERISTICS

Marco and Chaco viruses are sensitive to the effects of lipid solvents (sodium deoxycholate).[1] Co-electrophoretic analysis[5] of the proteins of 3_H-labeled Marco virus and 14_C-labeled VSV New Jersey in discontinuous-SDS polyacrylamide gels revealed differences in the mobilities of the G, N, and M viral proteins (Figure 3). The G and M proteins of Marco virus migrated faster than the corresponding VSV proteins, whereas the Marco viral nucleocapsid protein was slightly slower than that of VSV. The molecular weights of Marco viral proteins (estimated by comparison with mobilities of proteins of known molecular weight) are shown in Table 2. Comparative studies on Timbo and Chaco viruses have not been reported.

V. ANTIGENS AND ANTIGENIC RELATIONSHIPS

Hemagglutinins have not been produced from infected mouse brain extracted by sucrose acetone, but high-titered complement-fixing (CF) antigens are readily prepared by this method. Antigens and hyperimmune mouse ascitic fluids to Marco, Chaco,

TABLE 2

Molecular Weight (× 10³) Estimations of Marco and VSV Structural Proteins Separated in a Discontinuous—SDS—Polyacrylamide Gel System

Protein	Marco	VSV-New Jersey
L	149	148
G	49	61
N	42	50
M	32	24

TABLE 3

Cross-Complement Fixation Test Showing Relationship Between Timbo and Chaco Viruses and Lack of Relationship With Marco Virus[3]

	Hyperimmune mouse ascitic fluid		
Antigen	Marco	Timbo	Chaco
Marco	32/256[a]	0[b]	0
Timbo	0	512/256	16/64
Chaco	0	8/16	32/64

Note: No relationship between the three viruses and other rhabdoviruses has been found.

[a] Ab/Ag titer.
[b] <4/<4.

From Monath, T. P., Cropp, C. B., Murphy, F. A., and Frazier, C. L., *Arch. Virol.*, 60, 1—12, 1979. With permission.

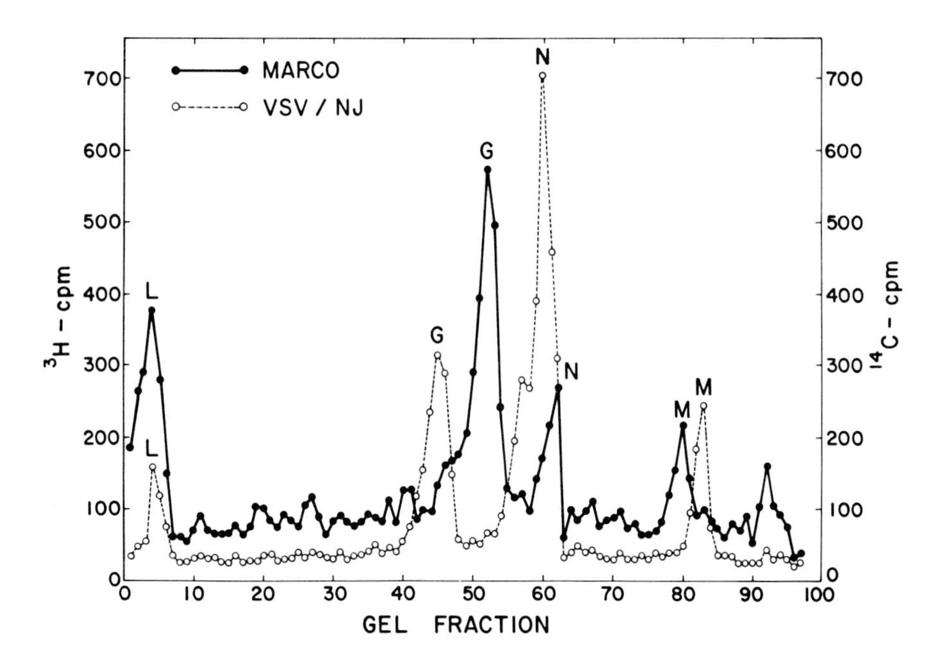

FIGURE 3. Co-electrophoresis in discontinuous-SDS-polyacrylamide gel of 3_H-Marco and 14_C-VSV proteins, showing different mobilities of the G, N, and M proteins.

and Timbo viruses have been crosstested against 30 other rhabdoviruses isolated from athropods and vertebrates (exclusive of fish), with negative results.[3] The CF relationship between Timbo and Chaco viruses has been repeatedly shown (Table 3).[2,3]

VI. BIOLOGIC PROPERTIES

All three viruses cause lethal infection of suckling mice[1-3] and hamsters[6] inoculated IC. Timbo virus is significantly less pathogenic for infant mice than Marco or Chaco

TABLE 4

Pathogenicity of Reptile Rhabdoviruses for Suckling Mice

Virus[a]	$ICLD_{50}/m\ell$	Suckling mice AST[b]	Pathogeni-city LD_{50}/ Vero cell PFU
Marco	$10^{7.4}$	3.2	5.0
Timbo	$10^{3.5}$	8.5	0.0005
Chaco	$10^{7.0}$	7.6	5.0

[a] Passage levels: Marco SM7; Timbo SM? + 2; Chaco SM7.

[b] Average survival time (days) at a dose of 100 $ICLD_{50}$.

From Monath, T. P., Cropp, C. B., Murphy, F. A., and Frazier, C. L., *Arch. Virol.*, 60, 1—12, 1979. With permission.

viruses Table 4). Infant mice inoculated by the intraperitoneal (IP) route and weaned mice inoculated IC or IP show no signs of illness, but weaned mice develop CF serum antibodies.[1,2] The illness in infant mice is characterized by signs of central nervous system dysfunction, including ataxia, lethargy, tremors, paraspinal muscle contraction, and coma. The encephalitic process is not rapidly progressive (or does not affect critical centers), and mice with signs of profound illness often survive for several days. Histopathologic studies have shown only encephalitic lesions.[2]

Viral growth in continuous cell lines derived from various classes of vetebrates has been investigated (Table 5).[3] Marco virus replicated to approximately equal titers (5 to 6 dex/mℓ) in mammalian (Vero, BHK-21) cells and three reptilian cell lines (VSW, IgH-2, and TH-1). Growth in amphibian cells was less than in mammalian or reptilian systems, and no detectable virus was released from cells of fish origin. Cytopathic effect (CPE) was marked in the mammalian cells, but in reptile cells it was absent (VSW and IgH-2) or minimal and subtle (TH-1). Chaco virus grew to high titer in one of four reptile cell lines (VSW) and in BHK-21 cells. Growth was observed in Vero cells and one amphibian cell line, but not in other reptile cells or in fish cells. CPE was present only in mammalian cells. Replication of Timbo virus was demonstrated in mammalian cells only.

In another study[7] Chaco virus did not produce CPE in HeLa cells, whereas Marco virus produced 2 to 3+ CPE to a titer of 4.5 $TCD_{50}/m\ell$. No CPE was observed in primary chick embryo cells inoculated with Marco virus 2 or in primary duck embryo cultures exposed to all three agents.[6]

Infection of reptilian cells with Marco and Chaco viruses was shown to protect against CPE after VSV challenge.[3] This technique was useful for virus assay, but the mechanism of heterologous interference has not been elucidated.

Timbo virus has been reported to produce plaques in BHK-21 cells.[2] All three viruses cause 0.5 to 2.0 mm plaques in Vero cells. Optimal plaquing conditions have been only partially investigated,[3] but a double agar overlay technique with addition of DEAE-dextran to the overlay medium gave satisfactory results.

The temperature optima for growth of all three viruses were 30°C. Reduced rates of replication were observed at 37°C in cells of both mammalian and poikilothermic origin. The data indicated that temperature sensitivity was determined both by the viral

TABLE 5

Growth of Reptile Rhabdoviruses in Cell Lines of Homeothermic and Poikilothermic Origin

	Marco			Timbo			Chaco		
Cell line[a]	CPE[b]	Maximum yield[c]	Day	CPE	Maximum yield	Day	CPE	Maximum yield	Day
Mammalian									
Vero	4+	5.3	6	4+	3.8	6	4+	3.5	6
BHK	3+	5.9	4—5	3—4+	5.0	3	3+	5.7	5
Reptilian									
IgH-2	0	5.7	4	0	Trace		0	3.2	3
VSW	0	5.4	4	0	0		0	6.0	6
TH-1	1+	5.1	3	0	0		0	0	
VH-2	0	2.3	9	0	0		0	0	
8625	0	2.3	8	0	0		0	Trace	
Amphibian									
A6	0	3.9	4	0	0		0	4.3	4—5
ICR-2A	0	0		0	0		0	0	
Piscine									
FHM	0	0		0	0		0	0	
RTG	0	0		0	0		0	0	

[a] IgH-2 = Iguana (*Iguana iguana*) heart. VSW = Russell's viper (*Viper russelli*) spleen. TH-1 = Turtle (*Terrapene carolina*) heart. VH-2 = Russell's viper heart. 8625 = Rattlesnake fibroma. A6 = South African clawed toad (*Xenopus laevis*) kidney. ICR-2A = Haploid frog (*Rana pipiens*) embryo. FHM = Fat Head minnow (*Pimephales promelas*). RTG = Rainbow trout (*Salmo gairdnesi*) gonad.

[b] Cytopathic effect.

[c] Log suckling mouse $ICLD_{50}$ or Vero cell PFU per milliliter supernatant fluid.

Note: Maximal replication in all cell lines was at 30°C.

genome and host cell factors.[3] The low temperature optima of these viruses may represent an adaptation to replication in ectothermic vertebrates; however, *Ameiva* is a diurnal lizard which frequents the forest edge, and during its normal activities probably has a body temperature well in excess of the 30°C optima for in vitro viral growth. The true relationships between viral infection and host physiology will be understood only with further field and experimental infection studies.

ACKNOWLEDGMENT

The electron microscopy was kindly performed by Dr.F. A. Murphy and Ms. Sylvia G. Whitfield, Viral Pathology Branch, Virology Division, Center for Disease Control, Atlanta, Georgia.

REFERENCES

1. **Causey, O. R., Shope, R. E., and Bensabath, G.,** Marco, Timbo, and Chaco, newly recognized arboviruses from lizards in Brazil, *Am. J. Trop. Med. Hyg.,* 15, 239—243, 1966.
2. **Berge, T. O., Ed.,** *International Catalogue of Arboviruses,* Publ. No. (CDC) 75-8301, U.S. Department of Health, Education and Welfare, Atlanta, 1975.
3. **Monath, T. P., Cropp, C. B., Murphy, F. A., and Frazier, C. L.,** Viruses isolated from repitles: preliminary identification of three new members of the Family Rhabdoviridae, *Arch. Virol.,* 60, 1—12, 1979.
4. **Pinheiro, F.,** personal communication.
5. **Trent, D. W.,** unpublished study.
6. **Monath, T. P.,** unpublished study.
7. **Buckley, J. M.,** Applicability of the HeLa (Gey) strain of human malignant epithelial cells to the propagation of arboviruses, *Proc. Soc. Exp. Biol. Med.,* 116, 354—360, 1964.

INDEX

A

C

D

Dimers, see Covalent dimers
Diribonucleic acids, see DNA
Disassociation, II: 146
Disease incidence, plant viruses, III: 158
Distal sugars, II: 102
DNA
  bovine ephemeral fever virus studies, III: 177—178
  organisms vs. RNA, II: 121
  RNA replication, role in, II: 76
  ultraviolet irradiation of, II: 64
  ultraviolet lesions, II: 42—43
DNA-dependent RNA polymerase, II: 64
DNA provirus, III: 87—88

E

F

G

H

I

J

K

L

M

N

O

P

Q

R

S

T

U

V

W

X

Y

Z